Research on Pomegranate Germplasm, Breeding, Genetics and Multiomics

Research on Pomegranate Germplasm, Breeding, Genetics and Multiomics

Guest Editors

Zhaohe Yuan
Gaihua Qin
Julián Bartual

MDPI

Basel • Beijing • Wuhan • Barcelona • Belgrade • Novi Sad • Cluj • Manchester

Guest Editors
Zhaohe Yuan
Nanjing Forestry University
Nanjing
China

Gaihua Qin
Anhui Academy of
Agricultrual Science
Hefei
China

Julián Bartual
Regional Ministry of
Agriculture, Water, Livestock
and Fisheries
Elche (Alicante)
Spain

Editorial Office
MDPI AG
Grosspeteranlage 5
4052 Basel, Switzerland

This is a reprint of the Special Issue, published open access by the journal *Horticulturae* (ISSN 2311-7524), freely accessible at: https://www.mdpi.com/journal/horticulturae/special_issues/60R05F1EMG.

For citation purposes, cite each article independently as indicated on the article page online and as indicated below:

Lastname, A.A.; Lastname, B.B. Article Title. *Journal Name* **Year**, *Volume Number*, Page Range.

ISBN 978-3-7258-2687-2 (Hbk)
ISBN 978-3-7258-2688-9 (PDF)
https://doi.org/10.3390/books978-3-7258-2688-9

Cover image courtesy of Zhaohe Yuan

Contents

About the Editors

Zhaohe Yuan

Zhaohe Yuan, Ph.D., is a professor at the College of Forestry, Nanjing Forestry University. He was the chairman of the Working Group Pomegranate and Minor Mediterranean Fruits of the International Society for Horticultural Science (ISHS) from Sept. 2013 to Feb. 2022 and is the honorary chairman of the Pomegranate Section of the Chinese Society for Horticultural Science. He mainly engages in research on tree fruit germplasm resources and genetic breeding, molecular biology, and genomics, especially in the evaluation and innovative utilization of pomegranate and other tree fruit germplasm resources. From January 1994 to December 1996, he successively worked as a visiting scholar at the Irrigated Agricultural Research and Extension Center, Washington State University, and the Department of Horticulture, University of Maryland, United States. He has presided over more than 30 national and provincial research projects, including the National Natural Science Foundation of China, the National Key Basic Research Program, the National Science and Technology Support Program, and the Science and Technology Development Plan. He has published more than 230 articles in the *Plant Biotechnology Journal, Plant Physiology, Food Chemistry,* and other publications. He obtained more than 50 patents in China and abroad, and edited more than 10 books, including *China Fruit Science and Practice - Pomegranate, Acta Horticulturae 1089, Modern Fruit Cultivation in China, Integrative Management of Pomegranate Diseases and Pests in China*, and *Pomegranate Culture, Art and Functional Utilization*. He serves as a reviewer for journals such as the *Plant Biotechnology Journal, The Plant Journal*, the *Journal of Integrative Agriculture*, as well as being an expert reviewer for the National Natural Science Foundation of China. His research results have been widely applied in the national tree fruit industry and provided technological support for China's pomegranate industry upgrade.

Gaihua Qin

Gaihua Qin, Ph.D., is a professor at the Horticultural Research Institute, Anhui Academy of Agricultural Science, China. She is interested in the genetic improvement of pomegranate and fruit quality biology. Recently, her work has focused on pomegranate seed coat development, the flavor metabolic mechanism of pear fruit, and physiological diseases of pomegranate. She has set the standard "Descriptors and Date standard for pomegranate (*Punica granatum* L.)", one of Technical Specification Series of crop germplasm resources in China. She has obtained more than 10 patents on breeding and cultivation techniques in China and abroad, and bred five pomegranate varieties, among them, Jinghua Yushizi and Feihong, which have been widely planted in Anhui Province, China. She has mapped the whole genome of pomegranate for the first time worldwide, and worked on uncovering the genetic mechanism of seed coat development and the mechanisms of pomegranate pericarp sunburning and browning.

Julián Bartual

Julián Bartual has an MS degree in Agronomy from the Polytechnic University of Valencia (UPV) and a Doctorate in Plant Production and Biotechnology from the Miguel Hernandez University (UMH). Currently, he is the director of the Agricultural Experiment Station, Elche, Spain (EEA Elx-GVA), and Ext. Research Associate at the University Institute for Agro-food and Agro-environmental Research and Innovation (CIAGRO-UMH, Orihuela, Alicante). He is a researcher, extensionist, consultant and coordinator on projects related to breeding, irrigation, fertigation and physiological studies on pomegranate. He is also a breeder of newly registered pomegranate varieties belonging to the Valencian Research Institute (IVIA-GVA) and developed under his leadership. He was designated Spanish expert in the Tropical Agriculture Platform (TAP/AIMS/FAO) working group by the National Research Institute in Agriculture (INIA). He has coordinated projects granted by the European Commission such as the prestigious MED framework. His recent research focuses on the relationship between pomegranate and climate change. He is an advisor for the Protected Denomination of Origin (PDO) 'Granada Mollar de Elche/Granada de Elche'. He was also the convener of the "IV International Symposium on Pomegranate and Other Mediterranean Minor Fruits" (ISHS, 2017), and the organizer of National and International Courses and Seminars on Horticulture (Fruit and Nuts). In recent years, he has authored books and papers in major conferences and journals related to pomegranate. He has also served as editor in refereed journals and as scientific committee member of flagship conferences. He has been a speaker in numerous important significant seminars in India, China, Mexico, Turkey, etc. He is a member of the Spanish Society of Horticultural Sciences (SECH) and has received a Gold Medal Award from the International Society for Horticultural Science (ISHS). He has also received the MASTER 2020 award from ASFPLANT.

horticulturae

MDPI

Editorial

Research on Pomegranate Germplasm, Breeding, Genetics and Multiomics

Zhaohe Yuan [1,2,*], Gaihua Qin [3] and Julián Bartual [4]

1 Co-Innovation Center for Sustainable Forestry in Southern China, Nanjing Forestry University, Nanjing 210037, China
2 College of Forestry, Nanjing Forestry University, Nanjing 210037, China
3 Institute of Horticultural Research, Anhui Academy of Agricultural Sciences, Hefei 230001, China
4 Estación Experimental Agraria de Elche, Ctra. CV-855 Km. 1, 03290 Alicante, Spain; bartual_jul@gva.es
* Correspondence: zhyuan88@hotmail.com

Citation: Yuan, Z.; Qin, G.; Bartual, J. Research on Pomegranate Germplasm, Breeding, Genetics and Multiomics. *Horticulturae* **2024**, *10*, 1162. https://doi.org/10.3390/horticulturae10111162

Received: 8 October 2024
Revised: 25 October 2024
Accepted: 27 October 2024
Published: 1 November 2024

1. Introduction

Pomegranate (*Punica granatum* L.), a fruit-bearing shrub with a rich cultural history [1,2], has long been appreciated for its unique flavor, nutritional value, and ornamental beauty. Native to the region stretching from Iran to northern India, pomegranate has been cultivated for thousands of years and is now grown globally in various climates [3,4]. Its fruits are valued for their antioxidant-rich arils [2,4,5], while ornamental varieties are sought after for their vibrant flowers and esthetic appeal. The importance of pomegranate in both culinary and medicinal traditions, coupled with its increasing demand in the global market, has spurred significant research into improving its yield, quality, and resilience [6–8].

In recent years, the advent of advanced genomics, transcriptomics, and phenotyping tools has allowed researchers to delve deeper into the genetic architecture of pomegranate, leading to new insights into its complex biology. These breakthroughs have significantly enhanced our ability to explore traits of agronomic importance, such as fruit quality [4,9–11], disease resistance [12], and stress tolerance [13,14]. Despite this progress, several gaps in our knowledge have persisted. Key questions remain regarding the genetic regulation of traits like fruit morphology, color [15,16], and texture, as well as the molecular mechanisms that underpin disease resistance and environmental adaptability [12].

Traditional breeding approaches, while effective, are often slow and constrained by the complex genetics of pomegranate, particularly when dealing with polygenic traits [3]. Moreover, the pomegranate germplasm is highly diverse, with many landraces and wild relatives showing wide variability in important traits [17]. As a result, molecular breeding and biotechnological tools are increasingly seen as indispensable in accelerating pomegranate improvement efforts [18,19].

However, until recently, comprehensive genomic resources for pomegranate have been limited. A complete reference genome has only been available for a short time, and the functional characterization of key genes and regulatory elements remains in its early stages. Moreover, the integration of multiomics approaches—combining genomics, transcriptomics, proteomics, and metabolomics—has been underutilized in pomegranate research [20]. These limitations have hindered our ability to fully exploit the genetic diversity of pomegranate and make precise, targeted improvements in breeding programs.

The goal of this Special Issue, "Research on Pomegranate Germplasm, Breeding, Genetics, and Multiomics", is to address these gaps by compiling cutting-edge research that advances our understanding of pomegranate at the genetic, molecular, and phenotypic levels. This collection of studies highlights the diversity and depth of current research, ranging from gene family analyses and molecular marker identification to postharvest treatment studies and disease resistance. By presenting this research, the Special Issue aims to offer a more comprehensive view of the genetic underpinnings of important traits and provide new tools and resources for breeders, researchers, and the broader horticultural community.

Moreover, the recent focus on integrating multiomics approaches in plant research offers exciting new avenues for exploration. By connecting the dots between genomic data and observable traits, researchers can better understand how genetic variation influences phenotypic diversity and develop strategies to improve pomegranate cultivars in a more targeted manner. The growing availability of genomic resources, including genome sequences and gene expression data, opens the door to more sophisticated breeding techniques, such as marker-assisted selection (MAS), genomic selection (GS), and even CRISPR-based genome editing.

This Special Issue also underscores the importance of interdisciplinary research, combining classical breeding with modern genetic technologies to address practical challenges such as improving disease resistance, extending postharvest shelf life, and enhancing fruit quality for both fresh consumption and industrial processing. As the pomegranate industry continues to grow, driven by increasing consumer demand for nutritious and functional foods, there is an urgent need to develop new cultivars that are resilient, high-yielding, and tailored to specific market needs.

By fostering a deeper understanding of the genetics and molecular biology of pomegranate, this Special Issue not only fills crucial gaps in our current knowledge but also sets the stage for future research aimed at sustainable and efficient pomegranate breeding and cultivation practices. It is hoped that this collection of articles will inspire further investigation and collaboration within the scientific community, ultimately leading to the development of new technologies and innovations that will benefit pomegranate growers and consumers alike.

2. Overview of Published Articles

This Special Issue includes ten research articles (List of Contributions). The ten articles offer comprehensive insights into the pomegranate's genetic and phenotypic diversity, genomic resources, and molecular mechanisms underlying critical traits. The research presented spans from fundamental genomic studies to applied horticultural practices, reflecting the ongoing progress in pomegranate breeding, multiomics, and postharvest management. A summary of the contributions is provided below.

2.1. Genetic Variation and Phenotypic Traits

A study on ornamental pomegranate investigates the genetic basis of floral morphology, linking genomic variation to petal traits, providing critical markers for breeding efforts aimed at ornamental purposes (Contribution 1). Another article (Contribution 2) delves into the role of anthocyanidin synthase (ANS) genes in anthocyanin biosynthesis, which determines fruit color, a major determinant of marketability. The comprehensive genomic identification and characterization of the ANS gene family contribute to understanding how color can be genetically manipulated to enhance consumer appeal. Additionally, researchers explore the expansin gene family, which influences fruit growth and texture, another vital quality trait for commercial production (Contribution 3). Research continues with a detailed analysis of the UDP-glycosyltransferase (UGT) gene family, providing insights into its role in synthesizing key flavonoids and other secondary metabolites (Contribution 4). This work not only uncovers the genetic underpinnings of fruit flavor and nutritional composition but also highlights potential targets for improving these characteristics through molecular breeding. The in-depth exploration of transcription factors like R2R3-MYB, involved in regulating anthocyanin and flavonoid biosynthesis, further enriches our understanding of the regulatory networks controlling secondary metabolism (Contribution 5).

2.2. Postharvest Physiology

One of the studies examines the effects of 1-methylcyclopropene (1-MCP) on maintaining the quality of pomegranate fruits during low-temperature storage (Contribution 7). This research provides valuable insights into delaying senescence and preserving fruit quality, which is crucial for extending shelf life and reducing postharvest losses—a key concern for both growers and retailers.

2.3. Disease Resistance and Environmental Stress

A study focused on anthracnose fruit rot caused by *Colletotrichum gloeosporioides*, a serious issue in the southeastern United States, identifies cultivars exhibiting strong resistance to the pathogen (Contribution 8). This finding is particularly relevant for developing disease-resistant cultivars for commercial cultivation in regions prone to this and similar diseases. Studies investigating lignin biosynthesis and the associated laccase gene family shed light on structural aspects of the pomegranate, particularly regarding tree integrity and stress tolerance (Contribution 6). These findings have practical implications for developing cultivars better equipped to withstand environmental challenges, such as drought and pathogen attacks.

2.4. Reproductive Biology

In addition to gene families and stress resistance, this Special Issue presents research on reproductive biology through the analysis of microRNAs (miRNAs) associated with pistil development (Contribution 9). By identifying miRNAs that regulate key developmental processes, this research offers insights that could enhance fruit set and yield through targeted breeding approaches.

2.5. Phenotypic Diversity

A study on the phenotypic diversity of pomegranate cultivars highlights the discriminating power of morphological and chemical characteristics (Contribution 10). This phenotypic diversity is invaluable for both germplasm conservation and breeding, offering a solid foundation for developing superior cultivars tailored to specific market or environmental needs.

In conclusion, this Special Issue bridges significant knowledge gaps by combining molecular, genomic, and phenotypic approaches, paving the way for future research that could further enhance pomegranate breeding and cultivation strategies. The integration of multiomics approaches promises to continue advancing the field, ultimately contributing to the development of more resilient, high-quality pomegranate cultivars.

3. Summary and Future Outlook

The contributions in this Special Issue underscore the remarkable progress that has been made in the field of pomegranate genetics and breeding. However, they also highlight the remaining challenges that need to be addressed to fully harness the genetic potential of this species. Future research should prioritize the following areas:

3.1. Integrating Multiomics Approaches

While considerable strides have been made in genomic and transcriptomic research, integrating these data with proteomic, metabolomic, and phenomic studies will provide a more holistic understanding of how genetic variation translates into phenotypic traits.

3.2. Functional Genomics and Gene Editing

The identification of key genes involved in traits like fruit quality, disease resistance, and stress tolerance is only the first step. Future efforts should focus on functional validation of these genes, utilizing CRISPR/Cas9 and other gene-editing technologies to accelerate the development of improved cultivars.

3.3. Climate Resilience

As global climates continue to change, developing pomegranate varieties that can withstand abiotic stresses such as drought will be critical. The role of transcription factors, such as the MYB family, in abiotic stress responses presents a promising avenue for future research.

3.4. Postharvest Biology

While this Special Issue has advanced our understanding of postharvest treatments, more research is needed to optimize storage conditions and explore natural alternatives to chemical treatments, ensuring the sustainability of the pomegranate industry.

3.5. Breeding Meeting Consumer Preferences and Marketability

As the pomegranate market grows, consumer-driven traits such as fruit flavor, texture and nutritional content will increasingly influence breeding priorities. Understanding the genetic basis for these traits and translating that knowledge into actionable breeding strategies should be a key focus.

In conclusion, this Special Issue has laid a solid foundation for the future of pomegrana research, presenting key discoveries and offering a roadmap for addressing the remaining challenges. We hope that the findings presented in this edition will inspire further investi gation and innovation, ultimately contributing to the sustainable cultivation and global appreciation of this remarkable fruit species.

Acknowledgments: The authors thank all the contributors and reviewers for their valuable contribu tions and support from the section editors of this Special Issue.

Conflicts of Interest: The authors declare no conflicts of interest.

List of Contributions

1. Huo, Y.; Yang, H.; Ding, W.; Yuan, Z.; Zhu, Z. Exploring the Relationship between Genomic Variation and Phenotype in Ornamental Pomegranate: A Study of Single and Double-Peta Varieties. *Horticulturae* **2023**, *9*, 361. https://doi.org/10.3390/horticulturae9030361.
2. Ni, H.; Suo, H.; Zhang, X.; Hu, L.; Yuan, F.; Zhang, M.; Zhang, S. Genome-Wide Identifi cation and Characterization of the ANS Gene Family in Pomegranate (*Punica granatum* L. *Horticulturae* **2023**, *9*, 468. https://doi.org/10.3390/horticulturae9040468.
3. Xu, X.; Wang, Y.; Zhao, X.; Yuan, Z. Uncovering the Expansin Gene Family in Pomegranate (*Punica granatum* L.): Genomic Identification and Expression Analysis. *Horticulturae* **2023**, *9*, 539. https://doi.org/10.3390/horticulturae9050539.
4. Zhao, X.; Feng, Y.; Ke, D.; Teng, Y.; Chen, Y.; Langjia, R. Molecular Identification and Characteri zation of UDP-glycosyltransferase (UGT) Multigene Family in Pomegranate. *Horticulturae* **2023**, *9*, 540. https://doi.org/10.3390/horticulturae9050540.
5. Suo, H.; Zhang, X.; Hu, L.; Ni, H.; Langjia, R.; Yuan, F.; Zhang, M.; Zhang, S. Unraveling the Pomegranate Genome: Comprehensive Analysis of R2R3-MYB Transcription Factors. *Horticul turae* **2023**, *9*, 779. https://doi.org/10.3390/horticulturae9070779.
6. Shi, J.; Yao, J.; Tong, R.; Wang, S.; Li, M.; Song, C.; Wan, R.; Jiao, J.; Zheng, X. Genome-Wid Identification of Laccase Gene Family from *Punica granatum* and Functional Analysis toward Potential Involvement in Lignin Biosynthesis. *Horticulturae* **2023**, *9*, 918. https://doi.org/10.33 0/horticulturae9080918.
7. Wan, R.; Song, J.; Lv, Z.; Qi, X.; Feng, Z.; Yang, Z.; Cao, X.; Shi, J.; Jian, Z.; Tong, R.; et al. Effects of 1-MCP Treatment on Postharvest Fruit of Five Pomegranate Varieties during Low-Temperatur Storage. *Horticulturae* **2023**, *9*, 1031. https://doi.org/10.3390/horticulturae9091031.
8. Schaller, A.; Chater, J.M.; Vallad, G.E.; Moersfelder, J.; Heinitz, C.; Deng, Z. Pomegranat Cultivars with Diverse Origins Exhibit Strong Resistance to Anthracnose Fruit Rot Caused b *Colletotrichum gloeosporioides*, a Major Disease in Southeast United States. *Horticulturae* **2023**, 1097. https://doi.org/10.3390/horticulturae9101097.
9. Zhao, Y.; Huang, J.; Li, M.; Ren, H.; Jiao, J.; Wan, R.; Liu, Y.; Wang, M.; Shi, J.; Zhang, K.; et a Exploring MicroRNAs Associated with Pomegranate Pistil Development: An Identification an Analysis Study. *Horticulturae* **2024**, *10*, 85. https://doi.org/10.3390/horticulturae10010085.

10. Radunić, M.; Jukić Špika, M.; Gadže, J. Phenotypic Diversity of Pomegranate Cultivars: Discriminating Power of Some Morphological and Fruit Chemical Characteristics. *Horticulturae* **2024**, *10*, 563. https://doi.org/10.3390/horticulturae10060563.

References

1. Levin, G.M. *Pomegranate Roads: A Soviet Botanist's Exile from Eden*; Floreant Press: Forestville, CA, USA, 2006; pp. 27–35.
2. Zohary, D.; Hopf, M.; Weiss, E. *The Origin and Spread of Domesticated Plants in South-West Asia, Europe, and the Mediterranean Basin*; Oxford University Press: Oxford, UK, 2013; pp. 134–135.
3. Holland, D.; Bar-Ya'akov, I. Pomegranate (*Punica granatum* L.) Breeding. In *Advances in Plant Breeding Strategies: Fruits*; Al-Khayri, J., Jain, S., Johnson, D., Eds.; Springer: Cham, Switzerland, 2018; pp. 601–647. [CrossRef]
4. Yuan, Z.; Fang, Y.; Zhang, T.; Fei, Z.; Han, F.; Liu, C.; Liu, M.; Xiao, W.; Zhang, W.; Wu, S.; et al. The pomegranate (*Punica granatum* L.) genome provides insights into fruit quality and ovule developmental biology. *Plant Biotechnol. J.* **2018**, *16*, 1363–1374. [CrossRef] [PubMed]
5. Saparbekova, A.A.; Kantureyeva, G.O.; Kudasova, D.E.; Konarbayeva, Z.K.; Latif, A.S. Potential of phenolic compounds from pomegranate (*Punica granatum* L.) by-product with significant antioxidant and therapeutic effects: A narrative review. *Saudi J. Biol. Sci.* **2023**, *30*, 103553. [CrossRef] [PubMed]
6. Shi, J.L.; Yao, J.N.; Tong, R.R.; Wang, S.; Li, M.; Song, C.H.; Wan, R.; Zheng, X.B. Genome-Wide Identification of Laccase Gene Family from *Punica granatum* and Functional Analysis towards Potential Involvement in Lignin Biosynthesis. *Horticulturae* **2023**, *9*, 918. [CrossRef]
7. Hernández, F.; Legua, P.; Martínez, R.; Melgarejo, P.; Martínez, J.J. Fruit quality characterization of seven pomegranate accessions (*Punica granatum* L.) grown in Southeast of Spain. *Sci. Hortic.* **2014**, *175*, 174–180. [CrossRef]
8. Hooks, T.; Niu, G.; Masabni, J.; Sun, Y.; Ganjegunte, G. Performance and Phytochemical Content of 22 Pomegranate (*Punica granatum*) Varieties. *Hortscience* **2021**, *56*, 217–225. [CrossRef]
9. Qin, G.H.; Xu, C.Y.; Ming, R.; Tang, H.B.; Guyot, R.; Kramer, E.M.; Hu, Y.D.; Yi, X.K.; Qi, Y.J.; Xu, Y.L.; et al. The pomegranate (*Punica granatum* L.) genome and the genomics of punicalagin biosynthesis. *Plant J.* **2017**, *91*, 1108–1128. [CrossRef] [PubMed]
10. Luo, X.; Li, H.X.; Wu, Z.K.; Yao, W.; Zhao, P.; Cao, D.; Yu, H.Y.; Li, K.D.; Cao, S.Y. The pomegranate (*Punica granatum* L.) draft genome dissects genetic divergence between soft- and hard-seeded cultivars. *Plant. Biotechnol. J.* **2018**, *18*, 955–968. [CrossRef] [PubMed]
11. Qin, G.H.; Liu, C.Y.; Li, J.Y.; Qi, Y.J.; Gao, Z.H.; Zhang, X.L.; Yi, X.K.; Pan, H.F.; Ming, R.; Xu, Y.L. Diversity of metabolite accumulation patterns in inner and outer seed coats of pomegranate: Exploring their relationship with genetic mechanisms of seed coat development. *Hortic. Res.* **2020**, *7*, 10. [CrossRef] [PubMed]
12. Yu, X.; Xavier, K.V.; Vallad, G.E.; Deng, Z. Diseases resistance in pomegranates: Importance, sources, breeding approaches, and progress. *Proc. Fla. State Hort. Soc.* **2018**, *131*, 1–5.
13. Rugienius, R.; Vinskienė, J.; Andriūnaitė, E.; Mcrkūnaitė-Haimi, Š.; Juhani-Haimi, P.; Graham, J. Genomic Design of Abiotic Stress-Resistant Berries. In *Genomic Designing for Abiotic Stress Resistant Fruit Crops*; Kole, C., Ed.; Springer: Cham, Switzerland, 2022; pp. 197–249. [CrossRef]
14. Pourghayoumi, M.; Bakhshi, D.; Rahemi, M.; Kamgar-Haghighi, A.A.; Aalami, A. The physiological responses of various pomegranate cultivars to drought stress and recovery in order to screen for drought tolerance. *Sci. Hortic.* **2017**, *217*, 164–172. [CrossRef]
15. Ben-Simhon, Z.; Judeinstein, S.; Trainin, T.; Harel-Beja, R.; Bar-Ya'akov, I.; Borochov-Neori, H.; Holland, D. A "White" Anthocyanin-less Pomegranate (*Punica granatum* L.) Caused by an Insertion in the Coding Region of the Leucoanthocyanidin Dioxygenase (LDOX.; ANS) Gene. *PLoS ONE* **2015**, *10*, e0142777. [CrossRef] [PubMed]
16. Zhao, X.Q.; Feng, Y.; Ke, D.; Teng, Y.; Yuan, Z.H. Comparative transcriptomic and metabolomic profiles reveal fruit peel color variation in two red pomegranate cultivars. *Plant Mol. Biol.* **2024**, *114*, 51. [CrossRef] [PubMed]
17. Gunnaiah, R.; Jagadeesha, R.C.; Cholin, S.; Prabhuling, G.; Govindaswamy Babu, A.; Fakrudin, B.; Murthy, S.B.N. Genetic diversity assessment and population structure analysis of pomegranate cultivars from different countries and Himalayan wild accessions. *J. Hortic. Sci. Biotechnol.* **2021**, *96*, 614–623. [CrossRef]
18. Özgüven, A.I.; Dönmez, D.E.R.Y.A.; Zahid, G.; Şimşek, Ö.; Kaçar, Y.A. Breeding and plant improvement of pomegranate (*Punica granatum* L.). *ISHS Acta Hortic.* **2022**, *1349*, 27–38. [CrossRef]
19. da Silva, J.A.T.; Rana, T.S.; Narzary, D.; Verma, N.; Meshram, D.T.; Ranade, S.A. Pomegranate biology and biotechnology: A review. *Sci. Hortic.* **2013**, *160*, 85–107. [CrossRef]
20. Rout, G.R.; Peter, K.V. *Omics in Horticultural Crops*; Academic Press: Cambridge, MA, USA, 2022; pp. 193–203.

horticulturae

MDPI

Article

Exploring the Relationship between Genomic Variation and Phenotype in Ornamental Pomegranate: A Study of Single and Double-Petal Varieties

Yan Huo [1,2,3,4], Han Yang [5], Wenjie Ding [6], Zhaohe Yuan [2,4,*] and Zunling Zhu [1,2,3,*]

1 College of Landscape Architecture, Nanjing Forestry University, Nanjing 210037, China
2 Southern Modern Forestry Collaborative Innovation Center, Nanjing Forestry University, Nanjing 210037, China
3 College of Art and Design, Nanjing Forestry University, Nanjing 210037, China
4 College of Forestry, Nanjing Forestry University, Nanjing 210037, China
5 College of Traditional Chinese Medicine, Weifang Medical University, Weifang 261053, China
6 College of Landscape Engineering, Suzhou Polytechnic Institute of Agriculture, Suzhou 215008, China
* Correspondence: zhyuan88@hotmail.com (Z.Y.); zhuzunling@njfu.edu.cn (Z.Z.)

Abstract: The double-petal varieties of ornamental pomegranate have higher ornamental value and garden development potential than the single-petal varieties but there has been no study on the genomic variation between them. This study aimed to determine the genomic variation between the two kinds of varieties and the relationship between the variation and phenotype by identifying the DNA variation of three single-petal varieties and three double-petal varieties using re-sequencing technology. The results showed that the variation number of each variety was in the order of single nucleotide polymorphisms (SNPs) > insertions and deletions (InDels) > structural variations (SVs) > copy number variations (CNVs). The number of SNPs and InDels in the double-petal varieties was significantly higher than that in the single-petal varieties, and there was no significant difference in the number of SVs and CNVs. The number of non-synonymous SNPs in the coding region (Nonsyn_CDS_SNPs) and InDels with a 3X length in the coding region (3X_shiftMutation_CDS_InDel) was significantly higher in the double-petal varieties than that in the single-petal varieties. The number of the two variants was strongly positively correlated with each morphological index that was related to the phenotypic difference between the two varieties. Nonsyn_CDS_SNPs and 3X_shiftMutation_CDS_InDel were enriched in the cell membrane system, cell periphery, and signal transduction, from which 15 candidate genes were screened. Our results provide genomic data for the study of the formation mechanism of the double-petal flower and lay a theoretical foundation for new variety breeding of ornamental pomegranate.

Keywords: *Punica granatum*; SNP; InDel; CNV; SV; double-petal flower; single-petal flower

Citation: Huo, Y.; Yang, H.; Ding, W.; Yuan, Z.; Zhu, Z. Exploring the Relationship between Genomic Variation and Phenotype in Ornamental Pomegranate: A Study of Single and Double-Petal Varieties. *Horticulturae* **2023**, *9*, 361. https://doi.org/10.3390/horticulturae9030361

Academic Editor: Giuseppe Ferrara

Received: 28 January 2023
Revised: 4 March 2023
Accepted: 7 March 2023
Published: 9 March 2023

1. Introduction

Pomegranate has rich germplasm resources. At present, the research on the phenotypic and genetic diversity of pomegranate varieties mainly focuses on the shape, quality and related molecular markers of fruit [1–3], while the research on the main ornamental character—the flower—has received less attention. The single-petal varieties of ornamental pomegranate (*Punica granatum* L.) have a narrow flower shape and single-petal layer, while the double-petal varieties have a full flower shape, more petaloid stamens, and numerous petals [4]. Thus, the double-petal varieties have higher ornamental value and garden development potential. The phenotypic differences between the two kinds of varieties are mainly manifested in the petalization of the stamens and the growth of the petal transitional form. Currently, research on the mechanism of double-petal flower formation mainly focuses on the regulation of transcription factors, while there is little research on DNA variation [5–7]. For example, scholars have established the ABCDE flower development model to explain how transcription factors regulate flower organ morphogenesis [8–11]. In recent years, second-generation sequencing technology, which is fast, efficient, and low cost, has realized the whole genome sequencing of many plants, and, thus, provided technical support for genomic research on flower types [12–17]. For instance, Xing [18] screened out flowering-related genetic variation by comparing the genomes of two apple varieties using re-sequencing technology. Then, Huang [13] screened out genes that are related to lip petal development by comparing the genomes of two *Phalaenopsis aphrodite* varieties and the transcriptomes of 21 tissues, and Wu [14] screened out single nucleotide polymorphism (SNP) variations that are related to sterile flowers and continuous flowering in *Hydrangea macrophylla* using a genome-wide association study of 82 bigleaf hydrangea cultivars. Currently, the whole genome sequences of three pomegranate varieties, 'Taishanhong' [19], 'Dabenzi' [20], and 'Tunisia' [21], have been published. However, pomegranate genome-based research has mainly focused on the fruit character [21], stress resistance [22], and the chloroplast genome [23], and a comparative study on the genomic variation between the single- and double-petal varieties of ornamental pomegranate has yet to be carried out.

Using re-sequencing technology, this study determined the genomic variation between single and double-petal varieties and the relationship between the variation and phenotype, thus contributing to the molecular mechanism of pomegranate petalization and further providing a reference for new breeding varieties of pomegranate.

2. Materials and Methods

2.1. Plant Material

Six ornamental pomegranate varieties with similar plant types and ecological habits were collected from the Chinese pomegranate germplasm resource nursery (Yicheng; 34°49′49.195″ N, 117°21′18.701″ E) for genome re-sequencing. The six ornamental pomegranate varieties included three single-petal varieties ('Taiansanbaitian', 'Yichengdanbanfenhongtian', and 'Zipitian') and three double-petal varieties ('Luoyangbaimasi', 'Yichengfenhongmudan', and 'Taianhongmudan'; Table 1). Among them, 'Taiansanbaitian' and 'Luoyangbaimasi' are white flower varieties, 'Yichengdanbanfenhongtian' and 'Yichengfenhongmudan' are pink flower varieties, and 'Zipitian' and 'Taianhongmudan' are red flower varieties.

Table 1. Six ornamental pomegranate varieties.

Variety Type	White Flower Variety	Pink Flower Variety	Red Flower Variety
Single-petal variety	 'Taiansanbaitian'	 'Yichengdanbanfenhongtian'	 'Zipitian'
Double-petal variety	 'Luoyangbaimasi'	 'Yichengfenhongmudan'	 'Taianhongmudan'

2.2. Collection of the Morphological Parameters of Six Varieties

Three plants were randomly selected from each variety for the investigation of morphological parameters. The morphological parameters included: the flower length, flower width, flower width/flower length, calyx length, calyx width, calyx width/calyx length, sepal number, petal number, and petaloid stamen number. A ruler or vernier caliper was used to measure the length, and two significant figures were retained after the decimal point.

2.3. Sample Collection, Library Establishment, and Genome Re-Sequencing

A one-leaf sample from mixed shoots for each variety were collected and immediately frozen using liquid nitrogen and then stored in an ultra-low temperature refrigerator at −80 °C. The improved cetyltrimethylammonium bromide (CTAB) method was used to extract the total DNA of the samples [24]. Then, the DNA samples were randomly fragmented by Covaris and the fragments were collected by magnetic beads. Adenine was added to 3′ end of end-repaired DNA fragments before adaptor ligation. The ligation products were then cyclized and then amplified by linear isothermal Rolling-Circle Replication and DNA NanoBall technology. Then, agarose gel electrophoresis was used to screen the size of the fragments. A 200–300 bp small fragment library was established using polymerase chain reaction amplification. The qualified library was sequenced on the BGISEQ platform.

2.4. Data Filtering and Mapping

We used SOAPnuke (v1.4.0) to obtain clean data. The BWA [25] software was used to match the clean reads to the reference genome of 'Dabenzi'. Then, Picard tools (v1.118 http://broadinstitute.github.io/picard/, accessed on 5 March 2020) were used to sort the SAM files according to the reference genome and convert them into BAM files. The BAM files were used to detect genomic variation after repairing the mate-pair information adding the read group information, and labeling the repetitive reads.

2.5. Detection of the Single Nucleotide Polymorphisms, Insertions and Deletions, Structural Variation, and Copy Number Variation Polymorphisms

The GATK [26] software was used to detect the SNPs and insertions and deletions (InDels). The SNP filtering parameters were: "QD < 2.0 | | FS > 60.0 | | MQ < 40.0 | | MQRankSum < −12.5 | | ReadPosRankSum < −8.0". The InDel filtering parameters were: "QD < 2.0 | | FS > 200.0 | | ReadPosRankSum < −20.0". Additionally, Breakdancer [27] was used to detect the structural variations (SVs) using the parameters "–m 100 –x 1,000,000 –s 30 –d 5". Moreover, SOAPcnv [28] was used to detect the copy number variations (CNVs) using the parameters "–u 2 –z".

2.6. Data Processing and Bioinformatics Analysis

Microsoft Excel 2020 was used for the basic statistics and mapping of the morphological parameters and the number of SNPs, InDels, SVs, and CNVs. The SPSS 24.0 software (IBM, Armonk, New York, NY, USA) was used to conduct an analysis of variance (ANOVA by Duncan's multiple-range test and Pearson correlation analyses. The differences between the means were considered statistically significant at both $p < 0.05$ and $p < 0.01$. Nonsynonymous SNPs in the coding (CDS) region and InDels causing a frameshift mutation with a length of 3X in the CDS region between 'Taiansanbaitian' and 'Luoyangbaimasi' 'Yichengdanbanfenhongtian' and 'Yichengfenhongmudan', and 'Zipidian' and 'Taianhongmudan' were compared. The common genes in the three comparison groups with important genetic variation between the single- and double-petal varieties were identified, and ggVennDiagram in the R software (http://cran.r-project.org/web/packages/cluster/ accessed on 10 February 2021) was used to make a Venn diagram. Online software (https://www.omicshare.com/tools, accessed on 4 December 2022) was used to carry out Gene Ontology (GO) and Kyoto Encyclopedia of Genes and Genomes (KEGG) classification and enrichment analysis.

3. Results

3.1. Flower Morphological Parameters for the Two Types of Varieties

The results of the flower morphological parameters for the single and double-petal varieties of ornamental pomegranate are shown in Figure 1. The results of the variance analysis showed that the variety had significant effects on the flower length, flower width, flower width/flower length, calyx length, calyx width, calyx width/calyx length, sepal number, petal number, and petaloid stamen number. The flower length, flower width, flower width/flower length, calyx length, calyx width, calyx width/calyx length, sepal number, petal number, and petaloid stamen number of the double-petal varieties were significantly higher than those of the single-petal varieties. Flower color had no significant effect on the flower morphological parameters.

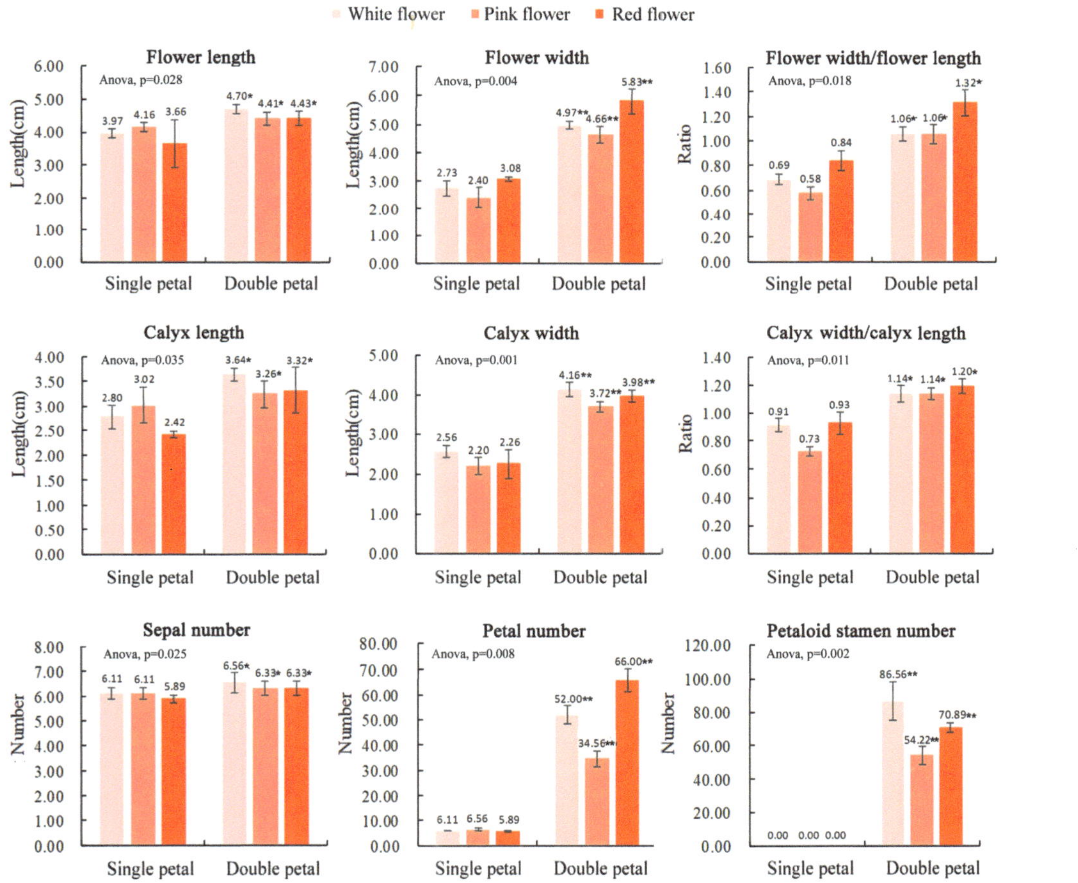

Figure 1. Morphological flower parameters of 6 ornamental pomegranate varieties. One asterisk and two asterisks indicated significant differences at $p < 0.05$ and $p < 0.01$, respectively, between the two types of flowers (single- and double-petals) according to the Duncan test. The standard deviation was also indicated.

3.2. Quality Evaluation and Mapping of the Sequencing Data

The re-sequencing data of the pomegranate varieties were filtered and quality control was conducted, and the results are shown in Table 2. A total of 98.40 GB of original sequencing data were obtained from the six ornamental pomegranate varieties, and the average ratio of the clean data to the original sequencing data was 92.72%. In addition, the average rate of high-quality (Q30) bases was 91.27%, and the average GC content was

40.96%. Moreover, 95.69% of the clean reads were aligned to the reference genome, with a mean sequencing depth of 51.38-fold and a mean coverage of 95.24%.

Table 2. Summary of the sequencing data of 6 ornamental pomegranate varieties.

Variety Name	Original Data (G)	Clean Data (G)	Q30 (%)	GC (%)	Mapped (%)	Depth	Coverage (%)
'Taiansanbaitian'	13.99	12.74	91.19	41.43	99.70	57.53	95.29
'Yichengdanbanfenhongtian'	17.49	16.37	91.43	40.67	93.45	55.25	95.06
'Zipidian'	18.36	17.26	91.70	40.09	81.81	58.23	95.36
'Luoyangbaimasi'	14.90	13.55	90.89	41.28	99.71	45.73	95.18
'Yichengfenhongmudan'	15.53	14.39	90.50	41.35	99.65	48.56	95.23
'Taianhongmudan'	18.13	17.05	91.88	40.95	99.83	42.97	95.29

3.3. *Basic Analysis of the Four Variation Types*

In this study, the SNPs, InDels, SVs, and CNVs of the ornamental pomegranate varieties were identified and counted. The results are shown in Figure 2. The average SNP number (463,840) in the double-petal varieties was significantly higher than that in the single-petal varieties (339,904). Furthermore, the average InDel number (111,995) in the double-petal varieties was significantly higher than that in the single-petal varieties (80,288). In addition, the average SV number in the single and double-petal varieties was 27,835 and 29,606, respectively, whereas the average CNV number in the single and double-petal varieties was 10,974 and 11,864, respectively. The variety had no significant effect on the SV and CNV numbers. The trend of the various types in the genome of the six ornamental pomegranate varieties was in the order of SNP > InDel > SV > CNV. Additionally, the variety had no significant effect on the number of SNPs, InDels, SVs, and CNVs. The trend of the various types in the coding sequences of the six ornamental pomegranate varieties was in the order of SNP_CDS > SV_CDS > CNV_CDS > InDel_CDS. On the other hand, there was no significant difference in the variable number of SNP, InDel, SV, CNV SNP_CDS, InDel_CDS, SV_CDS, and CNV_CDS among varieties of different colors.

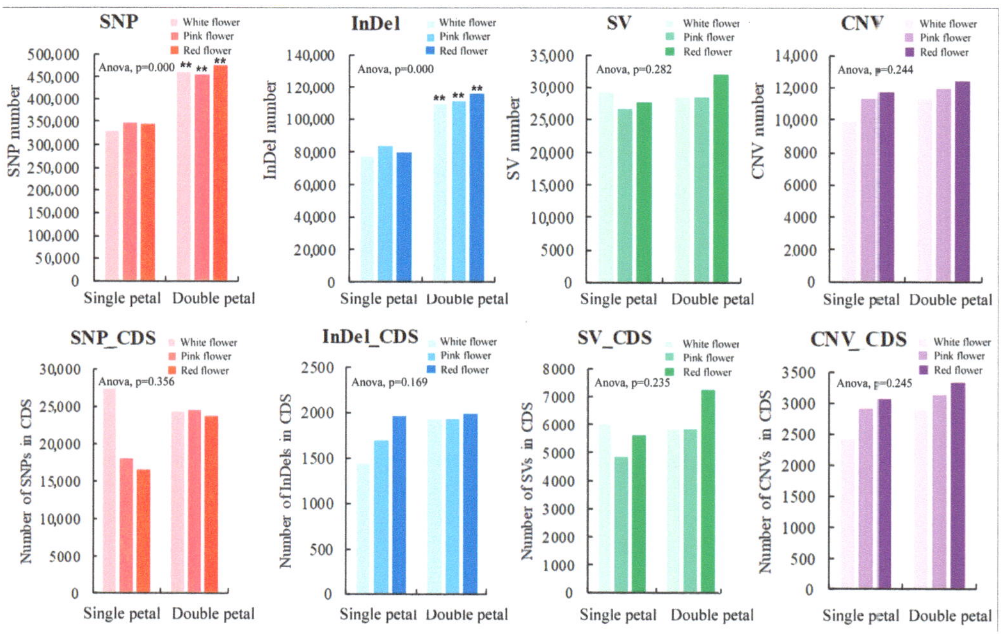

Figure 2. Quantitative statistics of 4 variation types of six ornamental pomegranate varieties. Two asterisks indicated significant differences at $p < 0.01$, between the two types of flowers (single- and double-petals) according to the Duncan test.

We conducted a Pearson correlation analysis between the number of different variation types identified in six varieties of genomes and the morphological parameters of 6 varieties in order to show the relationship between the variation and phenotype (Figure 3A). The SNP and InDel numbers were positively correlated with each morphological index, and the correlation coefficients ranged from 0.83 to 0.96 and 0.84 to 0.95, respectively. The correlation coefficients between the number of SNPs and InDels and the flower width, calyx width, petal number, and petaloid stamen number were higher than 0.90. The SV and CNV numbers were positively correlated with each morphological index, and the correlation coefficients ranged from 0.32 to 0.77 and 0.20 to 0.68, respectively. The trend of the correlation between the SNP number and each morphological index was similar to that between the InDel number and each morphological index, and the trend of the correlation between the SV number and each morphological index was similar to that between the CNV number and each morphological index.

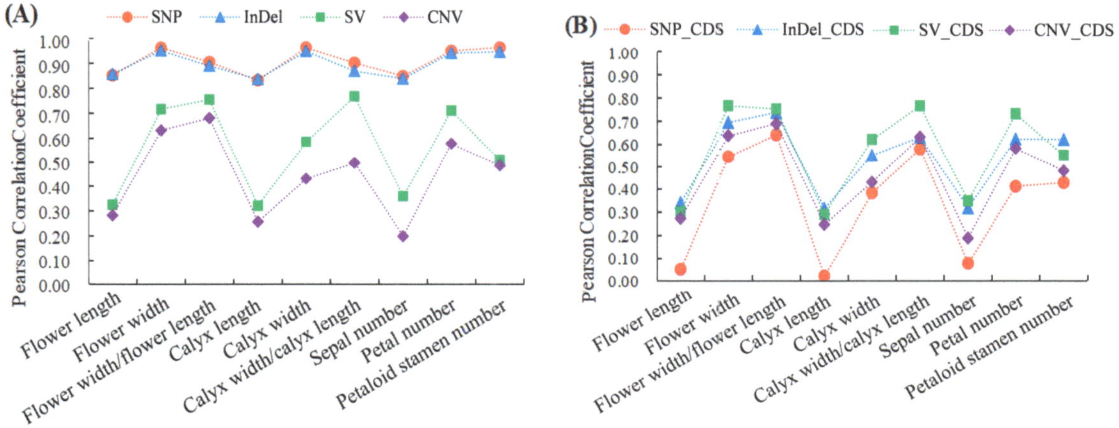

Figure 3. Correlation analysis between four variations of number and phenotypic parameters of ornamental pomegranate. (**A**) Correlation between the number of four types of variations in genome region and phenotypic parameters; (**B**) Correlation between the number of four types of variations in CDS region and phenotypic parameters.

The relationship between the number of the four variation types ©n the CDS and the morphological parameters was also analyzed, and the results are shown in Figure 3B. The SNP_CDS, InDel_CDS, SV_CDS, and CNV_CDS numbers were all positively correlated with each morphological index, and the correlation coefficients between them and the flower width, flower width/flower length, calyx width, calyx width/calyx length, petal number, and petaloid stamen number were higher than those between them and the flower length, calyx length, and sepal number. The correlation coefficients between the number of the four variation types in the intergenic region and the morphological indicators were also calculated (Table S1). The correlation coefficients between the SNP number in the intergenic region and the morphological parameters and between the InDel number in the intergenic region and the morphological parameters were much higher than those in the CDS region, indicating that a large number of SNPs and InDels that are closely related to the morphological parameters occur in the intergenic region.

3.4. Annotation Analysis of the Single Nucleotide Polymorphisms

Our results showed that the SNP heterozygosity of the single and double-petal varieties of ornamental pomegranate ranged from 55.36% to 75.38% and 67.46% to 72.38%, respectively (Table 3). The single and double-petal varieties had no significant effect on the SNP heterozygosity. Furthermore, the ranges of the number of synonymous SNPs in the CDS region (Syn_CDS_SNP) of the single- and double-petal varieties were 6340–16,142 and

9140–9416 respectively, and the variety had no significant influence on the Syn_CDS_SNP number (Figure 4A). The ranges of the number of non-synonymous SNPs in the CDS region (Nonsyn_CDS_SNP) of the single and double petal varieties were 10,118–11,254 and 14,566–15,162, respectively. The single-petal varieties had a very significant impact on the Nonsyn_CDS_SNP number ($p < 0.01$; Figure 4B). Moreover, the correlation analysis showed that the Syn_CDS_SNP number was negatively correlated with most of the morphological parameters, while the Nonsyn_CDS_SNP number was strongly positively correlated with all the morphological parameters (Figure 4C), indicating that the Nonsyn_CDS_SNP were closely related to all the morphological parameters and had a great impact on the phenotypic traits. Additionally, the variety had extremely significant effects on the SNP variations that involved the loss of a start codon, acquisition of a start codon, loss of a stop codon, and acquisition of a stop codon and the SNP variations that were located at the splicing site, in the region within 5K upstream/downstream of the gene, and in the gene region ($p < 0.01$). There were also positive correlations between these SNPs and each of the morphological indicators (Tables S2 and S3).

Table 3. SNPs statistics of six ornamental pomegranate varieties.

Variety Type	Variety Name	Total SNPs	Homozygous	Heterozygous	Heterozygosity Rate (%)
Single-petal variety	'Taiansanbaitian'	329,045	112,214	216,831	65.90
Single-petal variety	'Yichengdanbanfenhongtian'	347,440	155,103	192,337	55.36
Single-petal variety	'Zipidian'	343,226	84,488	258,738	75.38
Double-petal variety	'Luoyangbaimasi'	460,355	127,643	332,712	72.27
Double-petal variety	'Yichengfenhongmudan'	456,143	125,983	330,160	72.38
Double-petal variety	'Taianhongmudan'	475,022	154,553	320,469	67.46

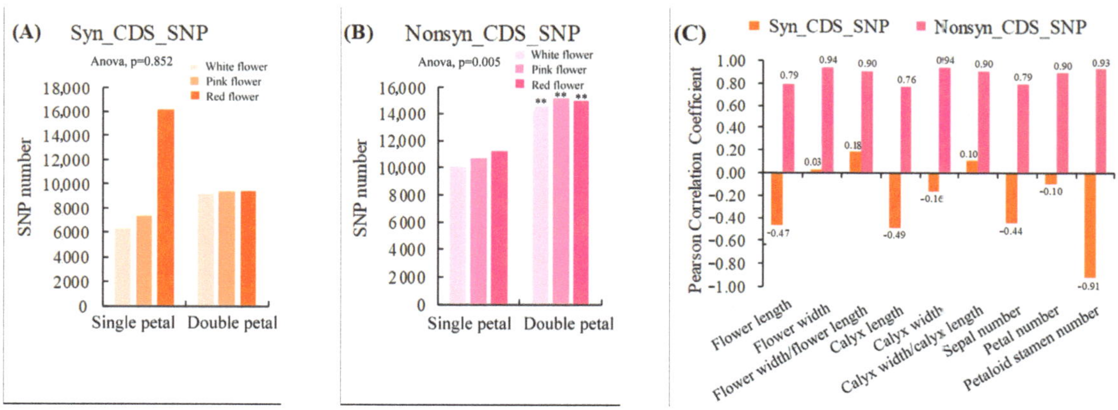

Figure 4. Statistics of synonymous and nonsynonymous SNPs in coding regions and their correlation analysis with morphological parameters. (**A**) Statistics of Syn_CDS_SNP number in six varieties; (**B**) Statistics of Nonsyn_CDS_SNP number in six varieties; (**C**) Correlation analysis between the number of Syn_CDS_SNP/Nonsyn_CDS_SNP and morphological indicators. Two asterisks indicated significant differences at $p < 0.01$, between the two types of flowers (single- and double-petals) according to the Duncan test.

3.5. Annotation Analysis of the Insertions and Deletions

The annotation statistics of the InDels in the six varieties are shown in Table 4. The results of the analysis of variance showed that the variety had a very significant impact on the number of InDels in the whole genome ($p < 0.01$) but there was no significant impact on the number of InDels in the CDS region. Additionally, the variety had a very significant effect on the InDels that were located in the region within 5K upstream/downstream of the gene, gene region, exon region, intron region, and pseudogene region ($p < 0.01$

Table S4), and there were positive correlations between these InDels and the morphological parameters (Table S5).

Table 4. Annotation statistics of InDels in six varieties.

Variety Type	Variety Name	Insertion in CDS/Genome	Deletion in CDS/Genome	InDel in CDS/Genome
Single petal variety	'Taiansanbaitian'	602/34,084	849/43,142	1451/77,226
Single petal variety	'Yichengdanbanfenhongtian'	722/37,739	976/46,177	1698/83,916
Single petal variety	'Zipidian'	892/35,653	1065/44,068	1957/79,721
Double petal variety	'Luoyangbaimasi'	807/50,503	1127/59,038	1934/109,541
Double petal variety	'Yichengfenhongmudan'	834/51,256	1092/59,718	1926/110,974
Double petal variety	'Taianhongmudan'	849/53,643	1137/61,826	1986/115,469

Figure 5A shows that the ranges of the number of InDels with a 3X length that caused a frameshift mutation in the CDS region (3X_shiftMutation_CDS_InDel) of the single and double-petal varieties were 517–593 and 697–738, respectively. The variety had a very significant impact on the number of 3X_shiftMutation_CDS_InDel ($p < 0.01$). In addition, the correlation analysis results showed that the number of 3X_shiftMutation_CDS_InDel was strongly positively correlated with the morphological parameters, and the correlation coefficients (0.72–0.95) between them were higher than those (0.32–0.74) between the InDels in the CDS region and the morphological parameters (Figure 5B), indicating that the 3X_shiftMutation_CDS_InDel had a great influence on the phenotypic characters of the ornamental pomegranates.

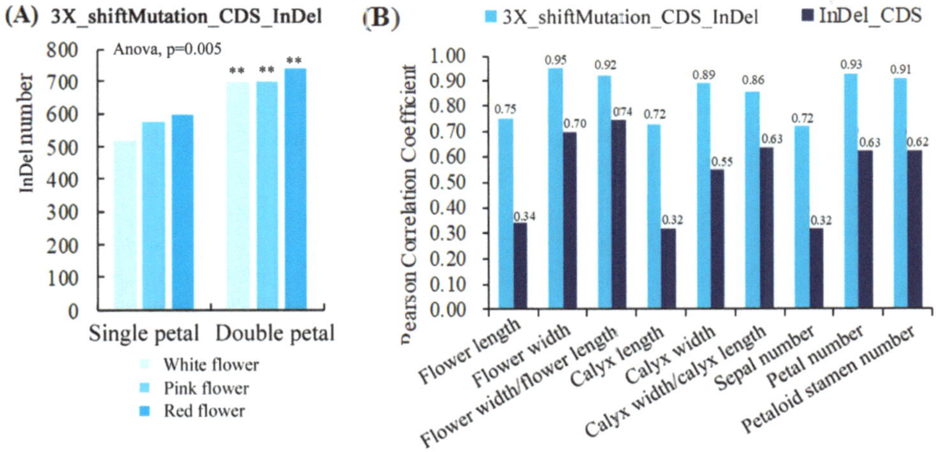

Figure 5. Quantitative statistics of 3X_shiftMutation_CDS_InDels and its correlation with phenotypic parameters. (**A**) Quantitative statistics of 3X_shiftMutation_CDS_InDels; (**B**) Correlation analysis between two types of InDels and morphological parameters. Two asterisks indicated significant differences at $p < 0.01$, between the two types of flowers (single- and double-petals) according to the Duncan test.

Figure 6 shows that the number of InDels of different lengths in the genome and CDS conforms to the normal distribution, with the largest number of one-base insertions or deletions. The analysis of variance showed that the variety had a significant impact on the number of InDels of different lengths in the genome ($p < 0.05$), while the variety had no significant impact in the CDS region. The number of InDels of each length in the single-petal varieties was significantly lower than that of the double-petal varieties. As shown in Figure 6A,B, within 15 bases that were inserted or deleted in the genome, the number of InDels of an even length was more than the number of adjacent InDels of an odd length (except for the one base InDels). As shown in Figure 6C,D, in the CDS region the number of InDels with a length of a multiple of three was more than the number of adjacent InDels with a length that was not a multiple of three.

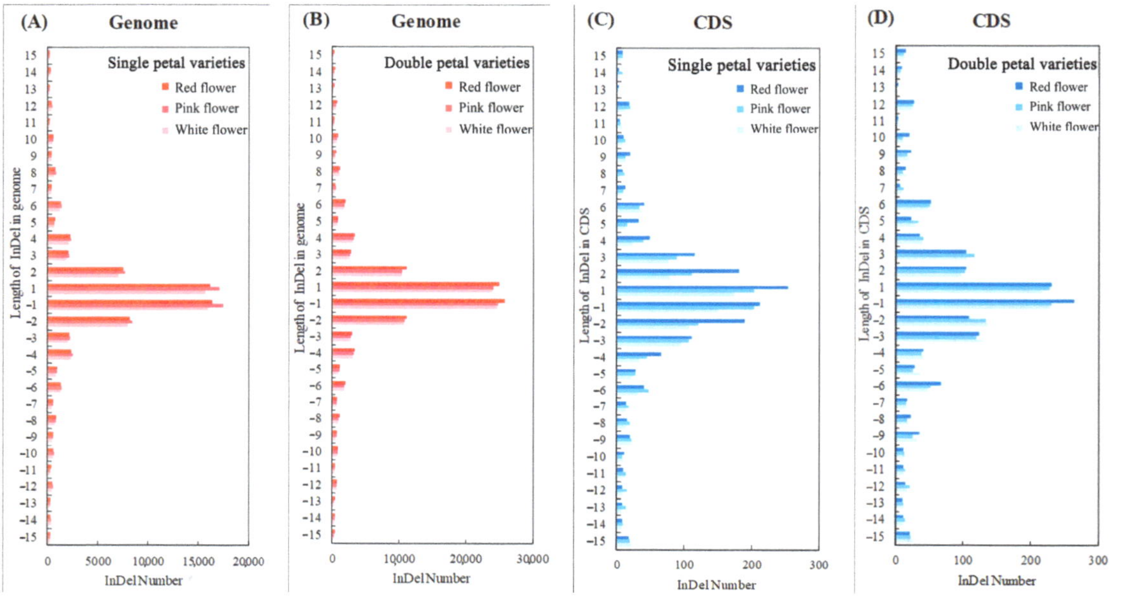

Figure 6. Length distribution statistics of insertion or deletion. (**A**) Length distribution map of insertion or deletion in genome region of single-petal varieties; (**B**) Length distribution map of insertion or deletion in genome region of double-petal varieties; (**C**) Length distribution map of insertion or deletion in CDS region of single-petal varieties; (**D**) Length distribution map of insertion or deletion in CDS region of double-petal varieties.

3.6. Annotation Analysis of the Structural Variations

A total of 172,324 SVs were identified in the six ornamental pomegranate varieties (Table S6). The type with the largest number was inter-chromosomal translocation (CTX) followed by inversion (INV), deletion (DEL), and intra-chromosomal translocation (ITX) and no insertions were detected. The variety had no significant effect on the total SVs DELs, INVs, and ITXs but it had a significant effect on the CTXs ($p < 0.05$). The CTX number in the single-petal varieties was significantly higher than that in the double-petal varieties. Moreover, the correlation analysis results showed that each morphological index was positively correlated with the number of DELs, INVs, and ITXs, while each morphological index was negatively correlated with the number of CTXs (Figure 7A)

3.7. Annotation Analysis of the Copy Number Variations

A total of 68,517 CNVs were identified in the six ornamental pomegranate varieties, of which the number of CNVs with an increased copy number was lower than that with a decreased copy number (Table S7). The analysis of variance showed that the variety had no significant effect on the number of total CNVs, CNVs with an increased copy number, and CNVs with a decreased copy number. Additionally, the correlation analysis results showed that the correlation coefficients between the morphological parameters and the number of CNVs with an increased copy number were lower than those between the morphological parameters and the number of CNVs with a decreased copy number (Figure 7B). In addition the number of CNVs with a decreased copy number was positively correlated with the flower width, flower width/flower length, calyx width, calyx width/calyx length, petal number, and petaloid stamen number (R > 0.56).

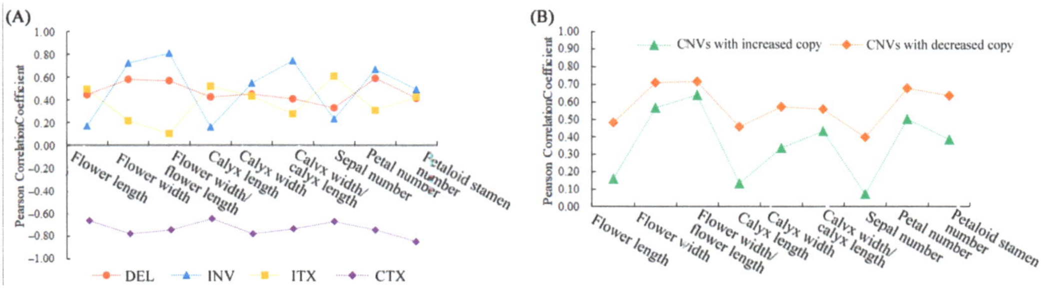

Figure 7. The correlation coefficient between SV/CNV types and morphological parameters. (**A**) The correlation coefficient between SV types and morphological parameters; (**B**) the correlation coefficient between CNV types and morphological parameters.

3.8. Variation Analysis between the Single- and Double-Petal Varieties

According to the above results, the variety had a significant effect on the number of Nonsyn_CDS_SNP and 3X_shiftMutation_CDS_InDel, so we compared these two variations in three groups and screened out the genes that were important for determining the single- or double-petal variety of ornamental pomegranate. Among them, there were 548 variations (belonging to 228 genes) between the genomes of 'Taiansanbaitian' and 'Luoyangbaimasi', 792 variations (belonging to 345 genes) between the genomes of 'Yichengdanbanfenhongtian' and 'Yichengfenhongmudan', and 314 variations (belonging to 181 genes) between the genomes of 'Zipidian' and 'Taianhongmudan'. In total, 37 common variants (belonging to 15 genes) were finally obtained from the three groups of genes. The GO enrichment analysis found (Figures S1–S9) that the enrichment of the variant genes among the three groups was similar. In terms of the cellular component GO terms, the three groups of mutant genes were enriched in the cell membrane system and cell periphery. Then, in terms of the biological process GO term, the three groups of mutant genes were enriched in signal transduction, and in terms of the molecular function GO terms, the three groups of mutant genes were enriched in transferase activity and GTP hydrolase activity. Finally, 15 mutant genes were screened from the three groups of comparison, and they were mainly involved in hormone pathways and stress responses, transcription and post-transcription regulation, translation, and post-translation regulation, and purine metabolism (Table 5).

Table 5. Classification of variant genes between single- and double-petal pomegranate.

Function Category	Gene Annotation	Gene ID
Hormone pathway and stress response	Indole-3-pyruvate monooxygenase, *YUC*	CDL15_Pgr015667
	CBS domain-containing protein, *CBSX6*	CDL15_Pgr010955
	Phosphate transporter, *PHO1*	CDL15_Pgr020531
	Receptor-like kinases, *RLK*	CDL15_Pgr015627
Transcription and post-transcriptional regulation	transcription factor, *TGA2.3*	CDL15_Pgr003069
	CCCH-type zinc finger protein, *C3H20*	CDL15_Pgr015624
	Werner Syndrome-like exonuclease, *WEX*	CDL15_Pgr003048
	Pentatricopeptide repeat protein, *PPR*	CDL15_Pgr003067
Translation and post-translation regulation	Tryptophan-tRNA ligase, *TrpS*	CDL15_Pgr015628
	Valine-tRNA ligase, *ValS*	CDL15_Pgr020557
	Ribosomal protein, *L23*	CDL15_Pgr004206
	Serine/threonine-protein phosphatase, *STPK*	CDL15_Pgr005655
	Palmitoyltransferase, *PAT*	CDL15_Pgr003037
	Charged multivesicular body protein, *CHMP*	CDL15_Pgr015621
Purine metabolism	phosphoribosylaminoimidazole-succinocarboxamide synthase	CDL15_Pgr015078

4. Discussion

The rapid development of bioinformatics and sequencing technology makes it possible to sequence the whole genome of many ornamental plants, which provides a starting point for revealing genetic variation at the genome level and exploring the relationship between genetic variation and phenotypic diversity [12,29,30]. When compared with the single-petal varieties, the double-petal varieties of ornamental pomegranates have higher ornamental value and wider garden application potential. Therefore, the genetic variation characteristics between the two varieties of ornamental pomegranates deserved further study. In this study, we found that the number of SNPs and InDels in the double-petal varieties was significantly higher than that in the single-petal varieties using whole genome re-sequencing technology, but there was no significant difference in the number of SNPs and InDels between varieties of different flower colors, indicating that the genetic variations among varieties of different flower types of ornamental pomegranates were more abundant than that among varieties of different flower colors. This also supported the previous researchers' finding that the single-petal varieties and the double-petal varieties were clustered in different branches through quantitative classification and molecular markers. In cluster analysis and principal component analysis, pomegranates were first classified according to flower type, and flower color was the second classification standard [18]. We also found that the number of genomic SNPs and InDels had a strong positive correlation with the morphological parameters, and the correlation coefficient was higher than that of SNPs and InDels located in the coding regions. These results indicated that the genetic differences between the two kinds of varieties were closely related to the flower type phenotype in the whole genome regions, but the variations that were related to the flower type phenotype might be only a small part of the differences in the coding regions. Further research also showed that SNPs and InDels closely related to flower phenotypic traits were mainly located in the intergenic region. This result supported the previous findings that miRNAs targeting transcription factors and hormone-related regulatory factors involved in pomegranate fruit development were located in the intergenic region [31], indicating that the intergenic region has an important role in pomegranate development. In addition we found that the variety had no significant effect on the number of CNVs in the genome or coding region, but the numbers of CNVs in the red flower varieties were the largest followed by the pink flower varieties, and the white flower varieties were the least, indicating that CNVs affecting gene expressions by disturbing gene activities and changing gene dosages might not participate in the flower development, but in the anthocyanin accumulation process.

Single nucleotide polymorphism heterozygosity is related to the abundance of parental resources, and SNP heterozygosity in citrus, soybean, and other plants is significantly related to phenotypic characteristics [18,32,33]. However, this study found that the single or double-petal variety had no significant effect on the SNP heterozygosity in ornamental pomegranate, indicating that ornamental pomegranate varieties may have undergone a complex natural selection and artificial breeding. Further analysis showed that the variety had no significant impact on the number of synonymous SNPs in the coding region but had a very significant impact on the number of non-synonymous SNPs in the coding region. Furthermore, it was found that Nonsyn_CDS_SNP were strongly positively correlated with the morphological parameters, indicating that Nonsyn_CDS_SNP was important to distinguish between single- and double petal-varieties.

According to genome variation studies of tomato, rice, apple, and other plants, it was found that the InDels number was generally less than the SNP number [24,34,35], which was consistent with our results. As with the SNPs, this study found that single- or double-petal variety had a very significant impact on the InDel number in the whole genome but had no significant impact on the InDels number in the CDS region. With further investigation, we found that the variety had a significant impact on the 3X_shiftMutation_CDS_InDel number and the 3X_shiftMutation_CDS_InDel number was strongly positively correlated with the phenotypic parameters, indicating that 3X_shiftMutation_CDS_InDel were important for distinguishing between the single-petal and double-petal varieties. Interestingly, within 15 bases that were inserted or deleted in the genome, the number of InDels of an even length was more than the number of InDels that were adjacent to them of an odd length. However, the law between the length and number of InDels in apples, millet, and other plants is different from this study. They generally follow the law that the longer the length is, the less the number of InDels [36,37]. Therefore, it is speculated that the findings in this study may be specific to pomegranates and very different from species such as grapes [38]. Additionally, we also found that the number of InDels with a length of three or a multiple of three was more than the number of adjacent InDels with a length that was not a multiple of three. This is because DNA mutations with lengths of three or multiple of three will not cause a frameshift, thus, avoiding fatal damage.

This study screened 15 candidate genes that were related to petalization, most of which were reported to be involved in reproductive development. Auxin affects stamen development and petal growth and *YUC1* encodes an important rate-limiting enzyme in the auxin synthesis pathway [39,40]. Yan [41] found that the overexpression of *YUC1* led to the overproduction of auxin and the poor development of the stamens, while variation in *YUC1* led to serious defects in the flower type development [42]. The cystathionine β-synthase domain proteins have the function of maintaining the balance of the redox reaction in the cells. In mutant plants, the scavenging capacity of active oxygen is reduced, the anthers are short and white, and there are no pollen grains [43]. Then, the phosphorus transporter gene, *PHO1*, can regulate grain filling and phosphorus distribution in crops [44,45] but there have been no studies on flower development. Additionally, the receptor-like protein kinase gene, *RLK*, regulates a series of biological processes, such as plant development, stress resistance, and hormone perception. It also plays an important role in the development of petunia pollen [46]. Moreover, the CCCH-type zinc finger protein is a transcription factor with a typical zinc finger structure. Liu [47] found that the C3H gene had the highest expression in the stamens of *Chimonanthus praecox*, and transgenic *Arabidopsis thaliana* had early flowering and abnormal stamen. Then, the pentatricopeptide repeat protein is an RNA-binding protein that participates in many post-transcriptional regulatory processes, such as splicing, editing, stabilization, and translation, and it plays a key role in cytoplasmic male sterility [48,49]. Furthermore, the ribosomal protein L23 is involved in the secretion and folding of new proteins. Moreover, the *L23* gene is expressed in inflorescences and other tissues in *Arabidopsis thaliana*, and the reproductive organs in mutant plants are deformed [50]. Additionally, STPK catalyzes the phosphorylation of serine and threonine residues on proteins, which is negatively regulated by the flower development gene *AGAMOUS* in *Arabidopsis thaliana* [51] and determines the number of female flowers and spike length in maize [52]. Moreover, *AGAMOUS*, *APETALA1*, and *APETALA2* resulted in more expression in brebas than in the main crop as reported in a recent investigation on *Ficus carica* [53]. Palmitoyltransferase catalyzes the palmitoylation modification of proteins [54], and pollen tube growth is defective in mutant plants [55]. The effect of these mutant genes on the petalization of ornamental pomegranate is worthy of further study.

5. Conclusions

In this paper, the genomic variation between single- and double-petal varieties was identified. The results showed that the number of SNPs and InDels caused by the mutation was larger than the number of SVs and CNVs caused by the recombination. The number of SNPs and InDels in the double-petal varieties was significantly higher than that in the single-petal varieties, and there was no significant difference in the number of SNPs and InDels between varieties of different flower colors, supporting the previous classification of pomegranate according to flower type. The variety had no significant effect on the SV and CNV numbers. In addition, the number of Nonsyn_CDS_SNP and 3X_shiftMutation_CDS_InDel was strongly positively correlated with the morphological parameters, showing that these two kinds of variants have an important influence on the phenotypic difference between the single- and double-petal varieties. Lastly, fifteen mutant genes were screened out from Nonsyn_CDS_SNPs and 3X_shiftMutation_CDS_InDel among the three groups of varieties and they were mainly involved in hormone pathways and stress responses, transcription and post-transcription regulation, translation, and post-translation regulation, and purine metabolism. This paper provides genomic mutation data between the single- and double-petal varieties and lays a theoretical foundation for double-flower molecular breeding in ornamental pomegranate. Further functional verification of mutation genes will provide insight and enable a deeper understanding of genetic involvement in the regulation of floral organ development.

Supplementary Materials: The following supporting information can be downloaded at: https://www.mdpi.com/article/10.3390/horticulturae9030361/s1, Table S1: Correlation coefficient between SNP, InDel, SV, CNV and morphological parameters; Table S2: SNP annotations of six ornamental pomegranate varieties; Table S3: Correlation coefficient between different SNP types and morphological parameters; Table S4: InDel annotations of six ornamental pomegranate varieties; Table S5: Correlation coefficient between different InDel types and morphological parameters; Table S6: Statistics of SV type in six ornamental pomegranate varieties; Table S7: Statistics of CNV type in six ornamental pomegranate varieties; Figure S1: Enrichment map of GO cell components of variant genes between 'Taiansanbaitian' and 'Luoyangbaimasi'; Figure S2: Enrichment map of GO biological process of variant genes between 'Taiansanbaitian' and 'Luoyangbaimasi'; Figure S3: Enrichment map of GO molecular function of variant genes between 'Taiansanbaitian' and 'Luoyangbaimasi'; Figure S4: Enrichment map of GO cell components of variant genes between 'Yichengdanbanfenhongtian' and 'Yichengfenhongmudan'; Figure S5: Enrichment map of GO biological process of variant genes between 'Yichengdanbanfenhongtian' and 'Yichengfenhongmudan'; Figure S6: Enrichment map of GO molecular function of variant genes between 'Yichengdanbanfenhongtian' and 'Yichengfenhongmudan'; Figure S7: Enrichment map of GO cell components of variant genes between 'Zipitian' and 'Taianhongmudan'; Figure S8: Enrichment map of GO biological process of variant genes between 'Zipitian' and 'Taianhongmudan'; Figure S9: Enrichment map of GO molecular function of variant genes between 'Zipitian' and 'Taianhongmudan'.

Author Contributions: Conceptualization, Y.H., Z.Y. and Z.Z.; methodology, Y.H.; software, Y.H. and H.Y.; validation, Y.H.; formal analysis, Y.H.; investigation, Y.H.; resources, Z.Y.; data curation, Y.H.; writing—original draft preparation, Y.H.; writing—review and editing, Y.H., H.Y. and W.D.; visualization, Y.H. and H.Y.; supervision, Z.Y. and Z.Z.; project administration, Z.Y. and Z.Z.; funding acquisition, Z.Y. and Z.Z. All authors have read and agreed to the published version of the manuscript.

Funding: This research was funded by the National Natural Science Foundation of China (32101582, 31770752), the Natural Science Foundation of Jiangsu Province of China (BK20210613), the Natural Science Foundation of the Jiangsu Higher Education Institutions of China (21KJB220008), the Initiative Project for Talents of Nanjing Forestry University (GXL2014070), the Priority Academic Program Development of Jiangsu High Education Institutions (PAPD), China Scholarship Council (CSC, 202008320482).

Institutional Review Board Statement: Not applicable.

Informed Consent Statement: Not applicable.

Data Availability Statement: Not applicable.

Conflicts of Interest: The authors declare no conflict of interest.

References

1. Patil, P.G.; Jamma, S.M.; Singh, N.V.; Bohra, A.; Parashuram, S.; Injal, A.S.; Gargade, V.A.; Chakranarayan, M.G.; Salutgi, U.D.; Dhinesh, B.K.; et al. Assessment of genetic diversity and population structure in pomegranate (*Punica granatum* L.) using hypervariable SSR markers. *Physiol. Mol. Biol. Plants* **2020**, *26*, 1249–1261. [CrossRef] [PubMed]
2. Tarantino, A.; Frabboni, L.; Mazzeo, A.; Ferrara, G.; Disciglio, G. Comparative evaluation of yield and fruit physico-chemical characteristics of five commercial cultivars of pomegranate grown in southeastern Italy in two consecutive years. *Horticulturae* **2022**, *8*, 497. [CrossRef]

3. Giancaspro, A.; Mazzeo, A.; Giove, S.L.; Zito, D.; Marcotuli, A.; Gallotta, P.; Colasuonno, D.; Nigro, A.; Blanco, M.; Aradhya, A.; et al. Exploiting DNA-based molecular tools to assess genetic diversity in pomegranate (*Punica granatum* L.) selections and cultivars. *Fruits* **2017**, *72*, 292–305. [CrossRef]
4. Ferrara, G.; Porfido, C.; Terzano, R.; Sarkhosh, A.; Mazzeo, A. A Study on the Characteristics of Buds and Flowers in Pomegranate: Differences among Cultivars. *Horticulturae* **2023**, *9*, 117. [CrossRef]
5. Jing, D.; Chen, W.; Hu, R.; Zhang, Y.; Xia, Y.; Wang, S.; He, Q.; Guo, Q.; Liang, G. An integrative analysis of transcriptome, proteome and hormones reveals key differentially expressed genes and metabolic pathways involved in flower development in Loquat. *Int. J. Mol. Sci.* **2020**, *21*, 5107. [CrossRef]
6. Ambawat, S.; Sharma, P.; Yadav, N.R.; Yadav, R.C. *MYB* transcription factor genes as regulators for plant responses: An overview. *Physiol. Mol. Biol. Plants* **2013**, *19*, 307–321. [CrossRef]
7. Chen, L.; Han, J.; Deng, X.; Tan, S.; Li, L.; Li, L.; Zhou, J.; Peng, H.; Yang, G.; He, G.; et al. Expansion and stress responses of *AP2/EREBP* superfamily in *Brachypodium distachyon*. *Sci. Rep.* **2016**, *6*, 21623. [CrossRef]
8. Coen, E.S.; Meyerowitz, E.M. The war of the whorls: Genetic interactions controlling flower development. *Nature* **1991**, *353*, 31–37. [CrossRef]
9. Kim, S.; Koh, J.; Yoo, M.J.; Kong, H.; Hu, Y.; Ma, H.; Soltis, P.S.; Soltis, D.E. Expression of floral MADS-box genes in basal angiosperms: Implications for the evolution of floral regulators. *Plant J.* **2005**, *43*, 724–744. [CrossRef]
10. Favaro, R.; Pinyopich, A.; Battaglia, R.; Kooiker, M.; Borghi, L.; Ditta, G.; Yanofsky, M.F.; Kater, M.M.; Colombo, L. MADS-box protein complexes control carpel and ovule development in *Arabidopsis*. *Plant Cell* **2003**, *15*, 2603–2611. [CrossRef]
11. Smaczniak, C.; Immink, R.G.; Muioj, M.; Blanvillain, R.; Busscher, M.; Busscher-Lange, J.; Dinh, Q.D.; Liu, S.; Westphal, A.H.; Boeren, S.; et al. Characterization of MADS-domain transcription factor complexes in *Arabidopsis* flower development. *Proc. Natl. Acad. Sci. USA* **2012**, *109*, 1560–1565. [CrossRef] [PubMed]
12. Saint-Oyant, L.H.; Ruttink, T.; Hamama, L. A high-quality genome sequence of *Rosa chinensis* to elucidate ornamental traits. *Nat. Plants* **2018**, *4*, 473–484. [CrossRef]
13. Huang, J.Z.; Lin, C.P.; Cheng, T.C.; Huang, Y.W.; Tsai, Y.J.; Cheng, S.Y.; Chen, Y.W.; Lee, C.P.; Chung, W.C.; Chang, B.C.; et al. The genome and transcriptome of *Phalaenopsis* yield insights into floral organ development and flowering regulation. *PeerJ* **2016**, *4*, e2017. [CrossRef] [PubMed]
14. Wu, X.; Alexander, L.W. Genome-wide association studies for inflorescence type and remontancy in *Hydrangea macrophylla*. *Hortic. Res.* **2020**, *7*, 27. [CrossRef]
15. Mariette, S.; Wong Jun Tai, F.; Roch, G.; Barre, A.; Chague, A.; Decroocq, S.; Groppi, A.; Laizet, Y.; Lambert, P.; Tricon, D.; et al. Genome-wide association links candidate genes to resistance to Plum Pox Virus in apricot (*Prunus armeniaca*). *New Phytol.* **2016**, *209*, 773–784. [CrossRef]
16. Wang, L.; Li, Y.; Jin, X.; Liu, L.; Dai, X.; Liu, Y.; Zhao, L.; Zheng, P.; Wang, X.; Liu, Y.; et al. Floral transcriptomes reveal gene networks in pineapple floral growth and fruit development. *Commun. Biol.* **2020**, *3*, 500. [CrossRef]
17. Fan, L.; Chen, M.; Dong, B.; Wang, N.; Yu, Q.; Wang, X.; Xuan, L.; Wang, Y.; Zhang, S.; Shen, Y. Transcriptomic analysis of flower bud differentiation in *Magnolia sinostellata*. *Genes* **2018**, *9*, 212. [CrossRef]
18. Wang, X.F. Studies on the Cultivar Classifieation of *Punica granatum* L. Ph.D. Thesis, Nanjing Forestry University, Nanjing, China, June 2007.
19. Yuan, Z.; Fang, Y.; Zhang, T.; Fei, Z.; Han, F.; Liu, C.; Liu, M.; Xiao, W.; Zhang, W.; Wu, S.; et al. The pomegranate (*Punica granatum* L.) genome provides insights into fruit quality and ovule developmental biology. *Plant Biotechnol. J.* **2018**, *16*, 1363–1374. [CrossRef]
20. Qin, G.H.; Xu, C.Y.; Ming, R.; Tang, H.; Guyot, R.; Kramer, E.M.; Hu, Y.; Yi, X.; Qi, Y.; Xu, X.; et al. The pomegranate (*Punica granatum* L.) genome and the genomics of punicalagin biosynthesis. *Plant J.* **2017**, *91*, 1108–1128. [CrossRef]
21. Luo, X.; Li, H.; Wu, Z.; Yao, W.; Zhao, P.; Cao, D.; Yu, H.; Li, K.; Poudel, K.; Zhao, D.; et al. The pomegranate (*Punica granatum* L.) draft genome dissects genetic divergence between soft- and hard-seeded cultivars. *Plant Biotechnol. J.* **2020**, *18*, 955–968. [CrossRef]
22. Kumar, A.; Sharma, J.; Munjal, V.; Sakthivel, K.; Thalor, S.K.; Mondal, K.K.; Chinchure, S.; Gharate, R. Polyphasic phenotypic and genetic analysis reveals clonal nature of *Xanthomonas axonopodis* pv. punicae causing pomegranate bacterial blight. *Plant Pathol.* **2020**, *69*, 347–359. [CrossRef]
23. Yan, M.; Zhao, X.; Zhou, J.; Huo, Y.; Ding, Y.; Yuan, Z. The complete chloroplast genomes of *Punica granatum* and a comparison with other species in Lythraceae. *Int. J. Mol. Sci* **2019**, *20*, 2886. [CrossRef] [PubMed]
24. Doyle, J.J. A rapid DNA isolation procedure for small quantities of fresh leaf tissue. *Phytochem Bull.* **1987**, *19*, 11–15. [CrossRef]
25. Li, H.; Durbin, R. Fast and accurate short read alignment with Burrows-Wheeler transform. *Bioinformatics* **2009**, *25*, 1754–1760. [CrossRef]
26. Li, H.; Handsaker, B.; Wysoker, A.; Fennell, T.; Ruan, J.; Homer, N.; Marth, G.; Abecasis, G.; Durbin, R.; 1000 Genome Project Data Processing Subgroup. The sequence alignment/map format and SAMtools. *Bioinformatics* **2009**, *25*, 2078–2079. [CrossRef]
27. McKenna, A.; Hanna, M.; Banks, E.; Sivachenko, A.; Cibulskis, K.; Kernytsky, A.; Garimella, K.; Altshuler, D.; Gabriel, S.; Daly, M.; et al. The genome analysis toolkit: A mapreduce framework for analyzing next-generation DNA sequencing data. *Genome Res.* **2010**, *20*, 1297–1303. [CrossRef]

28. Chen, K.; Wallis, J.W.; McLellan, M.D.; Larson, D.E.; Kalicki, J.M.; Pohl, C.S.; McGrath, S.D.; Wendl, M.C.; Zhang, Q.; Locke, D.P.; et al. BreakDancer: An algorithm for high-resolution mapping of genomic structural variation. *Nat. Methods* **2009**, *6*, 677–681. [CrossRef]
29. Chalvin, C.; Drevensek, S.; Chollet, C.; Gilard, F.; Šolić, E.M.; Dron, M.; Bendahmane, A.; Boualem, A.; Cornille, A. Study of the genetic and phenotypic variation among wild and cultivated clary sages provides interesting avenues for breeding programs of a perfume, medicinal and aromatic plant. *PLoS ONE* **2021**, *16*, e0248954. [CrossRef]
30. Anderson, N.O.; Kávová, T.; Daa, B.; Urn, V.; Květ, J. Phenotypic and genotypic variation in Czech forage, ornamental and wild populations of reed *Canarygrass*. *Crop. Sci.* **2016**, *56*, 2421–2435. [CrossRef]
31. Saminathan, T.; Bodunrin, A.; Singh, N.V.; Devarajan, R.; Nimmakayala, P.; Jeff, M.; Aradhya, M.; Reddy, U.K. Genome-wide identification of microRNAs in pomegranate (*Punica granatum* L.) by high-throughput sequencing. *BMC Plant Biol.* **2016**, *16*, 12[illegible] [CrossRef] [PubMed]
32. Liu, T.J.; Li, Y.P.; Zhou, J.J.; Hu, C.G.; Zhang, J.Z. Genome-wide genetic variation and comparison of fruit-associated traits between kumquat (*Citrus japonica*) and Clementine mandarin (*Citrus clementina*). *Plant Mol. Biol.* **2018**, *96*, 493–507. [CrossRef]
33. Guo, D.D.; Yuan, F.J.; Yu, X.M. Genome-wide variation analysis of grain and vegetable soybeans based on re-sequencing. *Mol. Plant Breed.* **2019**, *17*, 7306–7312. [CrossRef]
34. Hideki, H.; Kenta, S.; Akio, O.; Hiroyuki, F.; Koh, A.; Christophe, R.; Shusei, S.; Sachiko, I.; Satoshi, T. Genome-Wide SNP genotyping to infer the effects on gene functions in tomato. *DNA Res.* **2013**, *20*, 221–233. [CrossRef]
35. Parida, S.K.; Mukerji, M.M.; Singh, A.K.; Singh, N.K.; Mohapatra, T. SNPs in stress-responsive rice genes: Validation, genotyping, functional relevance and population structure. *BMC Genom.* **2012**, *13*, 26–426. [CrossRef] [PubMed]
36. Zhang, S.; Chen, W.; Xin, L.; Gao, Z.; Hou, Y.; Yu, X.; Zhang, Z.; Qu, S. Genomic variants of genes associated with three horticultural traits in apple revealed by genome re-sequencing. *Hortic. Res.* **2014**, *1*, 14045. [CrossRef] [PubMed]
37. Bai, H.; Cao, Y.; Quan, J.; Dong, L.; Li, Z.; Zhu, Y.; Zhu, L.; Dong, Z.; Li, D. Identifying the genome-wide sequence variations and developing new molecular narkers for genetics research by re-sequencing a landrace cultivar of *Foxtail Millet*. *PLoS ONE* **2013**, *8*, e73514. [CrossRef]
38. Fanizza, G.; Colonna, G.; Resta, P.; Ferrara, G. The effect of the number of RAPD markers on the evaluation of genotypic distances in *Vitis vinifera*. *Euphytica* **1999**, *107*, 45–50. [CrossRef]
39. Robert, H.S.; Crhak, K.L.; Mroue, S.; Benková, E. The importance of localized auxin production for morphogenesis of reproductive organs and embryos in *Arabidopsis*. *J. Exp. Bot.* **2015**, *66*, 5029–5042. [CrossRef]
40. Wang, J.; Wang, H.; Ding, L.; Song, A.; Shen, F.; Jiang, J.; Chen, S.; Chen, F. Transcriptomic and hormone analyses reveal mechanisms underlying petal elongation in *Chrysanthemum morifolium* 'Jinba'. *Plant Mol. Biol.* **2017**, *93*, 593–606. [CrossRef]
41. Yan, S.; Che, G.; Ding, L.; Chen, Z.; Liu, X.; Wang, H.; Zhao, W.; Ning, K.; Zhao, J.; Tesfamichael, K.; et al. Different cucumber *CsYUC* genes regulate response to abiotic stresses and flower development. *Sci. Rep.* **2016**, *6*, 20760. [CrossRef]
42. Cheng, Y.; Dai, X.; Zhao, Y. Auxin biosynthesis by the *YUCCA* flavin monooxygenases controls the formation of floral organs and vascular tissues in *Arabidopsis*. *Genes Dev.* **2006**, *20*, 1790–1799. [CrossRef]
43. Zafar, S.A.; Patil, S.B.; Uzair, M.; Fang, J.; Zhao, J.; Guo, T.; Yuan, S.; Uzair, M.; Luo, Q.; Shi, J.; et al. *DEGENERATED PANICLE AND PARTIAL STERILITY 1* (*DPS1*) encodes a cystathionine β-synthase domain containing protein required for anther cuticle and panicle development in rice. *New Phytol.* **2020**, *225*, 356–375. [CrossRef] [PubMed]
44. Ma, B.; Zhang, L.; Gao, Q.; Wang, J.; Li, X.; Wang, H.; Liu, Y.; Lin, H.; Liu, J.; Wang, X.; et al. A plasma membrane transporter coordinates phosphate reallocation and grain filling in cereals. *Nat. Genet.* **2021**, *53*, 906–915. [CrossRef]
45. Xiao, X.; Zhang, J.; Satheesh, V.; Meng, F.; Gao, W.; Dong, J.; Zheng, Z.; An, G.Y.; Nussaume, L.; Liu, D.; et al. SHORT-ROOT stabilizes PHOSPHATE1 to regulate phosphate allocation in *Arabidopsis*. *Nat. Plants* **2022**, *8*, 1074–1081. [CrossRef]
46. Gómez-Gómez, L.; Bauer, Z.; Boller, T. Both the extracellular leucine-rich repeat domain and the kinase activity of *FLS2* are required for flagellin binding and signaling in *Arabidopsis*. *Plant Cell* **2001**, *13*, 1155–1163. [CrossRef]
47. Liu, H.; Huang, R.; Jing, M.; Sui, S.; Guo, Y.; Liu, D.; Li, Z.; Lin, Y.; Li, M. Two C3H type zinc finger protein genes, *CpCZF1* and *CpCZF2*, from *Chimonanthus praecox* affect stamen development in *Arabidopsis*. *Genes* **2017**, *8*, 199. [CrossRef]
48. Igarashi, K.; Kazama, T.; Toriyama, K. A gene encoding pentatricopeptide repeat protein partially restores fertility in RT98-Type cytoplasmic male-sterile rice. *Plant Cell Physiol.* **2016**, *57*, 2187. [CrossRef] [PubMed]
49. Bentolila, S.; Bentolila, S.; Alfonso, A.A.; Hanson, M.R. A pentatricopeptide repeat-containing gene restores fertility to cytoplasmic male-sterile plants. *Proc. Natl. Acad. Sci. USA* **2002**, *99*, 10887–10892. [CrossRef] [PubMed]
50. Mcintosh, K.B.; Bonham-Smith, P.C. The two ribosomal protein *L23A* genes are differentially transcribed in *Arabidopsis thaliana*. *Genome* **2005**, *48*, 443–454. [CrossRef] [PubMed]
51. Ito, T.; Takahashi, N.; Shimura, Y.; Okada, K. A serine/threonine protein kinase gene isolated by an in vivo binding procedure using the *Arabidopsis* floral homeotic gene product, AGAMOUS. *Plant Cell Physiol.* **1997**, *38*, 248–258. [CrossRef] [PubMed]
52. Jia, H.; Li, M.; Li, W.; Liu, L.; Jian, Y.; Yang, Z.; Shen, X.; Ning, Q.; Du, Y.; Zhao, R.; et al. A serine/threonine protein kinase encoding gene *KERNEL NUMBER PER ROW6* regulates maize grain yield. *Nat. Commun.* **2020**, *11*, 988. [CrossRef] [PubMed]
53. Marcotuli, I.; Mazzeo, A.; Colasuonno, P.; Terzano, R.; Nigro, D.; Porfido, C.; Tarantino, A.; Aiese Cigliano, R.; Sanseverino, W.; Gadaleta, A.; et al. Fruit development in *Ficus carica* L.: Morphological and genetic pproaches to fig buds for an evolution from monoecy toward dioecy. *Front. Plant Sci.* **2020**, *11*, 1208. [CrossRef]

54. Kozakov, D.; Hall, D.R.; Xia, B.; Porter, K.A.; Padhorny, D.; Yueh, C.; Beglov, D.; Vajda, S. The ClusPro web server for protein-protein docking. *Nat. Protoc.* **2017**, *12*, 255–278. [CrossRef] [PubMed]
55. Hemsley, P.A.; Kemp, A.C.; Grierson, C.S. The TIP GROWTH DEFECTIVE1 S-acyl transferase regulates plant cell growth in *Arabidopsis*. *Plant Cell.* **2005**, *17*, 2554–2563. [CrossRef] [PubMed]

horticulturae

MDPI

Article

Genome-Wide Identification and Characterization of the ANS Gene Family in Pomegranate (*Punica granatum* L.)

Huihui Ni, Heming Suo, Xuan Zhang, Lei Hu, Fangyu Yuan, Maowen Zhang and Shuiming Zhang *

Department of Ornamental Horticulture, School of Horticulture, Anhui Agricultural University, Hefei 230036, China
* Correspondence: zhangshuiming@ahau.edu.cn

Abstract: Anthocyanidin Synthase (ANS) is a key enzyme in the later stages of the anthocyanin biosynthetic pathway, and its role is to convert colorless leucoanthocyanidins to colored anthocyanidins. In this study, a total of 75 members of the pomegranate ANS family were identified and divided into four groups (Group I, Group II, Group III and Group IV) based on evolutionary relationships. The 75 ANS gene family members were unevenly distributed on seven of the eight chromosomes of pomegranate. The results of the physical and chemical property analysis showed that 93.33% of the proteins were acidic proteins, 6.67% were alkaline proteins, 28% of the proteins were stable proteins and 72% were unstable proteins. Protein secondary structure analysis showed that α-Spiral and irregular curl are the main structural elements. Analysis of the conserved structural domains of the proteins showed that all 75 ANS family members contained one DIOX -N subfamily structural domain and one 2OG-FeII_Oxy subfamily structural domain. The results of subcellular localization showed that all 75 ANS family members of pomegranate were localized in the cytoplasm. Analysis of the transcriptome data showed that the expression of the pomegranate ANS genes were variety-specific and period-specific.

Keywords: anthocyanin synthase; genetic identification; co-linearity analysis; bioinformatics; expression analysis

Citation: Ni, H.; Suo, H.; Zhang, X.; Hu, L.; Yuan, F.; Zhang, M.; Zhang, S. Genome-Wide Identification and Characterization of the ANS Gene Family in Pomegranate (*Punica granatum* L.). *Horticulturae* **2023**, *9*, 468. https://doi.org/10.3390/horticulturae9040468

Academic Editor: Daniele Bassi

Received: 3 March 2023
Revised: 21 March 2023
Accepted: 6 April 2023
Published: 7 April 2023

1. Introduction

Pomegranate (*Punica granatum* L.) belongs to the genus *Punica* in the family of *Pomegranate* and is an excellent fruit tree that combines ecological, economic and social benefits, ornamental value and health functions [1]. Pomegranates have a long history of cultivation and were one of the first known edible fruits [2]. The pomegranate originated in the Middle East [3] and gradually developed in China after Zhang Qian's mission to the west in the Han Dynasty. It has developed rapidly in China in recent years. Currently, Shandong Zaozhuang, Xinjiang Yecheng, Anhui Huaiyuan and Sichuan Huili are famous pomegranate cultivation areas in China [4]. Pomegranates are increasingly popular in the consumer market for their sweetness, high economic value, nutritional value, medicinal value and health functions [5]. The red color of pomegranate seeds is the result of the accumulation of anthocyanins, which have many physiological functions, not only for the plant itself to differentiate cells and prevent the occurrence of diseases, but also for human health, such as antioxidation and the prevention of cardiovascular diseases [6]. The results of Wang, D et al. [7] showed that pomegranate juice and peels have strong antioxidant effects, with pomegranate polyphenols and anthocyanins being the active substances that exert antioxidant effects. Therefore, it is important to clarify the mechanism of its regulation of pomegranate seed color.

The colors that appear in plants are determined by the presence of different pigment substances in the plant, which mainly include flavonoids, carotenoids, betaines and chlorophyll, among which anthocyanins are the most abundant type of flavonoid pigments [8].

Anthocyanins are secondary metabolites of plants and are water-soluble natural pigments of the flavonoid group. In their natural form, anthocyanins are extremely unstable and easily combine with sugar molecules to form anthocyanidins [9]. Anthocyanins are an important class of secondary metabolites in the flavonoid family, mainly found in the vesicles of epidermal cells; are responsible for the color development of leaves, flowers, fruits, seeds and other organs; giving flowers, fruits, seed coats and other organs of plants red, blue, purple and other colors; and are the main pigment substances for plant coloring [10,11]. They also play an important role in insect pollination, growth hormone transport, the protection of leaves from UV damage, pest and disease suppression and root tumor induction [12].

Anthocyanidin Synthase (ANS) is involved in a series of metabolic reactions and is a 2-ketoglutarate-dependent enzyme belonging to a family of glutamate-dependent oxygenases [13]. The main chain of ANS contains thirteen strands, of which eight form a jellyroll or double-stranded helix topology. The jellyroll forms a hydrophobic cavity, one end of which forms the active site [14]. Anthocyanidin Synthase is the key enzyme at the end of the anthocyanidin synthase pathway, and catalyzes the conversion of colorless leucoanthocyanidins to colored anthocyanidins. ANS belongs to the dioxygenase gene family in the flavonoid pathway and catalyzes a number of two-electron oxidations, such as hydroxylation, desaturations and oxidative ring closures [15]. Current studies have shown that ANS expression regulates the accumulation of anthocyanins and the color of fruit and flowers [16]. The ANS gene was first isolated from a mutant of maize by transposon tagging [17]. In addition, it has now been cloned in a variety of plants including *Arabidopsis thaliana* [14], *Litchi chinensis* [18], *Mangifera indica* [19], *Malus pumila* [20], *Camellia sinensis* [21] and *Vitis vinifera* [22]. Overexpression of ANS can increase anthocyanin accumulation, while reducing the expression of ANS significantly decreases the anthocyanin level in plants, and this leads to the production of white flowers [23,24].

Pomegranates are receiving more attention for their antioxidant and cardiovascular-disease-prevention properties, of which the main substance is anthocyanin. The ANS gene plays a crucial role in the synthesis of anthocyanins, and the identification and expression analysis of the ANS gene family in pomegranate has not been reported. In this study, members of the pomegranate ANS gene family were identified and characterized by bioinformatics tools, and their expression patterns in three different varieties of 'Hongyushizi', 'Baiyushizi' and 'Tunisia' and at different periods were analyzed to lay the foundation for studying the functions of the ANS gene family in pomegranate.

2. Materials and Methods

2.1. Plant Materials

The varieties of pomegranate tested were 'Hongyushizi', 'Baiyushizi' and 'Tunisia'. Samples were taken from pomegranate seeds at 40, 80, and 120 days after bloom, with three biological replicates for each variety and period. The seeds of pomegranates were quick-frozen in liquid nitrogen and stored at ultra-low temperatures at −80 °C. Samples were sent to Guangdong GENE DENOVO Company (Guangzhou, China) for transcriptome sequencing.

2.2. Identification of Members of the ANS Gene Family of Pomegranate

The whole genome sequence of pomegranate (ASM765513v2) was downloaded from NCBI (https://www.ncbi.nlm.nih.gov/, accessed on 20 October 2021). Searching on NCBI (https://www.ncbi.nlm.nih.gov/, accessed on 20 October 2021) yielded sequences of *Arabidopsis thaliana* containing the structural domain of Anthocyanidin Synthase (PF03171, PF14226). The obtained sequences were used as probes and homology searches were performed by local BLASTP (E-value less than 1×10^{-10}) against pomegranate's proteins in order to prevent the erroneous loss of sequences with low similarity to the probes but containing the structural domain of ANS [25]. We also used HMMER to validate our results. Then, we finally obtained all the members of the pomegranate ANS gene family.

2.3. Analysis of Physicochemical Properties and Prediction of Secondary Structure of Pomegranate ANS Proteins

The physicochemical properties, such as molecular weight and theoretical isoelectric point, of the pomegranate ANS proteins were predicted using the online tool Expasy (http://web.expasy.org/, accessed on 21 October 2021) [26].

Subcellular localization analysis of the pomegranate ANS family was conducted using the online tool Plant-mPLoc in Cell-Ploc 2.0 (http://Plant-mPLoc server (sjtu.edu.cn), accessed on 21 October 2021).

The online software SPOMA (https://npsa-prabi.ibcp.fr/cgi-bin/npsa_automat.pl?page=npsa_sopma.html, accessed on 22 October 2021) [27] was used to predict the secondary structure of the pomegranate ANS gene family proteins.

2.4. Construction of Phylogenetic Trees and Mapping of Gene Structures

The phylogenetic tree was constructed using MEGA X software for the pomegranate ANS gene family proteins according to the neighbor-joining (NJ) method, with the check parameter step (bootstrap) set to 1000 and all other parameters set to default values [28]. The gene structure was mapped by Gene Structure View in TBtools software (version number: v1.098685) [29].

2.5. Structural and Conserved Motif Domain Analysis of the Pomegranate ANS Protein

CD search (NCBI Conserved Domain Search (nih.gov), accessed on 23 October 2021) [30] and SMART were used for protein conserved domain analysis. The conserved motifs of the pomegranate ANS family of proteins were analyzed using the online tool MEME (http://meme-suite.org, accessed on 23 October 2021) [31], where the parameters of MEME were set to a maximum number of motifs of 10 and occurrences of a single motif of zero, or one per sequence [32].

2.6. Chromosome Positioning and Co-Linearity Analysis

Chromosomal localization of pomegranate ANS family genes based on pomegranate whole genome annotation information was performed using Gene Location Visualization from GTF/GFF in TBtools software [30]. MCScanX in TBtools software was used to identify gene duplication patterns and collinearity analysis. The Simple Ka/Ks Calculator in TBtools software was used for calculating the Ka/Ks values [33].

2.7. RNA Extraction, Library Construction and Sequencing

Total RNA was extracted using Trizol reagent kit (Invitrogen, Carlsbad, CA, USA). RNA quality was assessed on an Agilent 2100 Bioanalyzer (Agilent Technologies, Palo Alto, CA, USA) and checked using RNase free agarose gel electrophoresis. After total RNA was extracted, mRNA was enriched by Oligo (dT) beads. Then, the enriched mRNA was fragmented into short fragments using fragmentation buffer and reverse transcripted into cDNA with random primers [34]. Second-strand cDNA were synthesized by DNA polymerase I, RNase H, dNTP and buffer. Then, the cDNA fragments were purified with a QiaQuick PCR extraction kit (Qiagen, Venlo, The Netherlands), end repaired, poly(A) added and ligated to Illumina sequencing adapters. The ligation products were size selected by agarose gel electrophoresis, PCR amplified and sequenced using Illumina HiSeq2500 by Gene Denovo Biotechnology Co. (Guangzhou, China).

2.8. Raw Data Filtering, GO Enrichment and Transcriptome Expression Analysis

To obtain high quality clean reads, reads were further filtered by fastp [35] (version 0.18.0). The parameters were as follows: (1) removing reads containing adapters; (2) removing reads containing more than 10% of unknown nucleotides (N); (3) removing low quality reads containing more than 50% of low quality (Q-value $\leq$ 20) bases.

To analyze the expression characteristics of pomegranate ANS genes, Log2 based on the Fragments Per kb per Million reads (FPKM) value was used to create a heat map with the HeatMap tool in TBtools software.

3. Results

3.1. Identification, Physicochemical Characterization and Subcellular Localization of Pomegranate ANS Gene Family Members

A total of 75 candidate ANS genes from pomegranate were screened in this study, and the physicochemical properties of ANS proteins were analyzed by the online tool ExPASy. The results (Table 1) showed that the number of amino acids in the protein ranged from 289 to 442. The molecular weight of the pomegranate ANS family proteins ranged from 32,101.26 Da (*PgANS59*) to 49,422.13 Da (*PgANS62*), with an average molecular weight of 40,507.81 Da. The theoretical isoelectric point ranged from 4.98 (*PgANS58*) to 9.11 (*PgANS36*), with an average theoretical isoelectric point of 5.83. Of these, 93.33% were acidic proteins (theoretical PI < 7) and 6.67% were basic proteins (theoretical PI > 7), indicating that most of the pomegranate ANS exhibited acidic proteins. Hydrophilicity analysis showed that the pomegranate ANS proteins were all hydrophilic. Based on the instability factor, 28% of these were stable proteins (instability coefficient < 40) and 72% of these were unstable proteins (instability coefficient > 40), indicating that the majority of the pomegranate ANS were unstable proteins. The subcellular localization of all pomegranate ANS gene family proteins were localized in the cytoplasm, indicating that pomegranate ANS proteins were non-secretory and carried out metabolic activities within the cell.

Table 1. Information on the members of the pomegranate ANS gene family proteins.

Protein ID	Gene ID	Gene Name	Number of Amino Acids	Molecular Weight/(D)	Isoelectric Point	Total Average Hydrophilicity	Instability Factor	Subcellular Localization
XP_031403964.1	XM_031548104	*PgANS1*	356	40,324.31	5.84	−0.392	53.57	cytoplasm
XP_031393652.1	XM_031537792	*PgANS2*	336	38,173.35	5.50	−0.563	46.13	cytoplasm
XP_031384984.1	XM_031529124	*PgANS3*	336	38,357.03	5.79	−0.405	44.57	cytoplasm
XP_031390545.1	XM_031534685	*PgANS4*	334	37,853.40	5.75	−0.344	50.46	cytoplasm
XP_031376432.1	XM_031520572	*PgANS5*	357	40,044.51	5.66	−0.382	50.46	cytoplasm
XP_031377848.1	XM_031521988	*PgANS6*	356	38,836.67	5.62	−0.305	52.91	cytoplasm
XP_031393854.1	XM_031537994	*PgANS7*	356	40,019.21	5.37	−0.192	44.10	cytoplasm
XP_031393852.1	XM_031537992	*PgANS8*	356	40,047.29	5.46	−0.172	44.14	cytoplasm
XP_031376880.1	XM_031521020	*PgANS9*	362	41,430.45	5.17	−0.413	40.53	cytoplasm
XP_031393856.1	XM_031537996	*PgANS10*	356	40,055.22	5.14	−0.197	43.63	cytoplasm
XP_031393853.1	XM_031537993	*PgANS11*	384	43,440.10	6.03	−0.295	46.39	cytoplasm
XP_031393851.1	XM_031537991	*PgANS12*	390	43,645.10	5.77	−0.254	44.39	cytoplasm
XP_031395560.1	XM_031539700	*PgANS13*	356	40,391.35	6.12	−0.366	40.64	cytoplasm
XP_031380350.1	XM_031524490	*PgANS14*	358	40,565.39	5.29	−0.377	40.89	cytoplasm
XP_031382717.1	XM_031526857	*PgANS15*	347	39,631.90	5.26	−0.390	40.16	cytoplasm
XP_031386650.1	XM_031530790	*PgANS16*	366	41,504.68	5.75	−0.352	35.46	cytoplasm
XP_031378923.1	XM_031523063	*PgANS17*	359	40,562.30	5.07	−0.338	36.23	cytoplasm
XP_031382330.1	XM_031526470	*PgANS18*	339	38,929.06	5.13	−0.481	48.51	cytoplasm
XP_031394087.1	XM_031538227	*PgANS19*	362	41,139.64	5.30	−0.152	48.30	cytoplasm
XP_031378476.1	XM_031522616	*PgANS20*	359	39,580.33	6.09	−0.187	38.09	cytoplasm
XP_031395559.1	XM_031539699	*PgANS21*	376	42,663.06	6.77	−0.366	38.50	cytoplasm
XP_031391993.1	XM_031536133	*PgANS22*	338	38,634.90	5.78	−0.453	44.89	cytoplasm
XP_031393855.1	XM_031537995	*PgANS23*	361	40,534.63	5.07	−0.265	45.11	cytoplasm
XP_031401502.1	XM_031545642	*PgANS24*	366	41,173.19	5.26	−0.373	40.11	cytoplasm
XP_031391176.1	XM_031535316	*PgANS26*	350	38,763.98	5.39	−0.277	54.46	cytoplasm
XP_031378903.1	XM_031523043	*PgANS27*	353	39,091.58	5.80	−0.244	39.59	cytoplasm
XP_031388526.1	XM_031532666	*PgANS28*	367	41,013.40	5.28	−0.487	44.63	cytoplasm
XP_031390878.1	XM_031535018	*PgANS29*	369	41,918.31	5.38	−0.226	49.33	cytoplasm
XP_031388423.1	XM_031532563	*PgANS30*	358	40,381.98	5.62	−0.295	42.66	cytoplasm
XP_031372937.1	XM_031517077	*PgANS31*	388	43,418.39	6.28	−0.314	55.54	cytoplasm
XP_031388151.1	XM_031532291	*PgANS32*	372	41,873.01	5.42	−0.402	43.37	cytoplasm
XP_031383472.1	XM_031527612	*PgANS33*	380	43,218.42	5.38	−0.396	36.43	cytoplasm
XP_031389012.1	XM_03153315	*PgANS34*	350	38,489.70	5.22	−0.169	36.97	cytoplasm
XP_031391174.1	XM_031535314	*PgANS35*	353	39,037.42	5.34	−0.220	43.80	cytoplasm
XP_031385121.1	XM_031529261	*PgANS36*	367	40,709.85	9.11	−0.195	41.41	cytoplasm
XP_031383168.1	XM_031527308	*PgANS37*	358	41,381.08	6.75	−0.524	35.12	cytoplasm
XP_031383192.1	XM_031527332	*PgANS38*	354	39,158.46	5.62	−0.272	35.71	cytoplasm
XP_031376436.1	XM_031520576	*PgANS39*	370	41,879.97	5.48	−0.252	41.88	cytoplasm
XP_031379827.1	XM_031523967	*PgANS40*	358	41,371.03	5.91	−0.486	40.20	cytoplasm

Table 1. *Cont.*

Protein ID	Gene ID	Gene Name	Number of Amino Acids	Molecular Weight/(D)	Isoelectric Point	Total Average Hydrophilicity	Instability Factor	Subcellular Localization
XP_031383167.1	XM_031527307	*PgANS41*	358	41,503.05	5.81	−0.542	43.97	cytoplasm
XP_031395634.1	XM_031539774	*PgANS42*	373	41,126.14	6.35	−0.136	43.77	cytoplasm
XP_031394103.1	XM_031538243	*PgANS43*	320	36,348.78	5.18	−0.379	34.24	cytoplasm
XP_031391175.1	XM_031535315	*PgANS44*	296	32,917.57	5.87	−0.233	42.64	cytoplasm
XP_031385605.1	XM_031529745	*PgANS45*	358	40,276.11	5.26	−0.264	49.52	cytoplasm
XP_031398105.1	XM_031542245	*PgANS46*	380	42,899.82	5.08	−0.312	52.48	cytoplasm
XP_031399235.1	XM_031543375	*PgANS47*	365	41,639.61	5.27	−0.390	33.34	cytoplasm
XP_031380799.1	XM_031524939	*PgANS48*	346	39,478.70	5.42	−0.471	38.38	cytoplasm
XP_031397740.1	XM_031541880	*PgANS49*	336	40,882.18	5.33	−0.391	44.01	cytoplasm
XP_031390650.1	XM_031534790	*PgANS50*	319	36,283.50	7.06	−0.546	38.19	cytoplasm
XP_031384128.1	XM_031528268	*PgANS51*	319	35,943.26	5.34	−0.300	30.71	cytoplasm
XP_031390887.1	XM_031535027	*PgANS52*	393	44,373.41	6.11	−0.328	47.95	cytoplasm
XP_031383134.1	XM_031527274	*PgANS53*	378	43,069.06	6.82	−0.443	32.77	cytoplasm
XP_031390764.1	XM_031534904	*PgANS54*	395	44,855.05	5.43	−0.434	40.97	cytoplasm
XP_031399302.1	XM_031543442	*PgANS55*	434	48,533.05	6.38	−0.026	36.22	cytoplasm
XP_031407498.1	XM_031551638	*PgANS56*	356	39,585.56	6.71	−0.185	55.07	cytoplasm
XP_031389388.1	XM_031533528	*PgANS57*	394	44,472.40	6.50	−0.383	49.59	cytoplasm
XP_031380299.1	XM_031524439	*PgANS58*	369	42,082.85	4.98	−0.291	46.91	cytoplasm
XP_031407500.1	XM_031551640	*PgANS59*	289	32,101.26	9.08	−0.104	49.69	cytoplasm
XP_031374700.1	XM_031518840	*PgANS60*	354	39,399.37	5.76	−0.018	51.91	cytoplasm
XP_031378238.1	XM_031522378	*PgANS61*	393	44,385.28	6.15	−0.334	36.41	cytoplasm
XP_031399301.1	XM_031543441	*PgANS62*	442	49,422.13	6.56	−0.030	35.95	cytoplasm
XP_031373129.1	XM_031517269	*PgANS63*	379	42,826.43	5.05	−0.374	46.89	cytoplasm
XP_031374648.1	XM_031518788	*PgANS64*	354	39,488.26	6.05	−0.159	51.08	cytoplasm
XP_031388742.1	XM_031532882	*PgANS65*	304	34,270.08	6.11	−0.509	42.64	cytoplasm
XP_031383847.1	XM_031527987	*PgANS66*	371	41,943.06	8.93	−0.481	39.13	cytoplasm
XP_031373149.1	XM_031517289	*PgANS67*	369	42,122.84	5.01	−0.282	46.38	cytoplasm
XP_031390843.1	XM_031534983	*PgANS68*	429	48,221.93	8.02	−0.373	44.95	cytoplasm
XP_031377515.1	XM_031521655	*PgANS69*	372	42,320.19	5.47	−0.326	43.29	cytoplasm
XP_031397862.1	XM_031542002	*PgANS70*	376	42,150.93	5.41	−0.348	38.47	cytoplasm
XP_031377516.1	XM_031521656	*PgANS71*	372	42,394.42	5.58	−0.323	45.43	cytoplasm
XP_031377517.1	XM_031521657	*PgANS72*	309	35,244.09	5.51	−0.376	44.64	cytoplasm
XP_031400740.1	XM_031544880	*PgANS73*	303	33,706.66	5.28	−0.166	45.19	cytoplasm
XP_031400741.1	XM_031544881	*PgANS74*	301	33,484.22	5.06	−0.182	41.88	cytoplasm
XP_031395563.1	XM_031539703	*PgANS75*	317	36,302.44	5.44	−0.409	35.63	cytoplasm

3.2. Predicted Secondary Structure of the Pomegranate ANS Gene Family Proteins

Structural predictions (Table 2) showed that all members of the pomegranate ANS proteins were composed of α-helices, irregular coils, extended chains and β-turns. The lowest proportion of these was β-turns, all below 10%, with most around 6%. The majority of proteins had around 35% alpha-helices, with extended chains accounting for around 20% and the highest proportion of irregular coils at around 40%. From the results, it appeared that irregular coiling and α-helix were the main constituent forms of the secondary structure of the pomegranate ANS proteins, with β-turns and extended chains being the secondary constituent forms.

Table 2. Predicted secondary structure of the pomegranate ANS gene family proteins.

Gene Name	α-Helix (%)	Extended Strand (%)	β-Turn (%)	Random Coil (%)
PgANS1	33.15	17.42	5.06	44.38
PgANS2	34.23	19.05	6.85	39.88
PgANS3	34.82	19.35	5.36	40.48
PgANS4	33.23	17.96	5.39	43.41
PgANS5	32.21	18.21	7.84	41.74
PgANS6	37.36	17.98	5.06	39.61
PgANS7	38.20	16.29	5.90	39.61
PgANS8	37.92	16.57	6.46	39.04
PgANS9	37.85	16.02	5.52	40.61
PgANS10	39.89	15.73	5.90	38.48
PgANS11	36.20	15.89	6.77	41.15

Table 2. *Cont.*

Gene Name	α-Helix (%)	Extended Strand (%)	β-Turn (%)	Random Coil (%)
PgANS12	33.85	16.41	4.10	45.64
PgANS13	35.67	15.73	6.46	42.13
PgANS14	34.08	16.20	5.03	44.69
PgANS15	35.16	17.29	5.48	42.07
PgANS16	33.61	18.03	7.38	40.98
PgANS17	42.90	15.60	6.96	34.54
PgANS18	38.05	16.52	5.31	40.12
PgANS19	38.40	15.19	6.35	40.06
PgANS20	30.92	19.78	6.41	42.90
PgANS21	36.44	16.49	6.12	40.96
PgANS22	34.91	17.75	6.21	41.12
PgANS23	36.57	16.07	5.26	42.11
PgANS24	37.98	18.58	5.74	37.70
PgANS25	34.39	15.61	6.35	43.65
PgANS26	35.43	16.57	6.29	41.71
PgANS27	34.28	17.56	5.38	42.78
PgANS28	35.97	14.99	5.45	43.60
PgANS29	38.21	15.18	7.05	39.57
PgANS30	38.55	17.04	5.87	38.55
PgANS31	34.54	16.49	5.93	43.04
PgANS32	36.83	17.47	5.65	40.05
PgANS33	40.53	16.32	7.11	36.05
PgANS34	35.14	18.00	6.00	40.86
PgANS35	34.56	19.55	5.10	40.79
PgANS36	32.43	16.35	6.27	44.96
PgANS37	33.52	17.60	5.59	43.30
PgANS38	35.88	18.08	5.08	40.96
PgANS39	37.84	16.49	5.95	39.73
PgANS40	30.73	19.27	4.47	45.53
PgANS41	31.01	17.60	5.59	45.81
PgANS42	30.29	16.62	4.83	48.26
PgANS43	40.62	17.19	6.88	35.31
PgANS44	37.84	17.57	6.76	37.84
PgANS45	36.31	19.27	6.98	37.43
PgANS46	40.53	16.32	6.05	37.11
PgANS47	37.81	16.99	5.48	39.73
PgANS48	35.26	16.76	4.62	43.35
PgANS49	39.34	16.39	6.28	37.98
PgANS50	40.44	17.87	7.21	34.48
PgANS51	43.26	17.55	7.52	31.66
PgANS52	33.59	14.76	5.34	46.31
PgANS53	34.39	18.25	6.61	40.74
PgANS54	32.41	17.47	4.30	45.82
PgANS55	38.94	20.97	5.99	34.10
PgANS56	34.55	14.51	5.06	45.79
PgANS57	31.98	18.53	5.58	43.91
PgANS58	36.04	17.62	5.69	40.65
PgANS59	38.75	18.69	5.88	36.68
PgANS60	36.16	14.41	5.65	43.79
PgANS61	36.64	16.28	6.36	40.71
PgANS62	38.69	20.81	4.75	35.75
PgANS63	37.73	17.68	5.54	39.05
PgANS64	37.29	15.25	5.93	41.53
PgANS65	37.83	17.76	7.57	36.84
PgANS66	38.27	17.25	6.20	38.27
PgANS67	37.40	18.97	5.42	38.21
PgANS68	35.43	16.32	5.13	43.12
PgANS69	38.98	16.40	6.18	38.44

Table 2. *Cont.*

Gene Name	α-Helix (%)	Extended Strand (%)	β-Turn (%)	Random Coil (%)
PgANS70	38.56	16.22	5.59	39.63
PgANS71	39.52	16.94	5.91	37.63
PgANS72	36.25	18.12	4.85	40.78
PgANS73	36.30	15.51	6.60	41.58
PgANS74	37.21	17.61	6.31	38.87
PgANS75	36.59	17.03	5.68	40.69

3.3. Phylogenetic and Genetic Structure Analysis of the Pomegranate ANS Family

To investigate the phylogenetic relationships of the pomegranate ANS proteins, a phylogenetic tree was constructed. The results (Figure 1a) show that the 75 ANS family members of pomegranate can be divided into four groups according to their distance of kinship, named Group I, Group II, Group III and Group IV, each containing 43, 13, 16 and 3 members, respectively. According to the analysis, the 75 pomegranate ANS proteins formed 29 paralogous proteins pairs. For example, *PgANS7* and *PgANS11*, *PgANS8* and *PgANS12*, *PgANS23* and *PgANS38*, *PgANS19* and *PgANS29* and *PgANS13* and *PgANS21* were paralogous proteins pairs. Of these, 21 pairs had 100% support from the 1000 bootstrapping test, except for 8 paralogous proteins pairs, including *PgANS8* and *PgANS12*, *PgANS23* and *PgANS39*, *PgANS17* 140 and *PgANS45*, *PgANS1* and *PgANS4*, *PgANS34* and *PgANS38*, *PgANS71* and *PgANS72*, *PgANS40* and *PgANS41*, *PgANS53* and *PgANS54*. This illustrates the robustness of the constructed phylogenetic tree, the similarity in protein sequences of the 21 pairs of pomegranate ANS proteins and their close affinity.

Structural mapping of the pomegranate ANS gene was carried out using Gene Structure View in TBtools software. The results (Figure 1b) showed that each member of the pomegranate ANS gene family had two–five CDSs (8 members had two CDSs, 29 members had three CDSs, 36 members had four CDSs and 2 members had five CDSs), with little variation in the number of CDSs between members and a relatively simple gene structure.

3.4. Analysis of Protein Structures and Conserved Motifs of Members of the Pomegranate ANS Family

Information on the location of the conserved structural domains of the pomegranate ANS proteins were analyzed using CD search and SMART online software and mapped using the "My Domains" function of the online software ProSite, in conjunction with a phylogenetic tree. The results (Figure 2a) showed that 75 pomegranate ANS protein sequences contained one DIOX -N [36] (non-haem dioxygenase in morphine synthesis N-terminal) subfamily structural domain and one 2OG-FeII_Oxy [37] (2-oxoglutarate Fe(II) oxygenase, 2-ketoglutarate-Fe^{2+} oxidase) subfamily with a characteristic polypeptide sequence of the Anthocyanidin Synthase family. The ANS enzyme oxidized colorless leucoanthocyanidins to colored anthocyanidins via Fe^{2+} and (2-oxoglutarate) ions and belonged to the family of dioxygenases.

Analysis of the conserved motifs of the pomegranate ANS proteins was conducted using the online tool MEME, which yielded 10 potentially conserved motifs (Figure 2b) with different motifs indicated by different colored boxes. The 10 obtained motifs were named as Motif 1–Motif 10. All members of the pomegranate ANS family contained Motif 1, Motif 3, Motif 4 and Motif 6 conserved motifs, and all protein motifs were highly conserved in their order of arrangement. Distribution of motifs in Group I: (1) all had Motif 8, excepting *PgANS44*; (2) all had Motif 3, Motif 6, Motif 4, Motif 2 and Motif 5 (3) Motif 7, except for *PgANS6*, *PgANS43*, *PgANS51*, *PgANS65* and *PgANS66*; (4) all had Motif 9, excepting *PgANS24*, *PgANS32*, *PgANS50*, *PgANS43*, *PgANS65* and *PgANS66* (5) all had Motif 10, excepting *PgANS29*, *PgANS24*, *PgANS32*, *PgANS43*, *PgANS65* and *PgANS66*. The distribution of motifs in Group II: (1) all had Motif 8, Motif 3, Motif 7 Motif 9, Motif 6, Motif 1 and Motif 4; (2) all had Motif 2, excepting *PgANS71* and *PgANS72* (3) all had Motif 5, excepting *PgANS49* and *PgANS72*; (4) all had Motif 10, excepting

PgANS55, *PgANS62* and *PgANS72*. Distribution of motifs in Group III: (1) all had Motif 8, excepting *PgANS59*; (2) all had Motif 3, Motif 9, Motif 6, Motif 1, Motif 4, Motif 2 and Motif 5; (3) all had Motif 7, excepting *PgANS57*; (4) all had Motif 10, excepting *PgANS52*, *PgANS68*, *PgANS53* and *PgANS54*. Distribution of motifs in Group IV: (1) all had Motif 8, excepting *PgANS75*; (2) all members of Group IV contained Motif 3, Motif 7, Motif 9, Motif 6, Motif 1, Motif 4, Motif 2, Motif 5 and Motif 10.

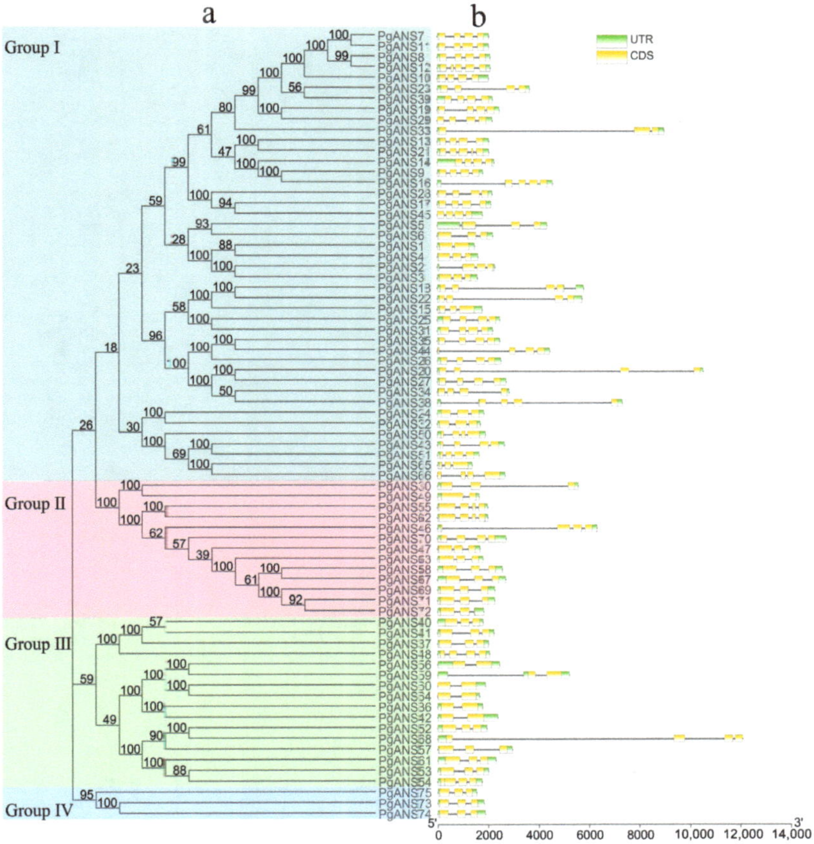

Figure 1. (**a**) Phylogenetic tree of the pomegranate ANS proteins and (**b**) genetic structure of the pomegranate ANS genes. UTR represents untranslated region, while CDS represents coding sequence.

3.5. Chromosome Positioning

Analysis of the chromosomal positioning of the 75 pomegranate ANS family genes based on the gene location file showed (Figure 3) that the 75 ANS genes were unevenly distributed on seven of the eight pomegranate chromosomes, Chr1, Chr2, Chr3, Chr4, Chr5, Chr6 and Chr8, with no genes being positioned on Chr7. The largest number of pomegranate ANS genes were distributed on Chr4, with 19 gene members; Chr2 had 15 gene members; Chr1 had 14 gene members; Chr5 had 10 gene members; Chr3 had 9 gene members; Chr8 had 5 gene members; and Chr6 had the lowest number of ANS genes distributed on it, with only 3 gene members.

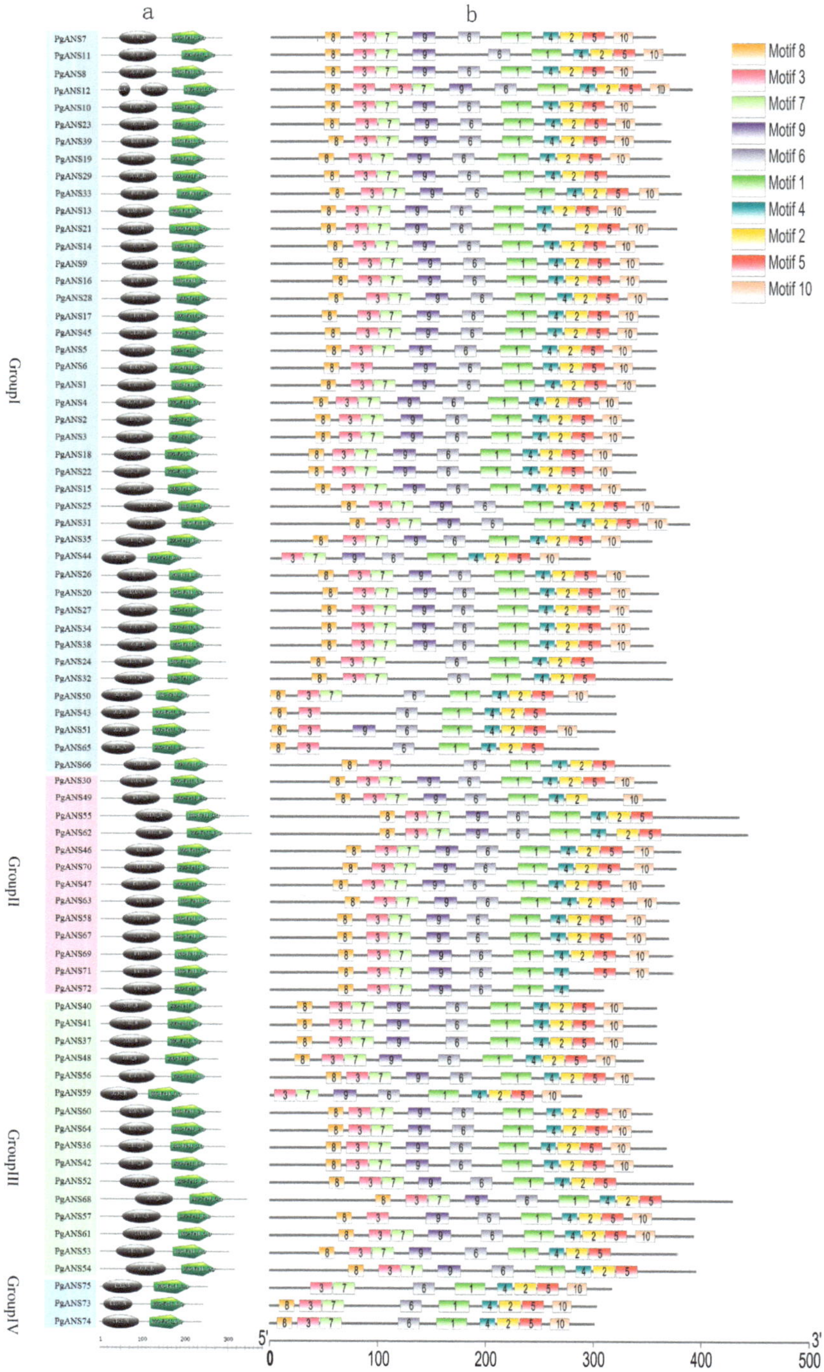

Figure 2. Conserved structural domains of the pomegranate ANS proteins (**a**) and conserved motifs of the pomegranate ANS proteins (**b**).

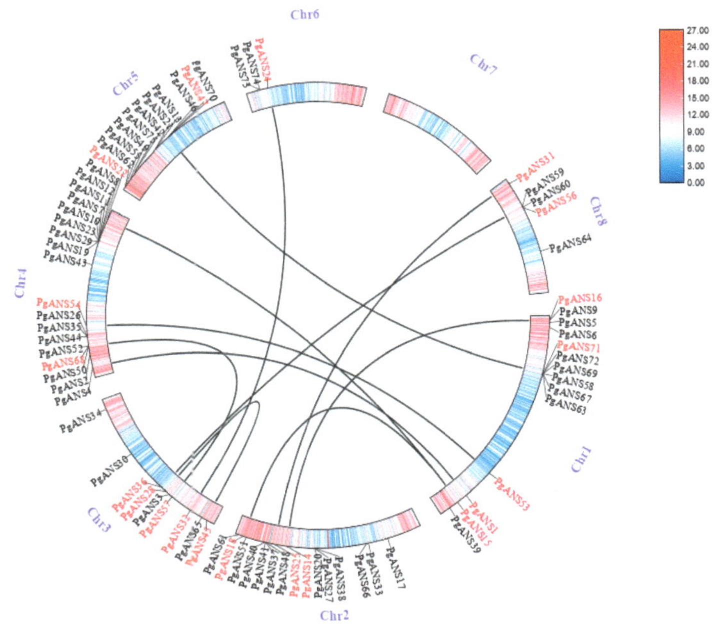

Figure 3. Location of the pomegranate ANS genes on the chromosome and gene duplication events. The red-colored names were the presence of co-linear genes.

Further analysis revealed tandem duplication between *PgANS45* and *PgANS28*, as well as fragment duplication. Ten pairs of fragment duplication genes were identified, which were *PgANS16* and *PgANS14*, *PgANS15* and *PgANS18*, *PgANS1* and *PgANS4*, *PgANS53* and *PgANS54*, *PgANS15* and *PgANS22*, *PgANS71* and *PgANS47*, *PgANS25* and *PgANS31*, *PgANS57* and *PgANS68*, *PgANS32* and *PgANS24* and *PgANS36* and *PgANS56*, suggesting possible functional similarities between some gene members of the pomegranate ANS family.

In addition, we calculated Ka/Ks ratios for 11 pomegranate ANS gene pairs to assess the selective pressure between duplicated pomegranate ANS genes. The results showed that the Ka values of the 11 pomegranate ANS gene pairs ranged from 0.10 to 0.51, and the Ks values ranged from 1.20 to 4.38. Additionally, all the values of Ka/Ks were less than one (Table 3).

Table 3. Estimated divergence period of the *PgANS* gene pairs. Ks, synonymous substitution rate; Ka, non-synonymous substitution rate.

Gene Pairs	Ka	Ks	Ka/Ks
PgANS1–PgANS4	0.48	1.33	0.36
PgANS45–PgANS28	0.51	1.73	0.29
PgANS16–PgANS14	0.47	4.38	0.11
PgANS15–PgANS18	0.36	3.87	0.09
PgANS53–PgANS54	0.20	2.90	0.07
PgANS15–PgANS22	0.33	1.70	0.19
PgANS71–PgANS47	0.38	2.18	0.17
PgANS25–PgANS31	0.28	3.12	0.09
PgANS57–PgANS68	0.38	2.40	0.16
PgANS32–PgANS24	0.10	1.20	0.08
PgANS36–PgANS56	0.38	1.66	0.23

3.6. Summary of RNA Sequencing Data

The output of RNA-seq date is shown in Table 4. After filtering low-quality, adapter polluted and high content of unknown base (N) reads, we acquired a total of 1,200,021,858 clean reads (from 35,721,370 to 53,391,468 for each sample) and 178,302,270,781 clean bases (from 5,322,204,950 to 7,882,028,946 for each sample) in 27 libraries. An average quality value of Q20 (sequencing base quality score > 20) was 96.89% for each library.

Table 4. Summary of RNA Sequencing Data, Clean Reads Q20—Sequencing base quality score > 20 and GC Content–GC ratio of sequence bases before filtering.

Sample	Raw Reads	Raw Bases (bp)	Clean Reads	Clean Bases (bp)	Clean Reads Q20 (%)	Clean Reads Ratio (%)	GC Content (%)
Baiyushizi-T1(1)	48,667,638	7,300,145,700	48,379,952	7,219,608,238	97.14%	99.41%	50.05%
Baiyushizi-T1(2)	47,486,554	7,122,983,100	47,186,944	7,050,368,799	96.74%	99.37%	49.77%
Baiyushizi-T1(3)	47,593,812	7,139,071,800	47,305,900	7,039,879,175	96.95%	99.40%	49.60%
Baiyushizi-T2(1)	36,016,766	5,402,514,900	35,721,370	5,322,204,950	96.42%	99.18%	50.36%
Baiyushizi-T2(2)	41,189,404	6,178,410,600	40,865,830	6,095,053,040	96.56%	99.21%	50.53%
Baiyushizi-T2(3)	39,571,150	5,935,672,500	39,300,102	5,862,500,285	97.00%	99.32%	50.59%
Baiyushizi-T3(1)	40,037,072	6,005,560,800	39,938,146	5,907,058,376	96.64%	99.75%	50.42%
Baiyushizi-T3(2)	42,861,892	6,429,283,800	42,742,380	6,337,181,505	96.83%	99.72%	50.31%
Baiyushizi-T3(3)	45,807,760	6,871,164,000	45,679,274	6,745,209,770	96.87%	99.72%	50.40%
Hongyushizi-T1(1)	48,234,574	7,235,186,100	47,911,350	7,148,996,855	96.80%	99.33%	49.68%
Hongyushizi-T1(2)	39,476,844	5,921,526,600	39,184,588	5,847,426,009	96.58%	99.26%	49.58%
Hongyushizi-T1(3)	41,389,884	6,208,482,600	41,102,138	6,137,884,644	96.64%	99.30%	49.66%
Hongyushizi-T2(1)	48,913,026	7,336,953,900	48,584,658	7,251,639,210	96.76%	99.33%	50.48%
Hongyushizi-T2(2)	38,788,642	5,818,296,300	38,486,532	5,744,610,574	96.52%	99.22%	50.08%
Hongyushizi-T2(3)	48,650,276	7,297,541,400	48,296,112	7,208,671,024	96.70%	99.27%	50.36%
Hongyushizi-T3(1)	43,736,804	6,560,520,600	43,660,662	6,324,509,067	97.56%	99.83%	49.25%
Hongyushizi-T3(2)	42,770,708	6,415,606,200	42,642,452	6,357,074,521	97.33%	99.70%	50.57%
Hongyushizi-T3(3)	43,698,610	6,554,791,500	43,576,138	6,466,042,710	97.44%	99.72%	50.76%
Tunisia-T1(1)	38,816,494	5,822,474,100	38,516,272	5,739,905,920	96.28%	99.23%	49.61%
Tunisia-T1(2)	47,502,710	7,125,406,500	47,163,990	7,039,556,375	96.39%	99.29%	49.70%
Tunisia-T1(3)	51,473,316	7,720,997,400	51,172,936	7,638,003,169	96.98%	99.42%	49.64%
Tunisia-T2(1)	48,979,242	7,346,886,300	48,609,708	7,242,609,081	96.62%	99.25%	50.15%
Tunisia-T2(2)	49,197,628	7,379,644,200	48,851,404	7,265,839,885	96.79%	99.30%	50.18%
Tunisia-T2(3)	49,308,242	7,396,236,300	48,974,856	7,303,980,121	96.86%	99.32%	50.22%
Tunisia-T3(1)	40,240,650	6,036,097,500	40,170,000	5,888,045,002	97.52%	99.82%	50.17%
Tunisia-T3(2)	53,530,512	8,029,576,800	53,391,468	7,882,028,946	97.56%	99.74%	49.20%
Tunisia-T3(3)	42,669,448	6,400,417,200	42,606,696	6,236,383,530	97.56%	99.85%	49.59%

3.7. Expression Analysis of the Pomegranate ANS Gene Family

The expression characteristics of pomegranate ANS genes were analyzed in three different varieties (Honhyushi, Baiyushizishi and Tunisia) 40, 80 and 120 days after the blooming period (Figure 4). Out of 75 aforementioned pomegranate ANS genes, we only detected the expression of 64 genes in the transcriptome datasets, while *PgANS5*, *PgANS7*, *PgANS8*, *PgANS13*, *PgANS18*, *PgANS38*, *PgANS44*, *PgANS55*, *PgANS59*, *PgANS69* and *PgANS72* were not detected in any samples, possibly due to special expression patterns that cannot be examined in our libraries. The 21 genes (*PgANS27*, *PgANS73*, *PgANS36*, *PgANS2*, *PgANS50*, *PgANS31*, *PgANS64*, *PgANS21*, *PgANS25*, *PgANS56*, *PgANS23*, *PgANS10*, *PgANS43*, *PgANS60*, *PgANS46*, *PgANS54*, *PgANS33*, *PgANS34*, *PgANS35*, *PgANS42* and *PgANS62*) had a similar expression pattern in the three different varieties; that is, the expression level of the genes tended to increase and then decrease as the growth period progressed, with the highest expression level at 80 days after the blooming period (T2). This suggested that most *PgANS* genes had a period-specific expression pattern.

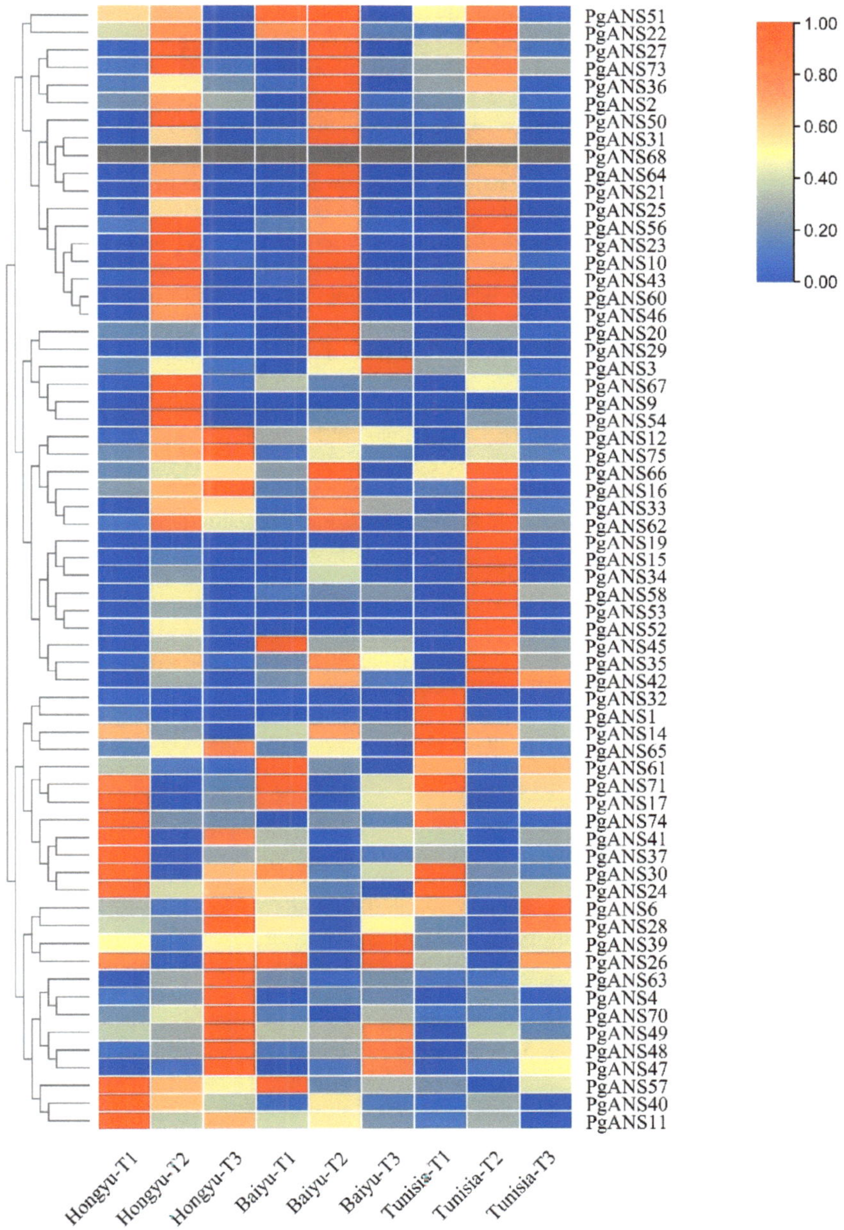

Figure 4. Heat map of the expression of pomegranate ANS genes in 3 species at different growth periods. T1—40 days after blooming period, T2—80 days after blooming period, T3—120 days after blooming period, 'Hongyu'—Hongyushizishizi, 'Baiyu'—Baiyushizi.

Analysis of the expression of ANS genes in the three varieties at T1 (40 days after the blooming period) showed that *PgANS17*, *PgANS74*, *PgANS41*, *PgANS37*, *PgANS57*, *PgANS40* and *PgANS11* were expressed at the highest levels in the red variety, 'Hongyushizi'. *PgANS51*, *PgANS22*, *PgANS45*, *PgANS61* and *PgANS26* were expressed at the highest levels in 'Baiyushizi'. Additionally, *PgANS32*, *PgANS1*, *PgANS14* and *PgANS65* had the highest expression levels in Tunisia.

Analysis of the expression of ANS genes in the three varieties at T2 (80 days after the blooming period) showed that *PgANS23, PgANS67, PgANS9, PgANS54, PgANS57* and *PgANS40* were expressed at the highest levels in 'Hongyushizi'. *PgANS51, PgANS36, PgANS2, PgANS31, PgANS64, PgANS21, PgANS20* and *PgANS29* were expressed at the highest levels in 'Baiyushizi'. Additionally, *PgANS42* and *PgANS71* were expressed at the highest levels in Tunisia.

Analysis of the expression of ANS genes in the three varieties at T3 (120 days after the blooming period) showed that *PgANS12, PgANS75, PgANS16, PgANS65, PgANS41, PgANS30, PgANS24, PgANS6, PgANS28, PgANS26, PgANS63, PgANS4, PgANS70, PgANS49, PgANS48, PgANS47, PgANS66, PgANS33* and *PgANS62* were expressed at the highest levels in the red variety, 'Hongyushizi'. Additionally, *PgANS42* was expressed at the highest levels in Tunisia. However, *PgANS3* and *PgAMS39* had the highest expression levels in the white variety, 'Baiyushizi'.

4. Discussion

The metabolism and regulation of plant anthocyanins have become a hot spot in scientific research in recent years. Anthocyanidin Synthase (ANS), a key enzyme in the later stages of the anthocyanidin synthesis pathway, has received widespread attention. The ANS gene has been cloned from most plants and has been shown to be an important structural gene for anthocyanin biosynthesis [38]. Anthocyanins readily combine with glycosides to form anthocyanidins, which, for the plant itself, not only give the plant a bright color and thus attract pollinators and foragers, facilitating pollination and seed dispersal [39], but also play a protective role in plant growth and development. In humans, anthocyanins have antioxidant, anti-aging and anti-vascular sclerosis effects [40,41]. ANS catalyzes the conversion of colorless anthocyanins to colored anthocyanins, which affects the accumulation of anthocyanins and determines the formation of flower and fruit coloration [42]. ANS is a family of iron- and 2-O-ketoglutarate-dependent dioxygenases, and the binding site for iron ions and 2-O-ketoglutarate in the conserved structural domain is the active central structure present in the cytochrome P450 family of genes [43]. In this study, a total of 75 members of the pomegranate ANS genes were screened by sequence alignment to predict and analyze the composition, structural characteristics and physicochemical properties of the proteins they encode. The results showed that members of the pomegranate ANS proteins contain conserved structural domains of the 2-ketoglutarate-Fe^{2+}-dioxygenase family, typical of the plant dioxygenase family of genes.

We performed covariance analysis of the members of the pomegranate ANS gene family and identified 11 co-linear gene pairs. Further analysis of these 11 pairs of genes showed that all gene pairs had Ka/Ks values less than one, indicating that they underwent strong purifying selection and played a key role in the evolution of the ANS genes [44].

The results of GO enrichment analysis showed that 64 genes were involved in metabolic processes, single organism processes and catalytic activities, and 3 genes were involved in the cell and cell part, respectively. It is conjectured that these genes may be involved in the growth and development of pomegranate fruits.

In this study, we examined the transcriptome data of 'Hongyushizi', 'Baiyushizi' and 'Tunisia' varieties at different fruit ripening periods using transcriptome sequencing technology. Studies have shown that the expression of the ANS gene was species-specific, with expression levels of dark > light > white/no color varieties; for example, in Saussurea medusa, ANS expression was higher in the red line than in the white and green lines of the healing tissue [45]. The expression level of purple kale (*Brassica oleracea var. Capitata Linnaeus*) ANS was significantly higher than that of green kale [46]. The expression of ANS was significantly higher in the pink petal line of peach (*Prunus persica*) than in the white petal line [47]. In this study, *PgANS12, PgANS75, PgANS16, PgANS65, PgANS41, PgANS30, PgANS24, PgANS6, PgANS28, PgANS26, PgANS63, PgANS4, PgANS70, PgANS49, PgANS48, PgANS47, PgANS66, PgANS33* and *PgANS62* were expressed at the highest levels in the red variety, 'Hongyushizi', confirming that the ANS gene was expressed at a higher level in the darker varieties than in the white/stainless varieties. We speculated that these genes played an important role in the formation of pomegranate seed color. However, *PgANS39* and *PgANS3* were expressed at the highest levels in 'Baiyushizi' during the T3 period. We speculated that the expression of these genes may have inhibited the accumulation of anthocyanins, thereby giving pomegranate seeds a white color.

A comparison of the expression pattern of the ANS genes in the same variety at different periods revealed that there was period of specificity in the expression pattern of the ANS genes in pomegranate. The 21 genes (*PgANS27*, *PgANS73*, *PgANS21*, et al.) showed a trend of increasing and then decreasing expression as the growth period progressed. The highest expression levels were found at 80 days after the bloom period, which is consistent with the study by Wang Chunhui et al. [48] on structural genes related to ANS in the fruit of the 'Hongyang' kiwi fruit mutant. It suggested that the regulatory mechanisms of the relevant structural genes differ in the synthesis of anthocyanosises and play different roles in the anthocyanidin synthase pathway. Further experiments are needed to verify the effect of *PgANS* expression on the color of pomegranate seeds.

5. Conclusions

In this study, a genome-wide analysis of the phylogenetic relationships, intron/exon structures, motif composition and expression characteristics of *PgANS* genes was performed. A total of 75 ANS members from pomegranate were identified and divided into four groups. The results of gene expression analysis indicated that the *PgANS* genes have both a variety-specific and a period-specific expression pattern. The results of this study further elucidated the expression pattern of pomegranate ANS genes in pomegranate seeds, and ultimately lay a certain theoretical foundation for the breeding of new anthocyanin-rich pomegranate varieties.

Author Contributions: Methodology, X.Z.; software, L.H. and M.Z.; validation, H.N. and H.S.; formal analysis, H.N.; data curation, H.N. and F.Y.; writing—original draft preparation, H.N. and H.S.; writing—review and editing, S.Z.; visualization, X.Z. and H.N.; supervision, S.Z.; funding acquisition, S.Z.; H.S was the co-first author of this article. All authors have read and agreed to the published version of the manuscript.

Funding: This research and the APC were funded by the Anhui Provincial Natural Science Foundation, grant number 2008085MC100, and the Anhui Provincial Natural Science Research Project Fund, grant number KJ2019ZD19.

Data Availability Statement: The original transcriptome data used in this study have been submitted to the NCBI, and the data are stored in the SRA database. the accession number is "PRJNA914887".

Conflicts of Interest: The authors declare no conflict of interest.

References

1. Yuan, Z.-H.; Yin, Y.-L.; Zhu, L.-Q.; Li, Y.; Hou, L.-F. Research progress of health functions of pomegranate. *Shandong For. Sci. Technol.* **2008**, *01*, 91–93, 59.
2. Viuda-Martos, M.; Fernández-López, J.; Pérez-Álvarez, J.A. Pomegranate and its many functional components as related to human health: A review. *Compr. Rev. Food Sci. Food Saf.* **2010**, *9*, 635–654. [CrossRef] [PubMed]
3. Julie, J. Therapeutic applications of pomegranate (*Punica granatum* L.): A review. *Altern. Med. Rev.* **2008**, *13*, 128–144.
4. Wang, X.-F. Classification of Pomegranate Varieties. Ph.D. Thesis, Nanjing Forestry University, Nanjing, China, 2007.
5. Schubert, S.Y.; Lansky, E.P.; Neeman, I. Antioxidant and eicosanoid enzyme inhibition properties of pomegranate seed oil and fermented juice flavonoids. *J. Ethnopharmacol.* **1999**, *66*, 11–17. [CrossRef]
6. Kähkönen, M.P.; Heinonen, M. Antioxidant activity of anthocyanins and their aglycons. *J. Agric. Food Chem.* **2003**, *51*, 628–633. [CrossRef]
7. Wang, D.; Özen, C.; Abu-Reidah, I.M.; Chigurupati, S.; Patra, J.K.; Horbanczuk, J.O.; Jóźwik, A.; Tzvetkov, N.T.; Uhrin, P.; Atanasov, A.G. Vasculoprotective effects of pomegranate (*Punica granatum* L.). *Front. Pharmacol.* **2018**, *9*, 544. [CrossRef]
8. Castañeda-Ovando, A.; de Lourdes Pacheco-Hernández, M.; Páez-Hernández, M.E.; Rodríguez, J.A.; Galán-Vidal, C.A. Chemical studies of anthocyanins: A review. *Food Chem.* **2009**, *113*, 859–871. [CrossRef]
9. Alvarez-Suarez, J.M.; Cuadrado, C.; Redondo, I.B.; Giampieri, F.; González-Paramás, A.M.; Santos-Buelga, C. Novel approaches in anthocyanin research—Plant fortification and bioavailability issues. *Trends Food Sci. Technol.* **2021**, *117*, 92–105. [CrossRef]

10. Hashimoto, M.; Suzuki, T.; Iwashina, T. New acylated anthocyanins and other flavonoids from the red flowers of Clematis cultivars. *Nat. Prod. Commun.* **2011**, *6*, 1631–1636. [CrossRef]
11. Behrens, C.E.; Smith, K.E.; Iancu, C.V.; Choe, J.-Y.; Dean, J.V. Transport of Anthocyanins and Other Flavonoids by the Arabidopsis ATP-Binding Cassette Transporter AtABCC2. *Sci. Rep.* **2019**, *9*, 437. [CrossRef]
12. Qi, Q.; Chu, M.; Yu, X.; Xie, Y.; Li, Y.; Du, Y.; Liu, X.; Zhang, Z.; Shi, J.; Yan, N. Anthocyanins and Proanthocyanidins: Chemical Structures, Food Sources, Bioactivities, and Product Development. *Food Rev. Int.* **2022**, 1–29. [CrossRef]
13. Schofield, C.J.; Zhang, Z. Structural and mechanistic studies on 2-oxoglutarate-dependent oxygenases and related enzymes. *Curr. Opin. Struct. Biol.* **1999**, *9*, 722–731. [CrossRef]
14. Wilmouth, R.C.; Turnbull, J.J.; Welford, R.W.; Clifton, I.J.; Prescott, A.G.; Schofield, C.J. Structure and mechanism of anthocyanidin synthase from *Arabidopsis thaliana*. *Structure* **2002**, *10*, 93–103. [CrossRef]
15. Saito, K.; Kobayashi, M.; Gong, Z.; Tanaka, Y.; Yamazaki, M. Direct evidence for anthocyanidin synthase as a 2-oxoglutarate-dependent oxygenase: Molecular cloning and functional expression of cDNA from a red forma of Perilla frutescens. *Plant J.* **1999**, *17*, 181–189. [CrossRef] [PubMed]
16. Gan, S.; Zheng, G.; Zhu, S.; Qian, J.; Liang, L. Integrative Analysis of Metabolome and Transcriptome Reveals the Mechanism of Color Formation in Liriope spicata Fruit. *Metabolites* **2022**, *12*, 144. [CrossRef] [PubMed]
17. Dellaporta, S.L.; Greenblatt, I.; Kermicle, J.L.; Hicks, J.B.; Wessler, S.R. Molecular cloning of the maize R-nj allele by transposon tagging with Ac. In *Chromosome Structure and Function*; Impact New Concepts; Springer: Boston, MA, USA, 1988; pp. 263–282.
18. Wei, Y.-Z.; Hu, F.-C.; Hu, G.-B.; Li, X.-J.; Huang, X.-M.; Wang, H.-C. Differential expression of anthocyanin biosynthetic genes in relation to anthocyanin accumulation in the pericarp of *Litchi chinensis* Sonn. *PLoS ONE* **2011**, *6*, e19455. [CrossRef]
19. Zhao, Z.; Chen, Y.; Gao, A.; Huang, J. Cloning and expression of anthocyanidin synthase (ANS) gene from peel of mango (*Mangifera indica Linn*). *Afr. J. Plant Sci.* **2014**, *8*, 147–152. [CrossRef]
20. Kim, S.H.; Lee, J.R.; Hong, S.T.; Yoo, Y.-K.; An, G.; Kim, S.-R. Molecular cloning and analysis of anthocyanin biosynthesis genes preferentially expressed in apple skin. *Plant Sci.* **2003**, *165*, 403–413. [CrossRef]
21. Jin, Q.-F.; Chen, Z.-D.; Sun, W.-J.; Lin, F.-M.; Xue, Z.-H.; Huang, Y.; Tang, X.-H. Cloning and bioinformatics analysis of tea tree CsANS gene and its promoter. *Tea Sci.* **2016**, *36*, 219–228.
22. Wang, H.; Wang, W.; Zhang, P.; Pan, Q.; Zhan, J.; Huang, W. Gene transcript accumulation, tissue and subcellular localization of anthocyanidin synthase (ANS) in developing grape berries. *Plant Sci.* **2010**, *179*, 103–113. [CrossRef]
23. Nakamura, N.; Fukuchi-Mizutani, M.; Miyazaki, K.; Suzuki, K.; Tanaka, Y. RNAi suppression of the anthocyanidin synthase gene in Torenia hybrida yields white flowers with higher frequency and better stability than antisense and sense suppression. *Plant Biotechnol.* **2006**, *23*, 13–17. [CrossRef]
24. Reddy, A.M.; Reddy, V.S.; Scheffler, B.E.; Wienand, U.; Reddy, A.R. Novel transgenic rice overexpressing anthocyanidin synthase accumulates a mixture of flavonoids leading to an increased antioxidant potential. *Metab. Eng.* **2007**, *9*, 95–111. [CrossRef] [PubMed]
25. Gai, J.-T.; Huang, J.-F.; Dang, Z.-G.; Zhu, M.; Chen, H.-R.; Wang, P.; Chen, Y.-Y. Identification of ANS genes in mango and comparative analysis with other plants. *Jiangsu Agric. Sci.* **2007**, *45*, 43–49.
26. Kumar, S.; Stecher, G.; Tamura, K. MEGA7: Molecular evolutionary genetics analysis version 7.0 for bigger datasets. *Mol. Biol. Evol.* **2016**, *33*, 1870–1874. [CrossRef] [PubMed]
27. Gasteiger, E.; Hoogland, C.; Gattiker, A.; Duvaud, S.; Wilkins, M.R.; Appel, R.D.; Bairoch, A. Protein identification and analysis tools on the ExPASy server. In *The Proteomics Protocols Handbook*; Humana: Louisville, KY, USA, 2005; pp. 571–607.
28. Sapay, N.; Guermeur, Y.; Deléage, G. Prediction of amphipathic in-plane membrane anchors in monotopic proteins using a SVM classifier. *BMC Bioinform.* **2006**, *7*, 1–11. [CrossRef] [PubMed]
29. Chen, C.J.; Chen, H.; Zhang, Y.; Thomas, H.R.; Frank, M.H.; He, Y.H.; Xia, R. TBtools: An Integrative Toolkit Developed for Interactive Analyses of Big Biological Data. *Mol. Plant* **2020**, *13*, 1194–1202. [CrossRef] [PubMed]
30. Marchler-Bauer, A.; Derbyshire, M.K.; Gonzales, N.R.; Lu, S.; Chitsaz, F.; Geer, L.Y.; Geer, R.C.; He, J.; Gwadz, M.; Hurwitz, D.I.; et al. CDD: NCBI's conserved domain database. *Nucleic Acids Res.* **2015**, *43*, D222–D226. [CrossRef]
31. Bailey, T.L.; Boden, M.; Buske, F.A.; Frith, M.; Grant, C.E.; Clementi, L.; Ren, J.; Li, W.W.; Noble, W.S. MEME SUITE: Tools for motif discovery and searching. *Nucleic Acids Res.* **2009**, *37* (Suppl. 2), W202–W208. [CrossRef]
32. Wang, J.; Wu, X.-Y.; Yang, L.; Duan, Q.-H.; Huang, J.-B. Genome-wide identification and expression analysis of the ACA gene family in Chinese cabbage. *Chin. Agric. Sci.* **2021**, *54*, 4851–4868.
33. Hurst, L.D. The Ka/Ks ratio: Diagnosing the form of sequence evolution. *Trends Genet.* **2002**, *18*, 486–487. [CrossRef]
34. Kukurba, K.R.; Montgomery, S.B. RNA sequencing and analysis. *Cold Spring Harb. Protoc.* **2015**, *2015*. 951–969. [CrossRef] [PubMed]
35. Chen, S.; Zhou, Y.; Chen, Y.; Gu, J. fastp: An ultra-fast all-in-one FASTQ preprocessor. *Bioinformatics* **2018**, *34*, i884–i890. [CrossRef] [PubMed]
36. Hagel, J.M.; Facchini, P.J. Dioxygenases catalyze the O-demethylation steps of morphine biosynthesis in opium poppy. *Nat. Chem. Biol.* **2010**, *6*, 273–275. [CrossRef] [PubMed]
37. Aravind, L.; Koonin, E.V. The DNA-repair protein AlkB, EGL-9, and leprecan define new families of 2-oxoglutarate-and iron-dependent dioxygenases. *Genome Biol.* **2001**, *2*, 1–8. [CrossRef]

38. Yang, H.-Z. Cloning of sweet potato Anthocyanidin Synthase (ANS) gene and analysis of its tissue expression pattern. *Shanxi Agric. Sci.* **2020**, *48*, 1718–1723.
39. Winkel-Shirley, B. Flavonoid biosynthesis. A colorful model for genetics, biochemistry, cell biology, and biotechnology. *Plant Physiol.* **2001**, *126*, 485–493. [CrossRef]
40. Wang, H.; Nair, M.G.; Strasburg, G.M.; Chang, Y.C.; Booren, A.M.; Gray, J.I.; DeWitt, D.L. Antioxidant and antiinflammatory activities of anthocyanins and their aglycon, cyanidin, from tart cherries. *J. Nat. Prod.* **1999**, *62*, 294–296. [CrossRef]
41. Liu, M.; Li, X.-Q.; Weber, C.; Lee, C.Y.; Brown, J.; Liu, R.H. Antioxidant and antiproliferative activities of raspberries. *J. Agric. Food Chem.* **2002**, *50*, 2926–2930. [CrossRef]
42. Zhang, H.; Zhao, X.; Zhang, J.; Yang, B.; Yu, Y.; Liu, T.; Nie, B.; Song, B. Functional analysis of an Anthocyanidin Synthase gene StANS in potato. *Sci. Hortic.* **2020**, *272*, 109569. [CrossRef]
43. Bu, X.-X.; Luo, X.-P.; Bai, Y.-C.; Li, C.-L.; Chen, H.; Wu, Q. Cloning of Anthocyanidin Synthase gene from golden buckwheat and correlation between its expression and anthocyanin amount. *Chin. Herb. Med.* **2014**, *45*, 985–989.
44. Tao, S.; Wang, D.; Jin, C.; Sun, W.; Liu, X.; Zhang, S.; Gao, F.; Khanizadeh, S. Cinnamate-4-hydroxylase gene is involved in the step of lignin biosynthesis in Chinese white pear. *J. Am. Soc. Hortic. Sci.* **2015**, *140*, 573–579. [CrossRef]
45. Fu, W.-Y. Flavonoid Composition and Gene Expression Analysis of Jellyfish Snowdrop. Master's Thesis, Northwest Agriculture and Forestry University, Xianyang, China, 2013.
46. Zhang, B. Study on the Regulatory Mechanism of Anthocyanin Biosynthesis Metabolic Pathway in Brassica Napus. Ph.D. Thesis, Chongqing University, Chongqing, China, 2011.
47. Hassani, D.; Liu, H.L.; Chen, Y.N.; Wan, Z.B.; Zhuge, Q.; Li, S.X. Analysis of biochemical compounds and differentially expressed genes of the anthocyanin biosynthetic pathway in variegated peach flowers. *Genet. Mol. Res.* **2015**, *14*, 13425–13436. [CrossRef] [PubMed]
48. Huang, C.-H.; Ge, C.-L.; Zhang, X.-H.; Wu, H.; Qu, X.-Y.; Xu, X.-B. Expression analysis of structural genes related to Anthocyanidin Synthase in fruits of 'Hongyang' kiwifruit mutant. *J. Fruit Trees* **2014**, *31*, 169–174+164.

horticulturae

Article

Uncovering the Expansin Gene Family in Pomegranate (*Punica granatum* L.): Genomic Identification and Expression Analysis

Xintong Xu [1,2], Yuying Wang [1,2], Xueqing Zhao [1,2] and Zhaohe Yuan [1,2,*]

1 Co-Innovation Center for Sustainable Forestry in Southern China, Nanjing Forestry University, Nanjing 210037, China; xuxintong@njfu.edu.cn (X.X.); wangyuying@njfu.edu.cn (Y.W.); zhaoxq402@163.com (X.Z.)
2 College of Forestry, Nanjing Forestry University, Nanjing 210037, China
* Correspondence: zhyuan88@hotmail.com

Abstract: Expansins, which are important components of plant cell walls, act as loosening factors to directly induce turgor-driven cell wall expansion, regulate the growth and development of roots, leaves, fruits, and other plant organs, and function essentially under environmental stresses. In multiple species, many expansin genes (*EXPs*) have been cloned and functionally validated but little is known in pomegranate. In this study, a total of 33 *PgEXPs* were screened from the whole genome data of 'Taishanhong' pomegranate, belonging to the EXPA(25), EXPB(5), EXLA(1), and EXLB(2) subfamilies. Subsequently, the composition and characteristics were analyzed. Members of the same branch shared similar motif compositions and gene structures, implying they had similar biological functions. According to *cis*-acting element analysis, *PgEXPs* contained many light and hormone response elements in promoter regions. Analysis of RNA-seq data and protein interaction network indicated that *PgEXP26* had relatively higher transcription levels in all pomegranate tissues and might be involved in pectin lyase protein synthesis, whilst *PgEXP5* and *PgEXP31* might be involved in the production of enzymes associated with cell wall formation. Quantitative real-time PCR (qRT-PCR) results revealed that *PgEXP* expression levels in fruit peels varied considerably across fruit developmental phases. *PgEXP23* was expressed highly in the later stages of fruit development, suggesting that *PgEXP23* was essential in fruit ripening. On the other hand, the *PgEXP28* expression level was minimal or non-detected. Our work laid a foundation for further investigation into pomegranate expansin gene functions.

Keywords: pomegranate; expansin gene family; bioinformatics; expression pattern

Citation: Xu, X.; Wang, Y.; Zhao, X.; Yuan, Z. Uncovering the Expansin Gene Family in Pomegranate (*Punica granatum* L.): Genomic Identification and Expression Analysis. *Horticulturae* **2023**, *9*, 539. https://doi.org/10.3390/horticulturae9050539

Academic Editor: Sergey V. Dolgov

Received: 17 March 2023
Revised: 26 April 2023
Accepted: 27 April 2023
Published: 28 April 2023

1. Introduction

In plant cells, the cell wall is an essential and distinct structure. It determines cell shape and size, provides mechanical support and stiffness, and is the cell's first barrier against pathogens [1]. Owing to the importance of cell wall enlargement in plant morphogenesis [2], it is becoming a hot focus to study the mechanism of cell wall extension. Previously, the 'acid growth phenomenon' showed that in an acidic environment, the cell wall can be extended without structural changes, and the extending characteristics can be induced or inhibited in a short time [3]. However, the acid growth hypothesis does not address the biochemical essence of cell wall relaxation, for which expansins provide a possible mechanism of action [4].

Expansin, a broad-spectrum protein, not only relaxes and irreversibly stretches cell walls in an acidic environment, but it also enhances cell extensibility. Its function is to regulate intercellular wall component relaxation and increase cell wall flexibility by breaking hydrogen bonds between cellulose microfibers and hemifibers [5]. In land plants, expansins are important regulators of turgor-driven cell wall expansion [6], while the expansin family was a highly ancient and conserved large gene family [7]. The analysis of gene structure and amino acid sequence showed that expansin genes were derived from a common ancestor and could be divided into four subfamilies: EXPA, EXPB, EXLA, and EXLB [8]. Moreover, studies found EXPA and EXPB subfamily genes mostly act upon plant cell wall extension and the processes of growth and development [9–11], while there was no proof that EXLA and EXLB were active on the cell wall, they play a major role in controlling plant stomata opening and closing [12,13]. Currently, with the advancement of genome sequencing and analysis technology, the expansin gene family has been comprehensively identified in many plants, including Arabidopsis [14], grape [15], apple [16], cotton [8], kiwi [17], and cannabis [18]. Although their sequence composition, structure –function, and number varied substantially among different species [19], expansin genes widely regulate plant growth, meristem growth, root hair emergence, pollen tube entry into stigma and ovary, fruit ripening, pericarp rupture, and other plant growth and development processes. For example, *ZmEXPB13*, an endosperm, specifically expressed genes that influenced seed germination [20]. In Arabidopsis, partial silencing of *AtEXPA7* led to shorter root hairs, and a point mutation in the rice gene *OsEXPA17* resulted in altered root hair emergence [21,22]. The *GgEXPA1* gene from gladiolus was shown to be highly expressed during the elongation stage of stamen filament cells [23]. Overexpressing the *FaEXP2* gene increased pectin content in the cell walls of transgentic lines while decreasing the expression level of genes encoding cell wall degrading enzymes, resulting in hard fruit or late ripening of fruit [24]. The expression of the expansin gene was significantly decreased in crack-prone longan pericarp, showing the gene plays a crucial role in crack resistance creation [25]. Moreover, expansin genes were involved in salt tolerance and drought resistance [26,27], among other functions.

Pomegranate (*Punica granatum* L.) is a species of economically important trees that is widely planted worldwide and is native to Central Asia, including Iran, Afghanistan, and the Caucasus [28,29]. It is well known for its vivid red skin and juicy seeds. Furthermore, the fruit peel and juice extracts are rich in antioxidants, such as polyphenols, and have been suggested to have positive effects in cardiovascular, tumors, diabetes, and other diseases [30–33]. In recent years, scholars have successively assembled several pomegranate genomes, including 'Taishanhong' [28], 'Dabenzi' [29], and 'Tunisia' [34], and acquired high-quality genome maps, offering an essential molecular biological basis for pomegranate genetic improvement. With advancements in molecular biology, we may not only study specific gene family functions bioinformatically but also analyze genes that regulate plant growth and development as well as environmental stress. What is more, these progresses play important roles in revealing the mechanism of development and stress tolerance [35,36]. The research to date on pomegranate expansins (*PgEXPs*) is still in its early stage. Thus, based on the 'Taishanhong' genome, we used bioinformatic approaches to identify members of the pomegranate expansin gene family and analyzed their physicochemical properties, conserved domain, evolutionary relationship, *cis*-acting element, and tissue organ expression. Our study will lay a foundation for further research on the functions of the expansin gene in pomegranate.

2. Materials and Methods

2.1. Plant Materials

The pomegranate variety for testing was 'Daqingpitian', and the sampling location was the Chinese Pomegranate Expo Park (34°77′ N, 117°48′ E) with sloppy management level. We selected three healthy, disease-free and uniformly growing adult fruit-bearing trees, and took a mixed sampling method to collect samples. A total of six different developmental periods (the dates were 25 August, 3 September, 12 September, 21 September, 30 September, and 9 October, designated as P1~P6, respectively) fruit samples were collected, finally.

2.2. Identification and Physicochemical Properties of PgEXP Family Genes

To identify the pomegranate expansin gene family, firstly, the Hidden Markov Model profile of the 'EXP domains (PF01357 and PF03330) was obtained from the Pfam database (http://pfam.xfam.org/, accessed on 3 February 2023), The pomegranate genome-wide data ('Taishanhong' ASM286412v1) were downloaded from the National Center for Biotechnology Information (NCBI) official website (http://www.ncbi.nlm.nih.gov/, accessed on 3 February 2023). Then, using HMMsearch, we compared sequences with all two conserved domains (E-value $\leq 10^{-10}$) in the pomegranate protein database to get the amino acid sequences of the originally screened pomegranate EXPs. Lastly, we removed redundant sequences manually by Excel, the rest were screened for conserved domains using the NCBI CDD (http://www.ncbi.nlm.nih.gov/cdd, accessed on 3 February 2023) and SMART (http://smart.embl-heidelberg.de, accessed on 3 February 2023) to exclude candidate sequences with missing conserved domains.

The physicochemical properties, which include the isoelectric point (pI), molecular weight (MW), and instability index of *PgEXP* members, were predicted using ExPaSy Protparam (http://web.expasy.org/protparam/, accessed on 4 February 2023) [37]. Signal peptides were predicted using SignalP 4.1 (http://www.cbs.dtu.dk/services/Signalp/index.php, accessed on 4 February 2023). Subcellular localization prediction was carried out by using Cell-PLoc 2.0 (http://www.csbio.sjtu.edu.cn/bioinf/Cell-PLoc-2/, accessed on 4 February 2023).

2.3. Phylogenetic Analysis

In order to classify PgEXPs based on phylogenic tree, the selected amino acid sequences of EXPs from pomegranate, *Arabidopsis*, grape, kiwi, and jujube were aligned by MUSCLE in MEGA 11 software with default settings [38], and the compared sequences were also trimmed with MEGA 11.A Maximum Likelihood (ML) in IQTree2 V2.1.3 and was then constructed under the best-fitting model with 1000 bootstrap replicates. EvolView (http://www.evolgenius.info/evolview/, accessed on 6 February 2023) was used to enhance the phylogenetic tree online. The phylogenetic position of PgEXPs in relation to reference EXPs was used to group them. We collected and summarized published records on the EXP family from other plant species to compare the phylogenetic groups of the EXPs in pomegranate and other species.

2.4. Analysis of Conserved Domains, Gene Structure and Protein Conserved Motif

According to the acquired PgEXP protein sequences and gene sequences, we used MUSCLE to compare them as shown by Jalview software. The online program MEME Suit (http://meme-suit.org/, accessed on 8 February 2023) was used to conduct motif analysis. To identify the exon–intron structure of 33 *PgEXP* genes, their annotation information was taken from pomegranate whole genome gff files and uploaded to the web application GSDS (http://gsds.cbi.pku.edu.cn/, accessed on 8 February 2023). Afterward, TBtools [39] was used to illustrate the phylogenetic tree, conserved motifs, and gene structure of *PgEXPs*.

2.5. Analysis of Cis-Acting Elements and Protein Interaction Networks

Promoter sequences (1500-bp upstream from the start codon) were extracted from the genome sequence of *PgEXPs*. Then, the online website PlantCARE (http://bioinformatics.psb.ugent.be/webtools/plantcare/html/, accessed on 8 February 2023) was used to analyze potential *cis*-acting elements, the findings visualized by TBtools software. To examine gene co-expression patterns, protein patterns with reasonably high specificity from the String (http://cn.string-db.org, accessed on 10 February 2023) were employed, and the model plant *Arabidopsis thaliana* was chosen as the species parameter.

2.6. RNA-Seq Analysis

To analyze expression patterns of *PgEXPs* in different pomegranate tissues and organs, the published transcriptome data of six pomegranate varieties ('Dabenzi', 'Tunisia', 'Baiyushizi', 'Black127', 'Nana', and 'Wonderful') were downloaded from NCBI (http://www.ncbi.nlm.nih.gov/, accessed on 10 February 2023), including outer seed coat, inner seed coat, pericarp, flower, root, leaf, and mixed samples of roots, leaf, flower, and fruit as shown in Table 1. Then, the transcriptomic data were calculated and analyzed with Kallisto v0.44.0 software (California, USA) [40], and the resulting values were transformed into Log_2(TPM+1) (Table S1) and finally, the expression heatmap was created by TBtools.

Table 1. Pomegranate transcript data.

Accession No.	Cultivar	Sample	ID	Reference
SRR5279388	Dabenzi	Outer seed coat	Dabenzi_OSC	[29]
SRR5279391	Dabenzi	Inner seed coat	Dabenzi_ISC	[29]
SRR5279394	Dabenzi	Pericarp	Dabenzi_pericarp	[29]
SRR5279395	Dabenzi	Flower	Dabenzi_flower	[29]
SRR5279396	Dabenzi	Root	Dabenzi_root	[29]
SRR5279397	Dabenzi	Leaf	Dabenzi_leaf	[29]
SRR5446592	Tunisia	Bisexual flowers (3.0–5.0 mm)	3–5 mm(B)	[34]
SRR5446595	Tunisia	Bisexual flowers (5.1–13.0 mm)	5.1–13 mm(B)	[34]
SRR5446598	Tunisia	Bisexual flowers (13.1–25.0 mm)	13.1–25 mm(B)	[34]
SRR5446601	Tunisia	Functional male flowers (3.0–5.0 mm)	3–5 mm(F)	[34]
SRR5446604	Tunisia	Functional male flowers (5.1–13.0 mm)	5.1–13 mm(F)	[34]
SRR5446607	Tunisia	Functional male flowers (13.1–25.0 mm)	13.1–25 mm(F)	[34]
SRR5678820	Tunisia	Inner seed coat	TNS_ISC	[29]
SRR5678819	Baiyushizi	Inner seed coat	BYSZ_ISC	[29]
SRR1054190	Black127	Mix of leaves, flowers, fruit and roots	Black127	[41]
SRR1055290	Nana	Mix of leaves, flowers, fruit and roots	Nana	[41]
SRR080723	Wonderful	Pericarp	Wonderful	[42]

2.7. RNA Isolation, Reverse Transcription and Quantitative Real-Time PCR (qRT-PCR)

qRT-PCR was performed to detect the expression of *PgEXP* genes. Total RNA was isolated from peels by RNA Extraction Kit (FastPure® Plant Total RNA Isolation Kit, Vazyme, Nanjing), and the quality was assessed by electrophoresis and A260/A280. The first-strand cDNA was synthesized from the total RNA by using a cDNA synthesis kit (HiScript III RT SuperMix for qPCR (+gDNA wiper), Vazyme, Nanjing). Specific quantification primers of *PgEXPs* (Table S2) were designed. Pomegranate *PgActin* served as an internal reference gene. The PCR system was 20 μL, including 10 μL Taq Pro Universal SYBR qPCR Master Mix (Vazyme, Nanjing), 0.4 μL upstream and downstream primers, 1 μL cDNA template (the concentration was around 200 ng/μL), and 8.2 μL ddH2O. Three biological replicates for each treatment were to be conducted. The PCR reaction protocol was as follows: 95 °C pre-denaturation for 30 s, 95 °C denaturation for 10 s, 60 °C for 30 s, 40 cycles; the melting curve procedure was as follows: 95 °C 15 s, 60 °C for 30 s, 40 cycles. The relative expression level was analyzed by $2^{-\Delta\Delta CT}$ method [43]. SPSS 23.0 (CA California, USA) and Origin 2018 software (MA Massachusetts, USA) were used to analyze and plotted the data, respectively.

3. Results

3.1. Identification and Physicochemical Properties of PgEXP Family Genes

In this study, 33 potential expansin genes were identified from the 'Taishanhong' pomegranate genome. They were named *PgEXP1-PgEXP33* in the order of gene ID to assist later investigation (Table S3). The analysis of physicochemical properties revealed that the coding region of *PgEXPs* ranged from 555 bp (*PgEXP5*) to 1005 bp (*PgEXP31*). The PgEXPs protein contained 185 (*PgEXP5*) to 335 (*PgEXP31*) amino acids, and the protein molecular mass was between 20495.67 kDa (*PgEXP5*) and 36742.26 kDa (*PgEXP31*). The theoretical isoelectric point ranged from 4.55 (*PgEXP15*) to 9.76 (*PgEXP21*) with 81.8% having pI values greater than 7, indicating basic, and the rest less than 7, indicating acidic. The PgEXP's protein instability indices ranged from 18.69 (*PgEXP22*) to 46.86 (*PgEXP7*) with 90.9% having good structural stability. The mean hydrophilic values of PgEXPs protein values ranged from −0.450 (*PgEXP16*) to 0.065 (*PgEXP14*), apart from *PgEXP14*, *PgEXP29*, and *PgEXP30* which were hydrophobic proteins, the others were hydrophilic. EXP usually had a signal peptide sequence at N-terminus. The analysis found that all members except

PgEXP2, *PgEXP5*, *PgEXP15*, *PgEXP18*, *PgEXP22*, *PgEXP23*, and *PgEXP33* contained the N-terminal signal peptides, and most signal peptide lengths were around 20 aa. According to the predicted subcellular localization results, all *PgEXPs* were localized in the cell wall.

3.2. Phylogenetic Analysis

To investigate the phylogenetic relationships, a phylogenetic tree was constructed using the selected protein sequences of EXPs from pomegranate, *Arabidopsis*, grape, kiwi, and jujube. In the phylogenetic tree (Figure 1), some members of the five species were clustered on one branch, concerning the phylogenetic relationships and naming rules of *AtEXPs*. *PgEXPs* were divided into four subfamilies, namely EXPA, EXPB, EXLA, and EXLB. The size of these four subfamilies varies slightly. The EXPA subfamily was the largest subfamily, with 25 members, while the EXLA subfamily had only one member. The EXPB subfamily had five members and the rest belonged to the EXLB subfamily. The analysis of the phylogenetic tree showed that all five species' EXPs were distributed in four subfamilies, indicating their functions differed. Some genes seemed to diverge early because of their long branches, and they might evolve differently during the long-term phylogeny. Moreover, there were some differences among these five species drawn from the phylogenetic tree, implying that the expansin gene family evolved separately. Through this analysis, we can infer the functions of *PgEXPs* with high similarity from other species' genes with verified functions.

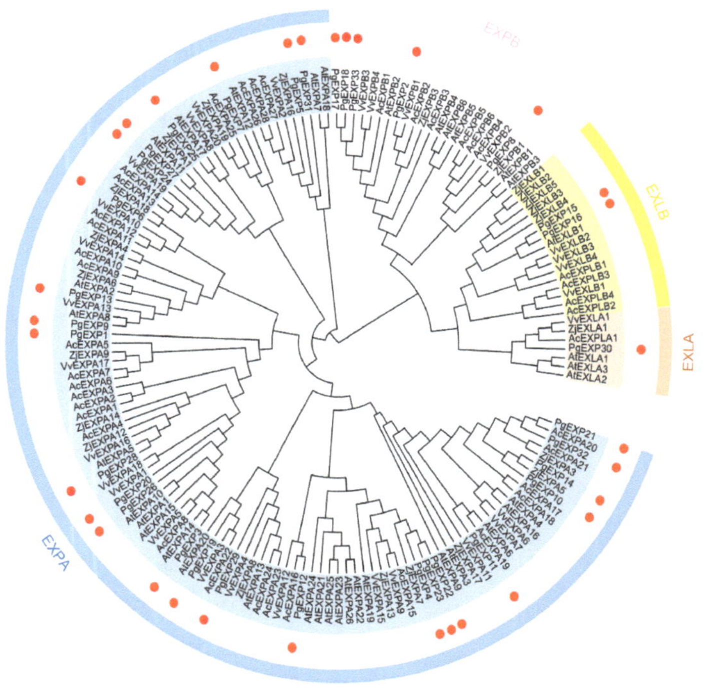

Figure 1. Phylogenetic trees in the expansin gene family in pomegranate, *Arabidopsis*, grape, kiwi, and jujube. The phylogenetic tree was constructed using the ML method with 1000 bootstrap replicates. The different colors indicated different subfamilies. The red dots represented *PgEXPs*.

To further compare the quantitative distribution of *EXPs* in different subfamilies, we summarized the number of *EXPs* of 21 species (Table 2), including monocotyledonous plants, dicotyledonous plants, and a nonvascular plant. The results showed that the EXPA subfamily had the largest number of *EXPs* in these species except that in corn (*Zea mays*). The number of EXPA and EXPB subfamilies members was less distinct in monocotyledonous plants, and the EXLB family members were extremely low. In addition, the number of monocotyledons EXPB subfamily members were significantly higher than that in dicots. The total number of *EXPs* in pomegranate was higher than that in cannabis (*Cannabis sativa*), ginkgo (*Ginkgo biloba*), grape (*Vitis vinifera*), and jujube (*Ziziphus zizyphus*), lower than that in other species within this table.

Table 2. Sizes of the four expansin subfamilies in different plants species.

Species	EXPA	EXPB	EXLA	EXLB	Total	Reference
Actinidia chinensis	28	6	1	4	39	[17]
Arabidopsis thaliana	26	6	3	1	36	[7]
Brassica napus	79	21	5	4	109	[44]
Brassica rapa	39	9	2	3	53	[45]
Cannabis sativa	19	7	1	5	32	[18]
Cucumis sativus	21	3	9	2	35	[46]
Glycine max	49	9	2	15	75	[47]
Ginkgo biloba	20	1	4	3	28	[48]
Gossypium hirsutum	67	12	15	1	93	[49]
Malus × Domestica	34	1	2	4	41	[15]
Nicotiana tabacum	36	6	3	7	52	[47]
Oryza sativa	34	19	4	1	58	[49]
Physcomitrella patens	32	0	6	0	38	[50]
Populus	27	3	2	4	36	[19]
Punica granatum	25	5	1	2	33	This study
Salix sinopurpurea	26	3	2	3	34	[51]
Solanum lycopersicum	25	8	1	4	38	[47]
Triticum aestivum	26	15	4	0	45	[12]
Vitis vinifera	20	4	1	4	29	[14]
Zea mays	36	48	4	0	88	[47]
Ziziphus zizyphus	19	3	1	7	30	[52]

3.3. Analysis of Conserved Domains, Gene Structure, and Protein Conserved Motif

To further verify the conserved structure of pomegranate *EXPs*, align 33 PgEXP protein sequences and then visualize them with Jalview software. As shown in Figure 2, the protein contained two conserved domains and a signal peptide. The conserved domains were both around 100 aa in length. Of these, domain 1 contained an 'HFD' structure which was part of the glycoside hydrolase-45 (GH45) protein's catalytic site. The 'HFD' structure was conserved in the EXPA and EXPB subfamilies but mutated to 'SFV' and 'DFI' in the EXLA and EXLB subfamilies. Except for *PgEXP28*, other EXPA members characterized the large insertion (α-Insertion) and deletion (α-Deletion). Of the three expansin-like protein sequences, one was classified into the EXLA subfamily based on the presence of a characteristic EXLA extension at the C-terminus with the remaining ones classified into EXLB subfamily. Additionally, *PgEXPs* had a BOX with the sequence signature 'GACGYG' which was missing or mutated in a few members. The variation in conserved sequence features revealed potential functional differences between subfamilies. Other members, with the exception of EXLA, EXLB subfamilies and some EXPA subfamily members (*PgEXP5*, *PgEXP10*, and *PgEXP23*), contained all eight cysteines (Cysteine, C) residues and four conserved tryptophan (tryptophan, W) residues at the hydroxyl terminus. It has been proposed that these sites were located in the expansin's glycosylation and catalytic regions, respectively, that cysteine might play a role in the formation of disulfide bonds, that HFD and conserved aspartate residues might act via a glycosylation mechanism, and that conserved tryptophan might work in the binding of protein and polysaccharide molecules [53].

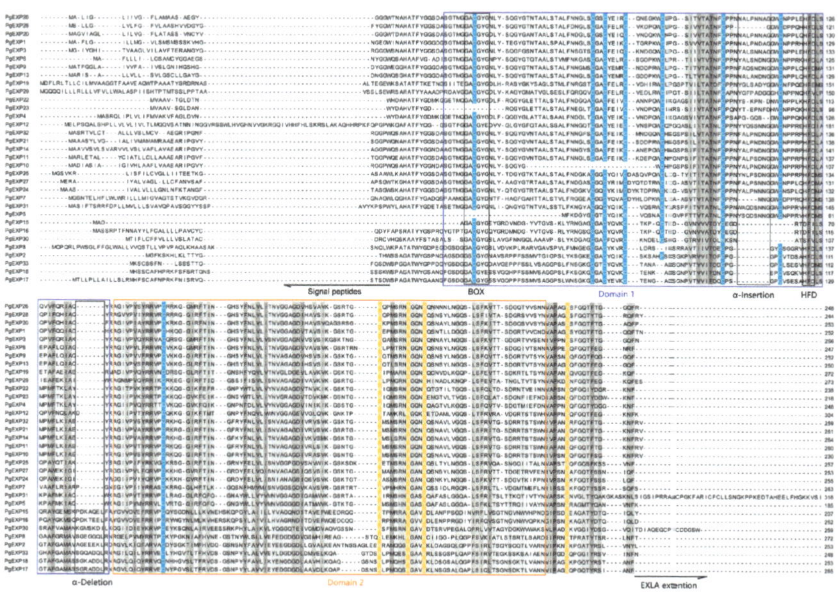

Figure 2. Multiple alignments of PgEXPs protein sequencing results. The blue square represented Domain 1. The orange square represented Domain 2. The blue background represented cysteine, and the orange background represented tryptophan.

A total of 10 conserved motifs were identified by the website MEME Suit (Figure 3a) When each motif was submitted to the Pfam server, it was discovered that motif1, motif4, and motif9 encoded conserved domain 1, motif2, motif3, and motif10 encoded the conserved domain 2. Then, the conserved motif distribution of *PgEXPs* was constructed (Figure 3b-B). The results showed that each gene contained 4–8 motifs. Furthermore, all members had motif5 and motif6, localized to essentially the same position, implying that they were the functional basis of pomegranate expansin genes. The others, with unknown functions, were distributed throughout the protein. Meanwhile, because of poor conservatism, motif deletions, additions, or substitutions in individual genes occurred.

To better understand the evolution of the expansin gene family in pomegranate, the exon–intron structures of all identified *PgEXPs* were analyzed. TBtools visualization results (Figure 3b-C) revealed that *PgEXPs* gene structure was relatively simple with 2–5 exons dividing the gene fragment into introns of varying lengths. The EXPA subfamily had three exon–intron structures: seventeen members had three exons and two introns, five members had two exons and one intron, and three members had four exons and three introns. The reason seemed to be the addition or deletion of introns in the structure. The EXPB subfamily had a very stable intron–exon structure as did the EXLB subfamily which had four exons and three introns. Additionally, the EXLA subfamily gene (*PgEXP30*) had the most introns and exons with a gene structure of five exons and four introns.

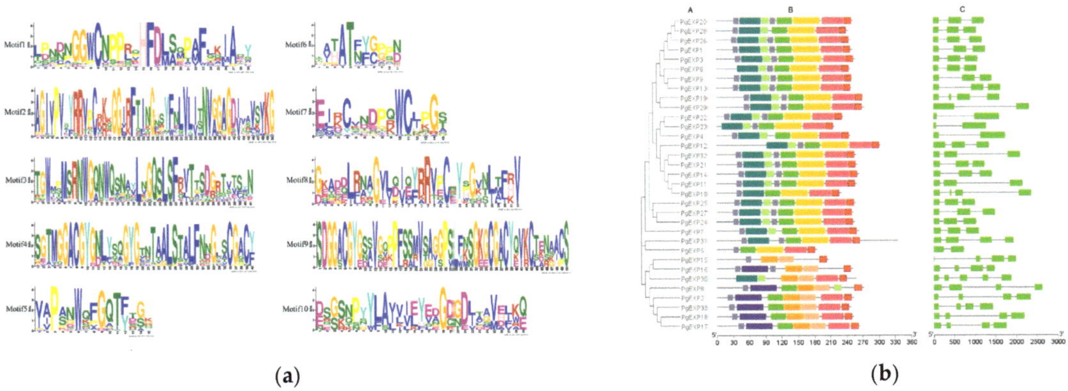

Figure 3. (**a**) *PgEXP* genes conserved motifs. The vertical coordinate represented the conserved amino acid and the height of the amino acid letter represented the frequency of occurrence. The horizontal coordinate represented the amino acid's position in the sequence. (**b**) The PgEXP gene family's phylogenetic tree (**A**), conserved motifs (**B**), and gene structure (**C**). Protein motifs in PgEXP members: colored boxes depict the various patterns. The results of phylogenetic analysis were used to perform clustering. Exons and introns were indicated by green boxes and black lines in the gene structure.

3.4. Analysis of Cis-Acting Elements

The *cis*-acting elements in the promoter were examined to preferably know the function of *PgEXPs* and the possible regulatory pathways involved. *PgEXPs* contained a total of 36 *cis*-acting elements which were broadly classified into three categories: biotic and abiotic stress responses, plant growth and development, and response elements related to hormone induction (Figure 4). For instance, some *PgEXP*s promoters contained MYB binding sites involved in drought inducibility (MBS), flavonoid biosynthesis genes regulation (MBSI), and light response (MRE). Besides MRE, *PgEXPs* promoter regions contained 11 light responsiveness elements, indicating these *PgEXPs* might be regulated by light. There were also 20 *PgEXPs* that could respond to plant biotic or abiotic stress. Of these, 13 contained the low-temperature responsiveness element LTR, 15 contained the defense and stress responsiveness element TC-rich repeats, and the rest contained the wound-responsive element WUN-motif. In addition, some *PgEXPs* contained anaerobic induction responsiveness elements (ARE) and anoxic specific inducibility responsiveness elements (GC-motif), implying they had a function in the oxygen shunt signaling response. Other regulatory elements were involved in zein metabolism regulation (O2-site), seed-specific regulation (RY-element), and palisade mesophyll cells (HD-Zip 1), suggesting *PgEXPs* related to them might be close to plant growth and development. Furthermore, there were 10 response elements associated with hormone induction. For example, CGTCA-motif and TGACG-motif were both involved in MeJA-responsiveness, but they had different binding sites, CGTCA in the former and TGACG in the latter. All other *PgEXPs*, except *PgEXP3*, *PgEXP29*, and *PgEXP11*, contained at least one phytohormone-responsive element.

3.5. Analysis of Protein Interaction Networks

The String protein interaction database was used to predict the co-expression of 33 PgEXP (Figure 5A) proteins (*Arabidopsis thaliana* was chosen as the model species and the AtEXP with the highest similarity was selected). The stronger the contact between the two proteins, the thicker the linkage line. The remaining PgEXPs had no direct contact, indicating that they were not directly controlled. Subsequently, the analysis of the protein interaction network was used to estimate the possible roles of each PgEXP. Among these, PgEXP1, PgEXP3, PgEXP20, and PgEXP26 were found to be identical to EXPA1 (Figure 5B) and were co-expressed with the gibberellin regulatory protein GASA6, the growth regulator At2g22840, and the pectin lyase protein AT5G04310. PgEXP5 and PgEXP31 were identical to EXPA18 (Figure 5C) and co-expressed with lignin synthesis proteins (RHS19, AT1G30870, and AT3G49960) as well as proteins in the cell wall metabolism-related enzymes production pathway (RHS12, XTH14, and AT5G04960).

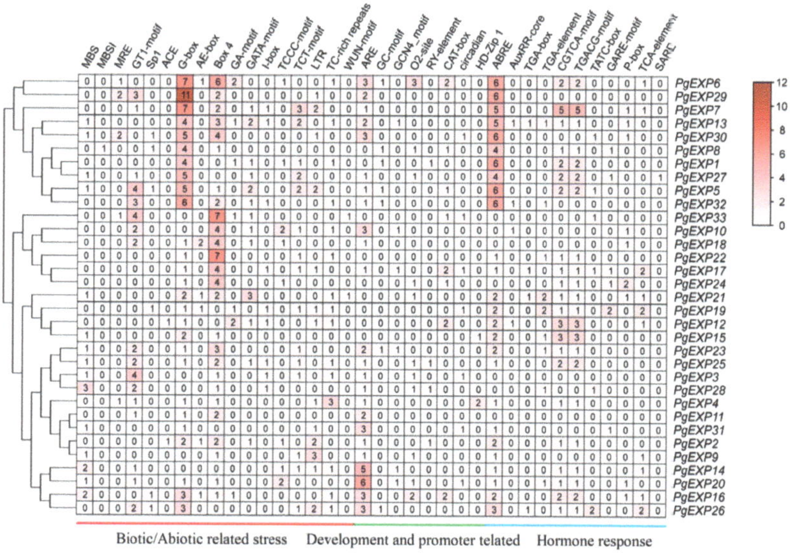

Figure 4. Pomegranate *PgEXPs* promoter predicted *cis*-acting elements. The numbers represented the number of *cis*-acting elements. The red line represented the biotic and abiotic related stress in plants. The green line represented plants development and promoter. The blue line represented hormone response related to plants.

3.6. Analysis of PgEXPs Gene Expression

To further investigate the gene expression divergence among different tissues, we downloaded RNA-seq data from NCBI for pomegranate (Table 1, Figure 6). Among the 33 *PgEXP* gene expressions in different tissues, ten genes (*PgEXP23, 20, 17, 25, 6, 4, 31, 18, 5,* and *22*) were not expressed or minimally expressed in various tissues. Four genes showed similar expression patterns, *PgEXP33* was expressed at the highest level in the mixed samples of 'Nana', *PgEXP15* was expressed at the highest level in the inner seed coat of 'Tunisia', *PgEXP1* was transcribed at the highest level in the pericarp and outer seed coat of 'Dabenzi', and *PgEXP27* was expressed at the highest level in the inner seed coat of 'Dabenzi'. There were 10 genes (*PgEXP26, 9, 13, 29, 32, 21, 14, 11, 16,* and *30*) that displayed expression in almost all tissues. Of these, *PgEXP26* expression in all tissues was all higher, indicating that it may play a complex function in pomegranate growth and development. Furthermore, we found all genes had low or no expression in two samples which were 5.1–13 mm hermaphrodite and 13.1–25 mm hermaphrodite.

3.7. qRT-PCR

To explore the particular roles of *PgEXPs* in fruit development, quantitative real-time PCR (qRT-PCR) was used to examine the expression patterns of *PgEXPs* in pomegranate pericarp. Finally, we chose 24 *PgEXP* and assessed their expression at 6 periods of 'Daqing-pitian' pomegranate pericarp (Figure 7). In general, *PgEXP1*, *PgEXP26*, *PgEXP28,* and *PgEXP32* expression levels decreased from P1 to P6; of these, *PgEXP28* was low or almost not expressed in P2-P6, and expression levels were considerably greater in P1 than in P2–P6. *PgEXP23* expression, on the other hand, rose gradually across all periods, peaking in P6. *PgEXP23* was found to be more abundant in the later phases of fruit growth as well, suggesting that it was important for fruit ripening. *PgEXP3* and *PgEXP13* had the highest expression levels at P5, *PgEXP15* had the highest expression levels at P2, *PgEXP9* had higher expression levels at both P2 and P5, and the expression levels of *PgEXPs* varied considerably at different phases of fruit development.

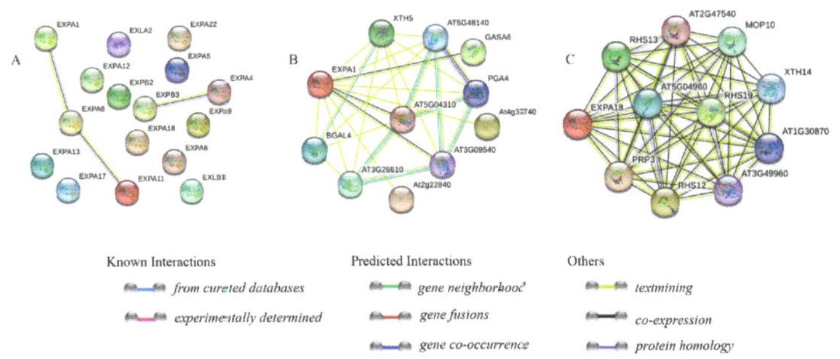

Figure 5. PgEXP proteins (**A**), PgEXP1, PgEXP3, PgEXP20, and PgEXP26 protein (**B**), PgEXP5 and PgEXP31 protein (**C**) functional interaction network and gene co-expression diagram. XTH5: Probable xyloglucan endotransglucosylase/hydrolase protein 5; BGAL4: Beta-galactosidase 4; AT5G48140: Galacturan 1,4-alpha-galacturonidase; AT5G04310: Pectin lyase-like superfamily protein; AT3G26610: Pectin lyase-like superfamily protein; GASA6: Gibberellin-regulated family protein; PGA4: Galacturan 1,4-alpha-galacturonidase; At4g37740: Growth-regulating factor 2; AT3G09540: Pectin lyase-like superfamily protein; At2g22840: Growth-regulating factor 1; RHS13: Root hair specific 13; PRP3: Arabidopsis thaliana proline-rich protein 3; AT5G04960: Plant invertase/pectin methylesterase inhibitor superfamily; MOP10 Pollen Ole e 1 allergen and extensin family protein; AT1G30870: Peroxidase superfamily protein; AT3G49960: Peroxidase superfamily protein.

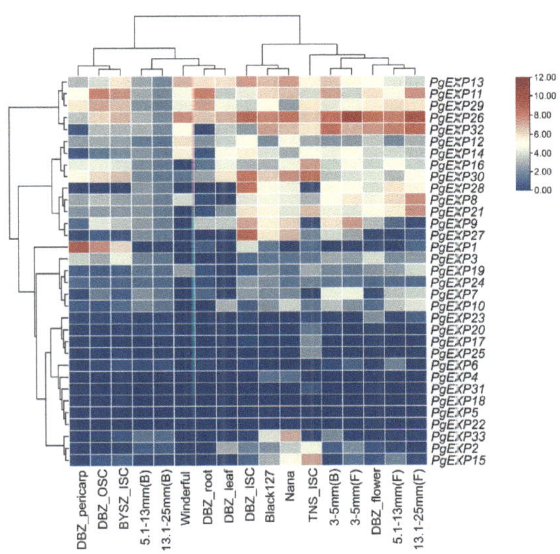

Figure 6. Heat map of *PgEXP* gene expression in pomegranate tissues. DBZ_pericarp: 'Dabenzi' Pericarp; DBZ_OSC: 'Dabenzi' Outer seed coat; BYSZ_ISC: 'Baiyushizi' Inner seed coat; 5.1–13 mm(B): 'Tunisia' Bisexual flowers (5.1–13.0 mm); 13.1–25 mm(B): 'Tunisia' Bisexual flowers (13.1–25.0 mm); Wonderful: 'Wonderful' Pericarp; DBZ_root: 'Dabenzi' Root; DBZ_leaf: 'Dabenzi' Leaf; DBZ_ISC: 'Dabenzi' Inner seed coat; Black127: 'Black127' Mix of leaves, flowers, fruit and roots; Nana: 'Nana' Mix of leaves, flowers, fruit and roots; TNS_ISC: 'Tunisia' Inner seed coat; 3–5 mm(B): 'Tunisia' Bisexual flowers (3–5 mm); 3–5 mm(F): 'Tunisia' Functional male flowers (3–5 mm); DBZ_flower: 'Dabenzi' Flower; 5.1–13 mm(F): 'Tunisia' Functional male flowers (5.1–13.0 mm); 13.1–25 mm(F): 'Tunisia' Functional male flowers (13.1–25.0 mm).

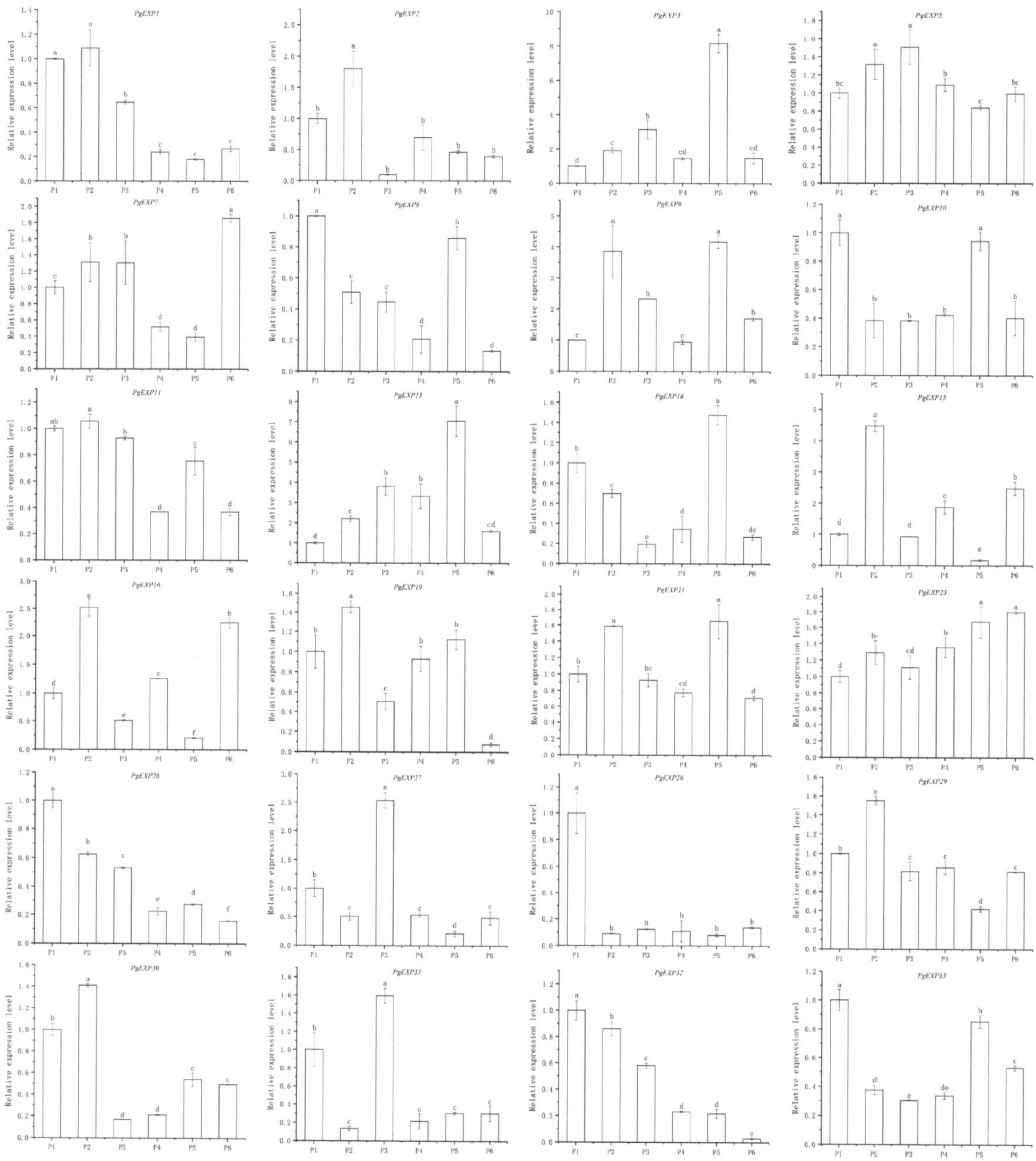

Figure 7. The expression patterns of 24 *PgEXPs* in pomegranate peel at 6 developmental periods obtained by qRT-PCR analysis. The dates were 25 August, 3 September, 12 September, 21 September, 30 September, and 9 October designated as P1~P6, respectively. The data shown represent an average of three independent experiments ± SD. The vertical bars show the standard error. Bars with different letters (a–f) indicate significant differences at $p < 0.05$ according to Duncan's test.

4. Discussion

Since cell wall enlargement played a key role in plant morphogenesis, the mechanism of cell wall extension had been a focus of investigation [2]. During turgor-mediated growth, cell wall stress relaxation occurred [54] which involved cell wall-loosening factors, such as EXPs [55]. Expansins might offer a mechanism for cell wall elongation by interfering with the binding of microfibrils to the cell wall [4]. The expansin gene family has been thoroughly discovered in numerous plants [17,18]. Previously, systematic analyses were carried out to identify and characterize expansin families in a variety of models, crop and fruit plant species, such as tomatoes [56], pepper [57], litchi [58], and apple [59]. Studies, subsequently, showed that a close relationship was established between expansins and fruit cracking. Fruit cracking tremendously damages the appearance of the fruit, easily leads to pathogen invasion, greatly reduces marketability, and causes immense economic losses. Currently, pomegranate fresh fruit, seedlings, and processing industry have a great promising future, while fruit cracking is a major cause of fruit loss in pomegranate [60]. Thus, it is considerable and urgent to explore the roles expansins play in pomegranates.

In this work, we first identified 33 pomegranate expansins through genome-wide analysis and then used web-based tools and resources to predict the properties of these proteins, such as isoelectric point and molecular weight. The prediction of subcellular localization showed that all genes localized in the cell wall. These helped us learn more about these genes.

Based on the genes of each ancestor, plant expansins experienced varying degrees of gene duplication and expansion [6]. The dicotyledonous plant expansin gene family, which included *Arabidopsis*, soybean [61], and watermelon [62], was mostly extended via tandem and segmental replication. According to the established EXPs classification of pomegranate, *Arabidopsis*, grape, kiwi, and jujube, *PgEXPs* were divided into four subfamilies, EXPA, EXPB, EXLA, and EXLB. In addition, we found that the large branches of the phylogenetic tree usually contained *EXPs* of different species, while on some small branches, there were usually only *EXPs* of the same species, suggesting that EXPs amplification had occurred before the differentiation of these species and after the differentiation of the species, EXPs amplification occurred again. This view was corroborated by the phylogenetic analysis of tobacco [47], soybean [47], and rice [49]. Analysis and comparison of the sizes of expansin subfamilies in 21 species revealed an uneven distribution of each gene subfamily among species. For example, the EXPB subfamily members were significantly more numerous in momocots than that in dicots, with 19 in rice, 15 in wheat, and 48 in maize, while around 10 in dicots (Table 2), it may be due to the greater expansion and retention of gene duplication events in monocots [61].

The architecture of various *PgEXP* genes varied with minimal variation in the number of exons and a substantial diversity in the length of introns. The analysis of gene structure revealed that *PgEXPs* had 2–5 exons which was compatible with the number of exons found in grapes [15] and land cotton [63], showing that the pomegranate expansin gene structure was largely conserved. According to the conserved motif analysis, all *PgEXPs* contained a total of 10 motifs with a similar distribution of motifs within the same subgroup. The high degree of sequence identity and similar gene structure of *PgEXPs* within each family indicated that the pomegranate expansin family had undergone gene duplication throughout evolution, resulting in multiple copies with a partial or complete overlap in function.

Cis-acting elements offer genes the ability to work in developmental or environmental regulation. Light responsiveness elements were the first class of enriched *cis*-acting elements in *PgEXPs*. Of these, G-box was the most abundant light responsiveness element. In *PgEXPs* promoter regions, several elements associated with plant growth, environmental stress, and hormone induction existed. The results were consistent with the findings of that in cannabis [18] and *Panax ginseng* [64], indicating the regulatory elements of the *EXPs* were more conserved across species. For instance, the promoter regions contained many *cis*-acting elements engaged in seed-specific regulation, zein metabolism regulation, and palisade mesophyll cell differentiation, suggesting that these *PgEXPs* worked in pomegranate seed germination and leaf development [65]. Additionally, MYB binding sites were found in the promoter regions of 20 *PgEXPs*. A total of fifteen *PgEXPs* had binding sites (MBS) implicated in drought-inducibility which might be related to plants' biotic and abiotic stress. Five *PgEXPs* had sites (MRE) involved in light responsiveness and there was also one site (MBSI) implicated in the regulation of flavonoid biosynthesis genes. These MYB protein binding sites could be detected in the sour cherry expansin gene family as well [66]. Furthermore, the findings of hormone-inducing *cis*-acting elements, such as abscisic acid, MeJA, gibberellin, and salicylic acid, implied that *PgEXPs* might be triggered by a variety of hormone-signaling molecules [67].

The protein interaction network of PgEXPs revealed that some genes had cocations. The prediction results revealed that PgEXP1, PgEXP3, PgEXP20, and PgEXP26 were co-expressed with pectin lyase proteins. Combined with the expression level of these genes, we found that PgEXP26 might be involved in pectin lyase protein production of flower, leaf, pericarp, and mixed tissues. In addition, PgEXP1 expression levels in the inner seed coat of 'Dabenzi' and 'Baiyushizi' were greater than that in 'Tunisia', it might be owing to genotype differences. According to further analysis of the expression pattern of *PgEXPs* in different tissues, *PgEXP30*, a member of the EXLA subfamily, was expressed in a higher level in flower, root, and mixed sample than other tissues. In contrast, two members of the EXLB subfamily, *PgEXP15* and *PgEXP16*, were exclusively expressed in 'Dabenzi' root tissues, implying that EXLB played a role in pomegranate root growth and development. Furthermore, some genes were not or weakly expressed in all tissues, indicating that they were not engaged in pomegranate growth or stress regulation. Finally, qRT-PCR was used to examine the expression pattern of *PgEXPs* in pomegranate pericarp. The expression level of *PgEXP23* rose consistently with time advancement, indicating that *PgEXP23* may play an important role in fruit ripening. *PgEXPs* expression levels varied considerably across fruit developmental phases, suggesting that *PgEXPs* genes may have diverse functions.

5. Conclusions

This work was the first comprehensive genome-wide analysis of the pomegranate expansin gene family. We identified 33 *PgEXPs* from the pomegranate 'Taishanhong' genome. Based on a phylogenetic tree of pomegranate, *Arabidopsis*, grape, kiwi, and jujube expansin genes, pomegranate genes were split into four subfamilies, EXPA (25 members), EXPB (5 members), EXLA (1 member), and EXLB (2 members). Members of subfamilies were highly conserved in motif and gene structure. Analysis of tissue-specific expression patterns of *PgEXPs* revealed that they may function differently in regulating organ/tissue morphology formation and development. Analysis of promoter cis-acting elements revealed that *PgEXPs* may respond to development and stress. qRT-PCR results played a role in exploring the function of *PgEXP* genes in pomegranate. The above analysis added to our knowledge of the expansin gene family in pomegranates and provided insights into the possible functional involvement of pomegranate expansin genes.

Supplementary Materials: The following supporting information can be downloaded at: https://www.mdpi.com/article/10.3390/horticulturae9050539/s1, Table S1: Referenced EXPs used to contruct phylogenetic tree, Table S2: Primers for qRT-PCR, Table S3: Basic information of pomegranate gene family, Table S4: The cis-acting elements of PgEXPs.

Author Contributions: Conceptualization, X.X. and Z.Y.; methodology, X.X. and Z.Y.; software, X.X. and Y.W.; visualization, X.X. formal analysis, X.X.; writing—original draft preparation, X.X.; writing—review and editing, Y.W., X.Z. and Z.Y.; supervision, Z.Y.; funding acquisition, X.Z. and Z.Y. All authors have read and agreed to the published version of the manuscript.

Funding: This work was supported by National Natural Science Foundation of China (31901341) and the Priority Academic Program Development of Jiangsu High Education Institutions [PAPD].

Data Availability Statement: The data presented in this work are available upon request from the corresponding author and the public pomegranate transcriptomes presented in this work are available in the insert article.

Conflicts of Interest: The authors declare no conflict of interest.

References

1. Liu, N.; Sun, Y.; Pei, Y.; Zhang, X.; Wang, P.; Li, X.; Li, F.; Hou, Y. A Pectin Methylesterase Inhibitor Enhances Resistance to *Verticillium* Wilt. *Plant Physiol.* **2018**, *176*, 2202–2220. [CrossRef] [PubMed]
2. Lin, C.; Choi, H.-S.; Cho, H.-T. Root hair-specific EXPANSIN A7 is required for root hair elongation in Arabidopsis. *Mol. Cells* **2011**, *31*, 393–397. [CrossRef] [PubMed]
3. Rayle, D.L.; Cleland, R. Enhancement of Wall Loosening and Elongation by Acid Solutions. *Plant Physiol.* **1970**, *46*, 250–253. [CrossRef] [PubMed]
4. Cosgrove, D.J. Loosening of plant cell walls by expansins. *Nature* **2000**, *407*, 321–326. [CrossRef]
5. Feng, X.; Xu, Y.Q.; Peng, L.N.; Yu, X.Y.; Zhao, Q.Q.; Feng, S.S.; Zhao, Z.Y.; Li, F.L.; Hu, B.Z. TaEXPB7-B, a β-expansin gene involved in low-temperature stress and abscisic acid responses, promotes growth and cold resistance in Arabidopsis thaliana. *Plant Physiol.* **2019**, *9*, 153004. [CrossRef]
6. Vannerum, K.; Huysman, M.J.; De Rycke, R.; Vuylsteke, M.; Leliaert, F.; Pollier, J.; Lütz-Meindl, U.; Gillard, J.; De Veylder, L.; Goossens, A.; et al. Transcriptional analysis of cell growth and morphogenesis in the unicellular green alga *Micrasterias* (Streptophyta), with emphasis on the role of expansin. *BMC Plant Biol.* **2011**, *11*, 128. [CrossRef]
7. Sampedro, J.; Lee, Y.; Carey, R.E.; DePamphilis, C.; Cosgrove, D.J. Use of genomic history to improve phylogeny and understanding of births and deaths in a gene family. *Plant J.* **2005**, *44*, 409–419. [CrossRef]
8. Lv, L.-M.; Zuo, D.-Y.; Wang, X.-F.; Cheng, H.-L.; Zhang, Y.-P.; Wang, Q.-L.; Song, G.-L.; Ma, Z.-Y. Genome-wide identification of the expansin gene family reveals that expansin genes are involved in fibre cell growth in cotton. *BMC Plant Biol.* **2020**, *20*, 223. [CrossRef]
9. Gasperini, D.; Greenland, A.; Hedden, P.; Dreos, R.; Harwood, W.; Griffiths, S. Genetic and physiological analysis of Rht8 in bread wheat: An alternative source of semi-dwarfism with a reduced sensitivity to brassinosteroids. *J. Exp. Bot.* **2012**, *63*, 4419–4436.
10. Yu, Z.M.; Kang, B.; He, X.W.; Lv, S.L.; Bai, Y.H.; Ding, W.N.; Chen, M.; Cho, H.T.; Wu, P. Root hair-specific expansin modulate root hair elongation in rice. *Plant J.* **2011**, *66*, 725–734.
11. Tabuchi, A.; Li, L.-C.; Cosgrove, D.J. Matrix solubilization and cell wall weakening by β-expansin (group-1 allergen) from maize pollen. *Plant J.* **2011**, *68*, 546–559. [CrossRef]
12. Han, Z.; Liu, Y.; Deng, X.; Liu, D.; Liu, Y.; Hu, Y.; Yan, Y. Genome-wide identification and expression analysis of expansin gene family in common wheat (*Triticum aestivum* L.). *BMC Genom.* **2019**, *20*, 101. [CrossRef]
13. Ludidi, N.; Heazlewood, J.; Seoighe, C.; Irving, H.; Gehring, C. Expansin-Like Molecules: Novel Functions Derived from Common Domains. *J. Mol. Evol.* **2002**, *54*, 587–594. [CrossRef]
14. Sampedro, J.; Carey, R.E.; Cosgrove, D.J. Genome histories clarify evolution of the expansin superfamily: New insights from the poplar genome and pine ESTs. *J. Plant Res.* **2006**, *119*, 11–21. [CrossRef]
15. Santo, S.D.; Vannozzi, A.; Tornielli, G.B.; Fasoli, M.; Venturini, L.; Pezzotti, M.; Zenoni, S. Genome-Wide Analysis of the Expansin Gene Superfamily Reveals Grapevine-Specific Structural and Functional Characteristics. *PLoS ONE* **2013**, *8*, e62206. [CrossRef]
16. Zhang, S.Z.; Xu, R.R.; Gao, Z.; Chen, C.T.; Jiang, Z.S.; Shu, H.R. A genome-wide analysis of the expansin genes in Malus Domestica. *Mol. Genet. Genom.* **2014**, *289*, 225–236. [CrossRef]
17. Li, J.; Yin, P.; Wang, H.R.; An, M.; Li, G.T. Identification of Actinidia Chinensis expansin gene family and expression under different stresses. *Mol. Plant Breed.* **2021**, 1–17. Available online: http://kns.cnki.net/kcms/detail/46.1068.S.20210820.1356.017.html (accessed on 3 February 2023).
18. Liang, Z.X.; Qi, H.Y.; Xu, H.G. Identification and bioinformatics analysis of hemp expansin gene family. *Mol. Plant Breed.* **2022**, 1–13. Available online: http://kns.cnki.net/kcms/detail/46.1068.S.20220630.1424.006.html (accessed on 3 February 2023).
19. Li, H.Y.; Shi, Y.; Ding, Y.N.; Xu, J.C. Bioinformatics analysis of expansin gene family in poplar genome. *Beijing Linye Daxue Xuebao* **2014**, *36*, 59–67.
20. Zhang, W. Identification of Endosperm-Specific Expression Genes and Functional Analysis of ZmEXPB13 in Maize EXPANSIN Family. Ph.D. Thesis, Anhui Agricultural University, Hefei, China, 2014.
21. Cho, H.-T.; Cosgrove, D.J. Altered expression of expansin modulates leaf growth and pedicel abscission in *Arabidopsis thaliana*. *Proc. Natl. Acad. Sci. USA* **2000**, *97*, 9783–9788. [CrossRef]
22. Won, S.-K.; Choi, S.-B.; Kumari, S.; Cho, M.; Lee, S.H.; Cho, H.-T. Root hair-specific EXPANSIN B genes have been selected for graminaceae root hairs. *Mol. Cells* **2010**, *30*, 369–376. [CrossRef] [PubMed]
23. Azeez, A.; Sane, A.P.; Tripathi, S.K.; Bhatnagar, D.; Nath, P. The gladiolus GgEXPA1 is a GA-responsive alpha-expansin gene expressed ubiquitously during expansion of all floral tissues and leaves but repressed during organ senescence. *Postharvest Biol. Technol.* **2010**, *58*, 48–56. [CrossRef]

24. Nardi, C.; Villarreal, N.M.; Rossi, F.; Martínez, S.; Martínez, G.A.; Civello, P.M. Overexpression of the carbohydrate binding module of strawberry expansin2 in Arabidopsis thaliana modifies plant growth and cell wall metabolism. *Plant Mol. Biol.* **2015**, *88*, 101–117. [CrossRef] [PubMed]
25. Yang, W.H.; Zeng, H.; Zou, M.H.; Lu, C.Z.; Huang, X.M. Progress of research on the relationship between fruit splitting occurrence and pericarp cell wall modification. *J. Trop. Crops* **2011**, *32*, 1995–1999.
26. Lu, P.; Kang, M.; Jiang, X.; Dai, F.; Gao, J.; Zhang, C. RhEXPA4, a rose expansin gene, modulates leaf growth and confers drought and salt tolerance to Arabidopsis. *Planta* **2013**, *237*, 1547–1559. [CrossRef]
27. Chen, Y.; Zhang, B.; Li, C.; Lei, C.; Kong, C.; Yang, Y.; Gong, M. A comprehensive expression analysis of the expansin gene family in potato (*Solanum tuberosum*) discloses stress-responsive expansin-like B genes for drought and heat tolerances. *PLoS ONE* **2019**, *14*, e0219837. [CrossRef]
28. Yuan, Z.; Fang, Y.; Zhang, T.; Fei, Z.; Han, F.; Liu, C.; Liu, M.; Xiao, W.; Zhang, W.; Wu, S.; et al. The pomegranate (*Punica granatum* L.) geneme provides insights into fruit quality and ovule developmental biology. *Plant Biotechnol J.* **2018**, *16*, 1363–1374. [CrossRef]
29. Qin, G.; Xu, C.; Ming, R.; Tang, H.; Guyot, R.; Kramer, E.M.; Hu, Y.; Yi, X.; Qi, Y.; Xu, X.; et al. The pomegranate (*Punica granatum* L.) geneme and the genomics of punicalagin biosynthesis. *Plant J.* **2017**, *91*, 1108–1128. [CrossRef]
30. Hassanen, E.I.; Tohamy, A.F.; Issa, M.Y.; Ibrahim, M.A.; Farroh, K.Y.; Hassan, A.M. Pomegranate juice diminishes the mitochondria-dependent cell death and NF-kB signaling pathway induced by copper oxide nanoparticles on liver and kidneys of rats. *Int. J Nanomed.* **2019**, *14*, 8905–8922. [CrossRef]
31. Pirzadeh, M.; Caporaso, N.; Rauf, A.; Shariati, M.A.; Yessimbekov, Z.; Khan, M.U.; Imran, M.; Mubarak, M.S. Pomegranate as a source of bioactive constituents: A review on their characterization, properties and applications. *Crit. Rev. Food Sci. Nutr.* **2021**, *61*, 982–999. [CrossRef]
32. Kandylis, P.; Kokkinomagoulos, E. Food Applications and Potential Health Benefits of Pomegranate and its Derivatives. *Foods* **2020**, *9*, 122. [CrossRef]
33. Akhtar, S.; Ismail, T.; Fraternale, D.; Sestili, P. Pomegranate peel and peel extracts: Chemistry and food features. *Food Chem.* **2015**, *174*, 417–425. [CrossRef]
34. Chen, L.; Zhang, J.; Li, H.; Niu, J.; Xue, H.; Liu, B.; Wang, Q.; Luo, X.; Zhang, F.; Zhao, D.; et al. Transcriptomic Analysis Reveals Candidate Genes for Female Sterility in Pomegranate Flowers. *Front. Plant Sci.* **2017**, *8*, 1430. [CrossRef]
35. Yuan, Z.; Yin, Y.; Qu, J.; Zhu, L.; Li, Y. Population Genetic Diversity in Chinese Pomegranate (*Punica granatum* L.) Cultivars Revealed by Fluorescent-AFLP Markers. *J. Genet. Genom.* **2007**, *34*, 1061–1071. [CrossRef]
36. Usha, T.; Goyal, A.K.; Lubna, S.; Prashanth, H.; Mohan, T.M.; Pande, V.; Middha, S.K. Identification of Anti-Cancer Targets of Eco-Friendly Waste Punica granatum Peel by Dual Reverse Virtual Screening and Binding Analysis. *Asian Pac. J. Cancer Prev.* **2014**, *15*, 10345–10350. [CrossRef]
37. Artimo, P.; Jonnalagedda, M.; Arnold, K.; Baratin, D.; Csardi, G.; de Castro, E.; Duvaud, S.; Flegel, V.; Fortier, A.; Gasteiger, E.; et al. ExPASy: SIB bioinformatics resource portal. *Nucleic Acids Res.* **2012**, *40*, W597–W603. [CrossRef]
38. Tamura, K.; Stecher, G.; Kumar, S. MEGA11: Molecular Evolutionary Genetics Analysis Version 11. *Mol. Biol. Evol.* **2021**, *38*, 3022–3027. [CrossRef]
39. Chen, C.; Chen, H.; Zhang, Y.; Thomas, H.R.; Frank, M.H.; He, Y.; Xia, R. TBtools: An Integrative Toolkit Developed for Interactive Analyses of Big Biological Data. *Mol. Plant* **2020**, *13*, 1194–1202. [CrossRef]
40. Bray, N.L.; Pimentel, H.; Melsted, P.; Pachter, L. Near-optimal probabilistic RNA-seq quantification. *Nat. Biotechnol.* **2016**, *34*, 525–527. [CrossRef]
41. Ophir, R.; Sherman, A.; Rubinstein, M.; Eshed, R.; Schwager, M.S.; Harel-Beja, R.; Bar-Ya'Akov, I.; Holland, D. Single-Nucleotide Polymorphism Markers from De-Novo Assembly of the Pomegranate Transcriptome Reveal Germplasm Genetic Diversity. *PLoS ONE* **2014**, *9*, e88998. [CrossRef]
42. Ono, N.N.; Britton, M.T.; Fass, J.N.; Nicolet, C.M.; Lin, D.; Tian, L. Exploring the Transcriptome Landscape of Pomegranate Fruit Peel for Natural Product Biosynthetic Gene and SSR Marker DiscoveryF. *J. Integr. Plant Biol.* **2011**, *53*, 800–813. [CrossRef] [PubMed]
43. Livak, K.; Schmittgen, T. Analysis of relative gene expression data using realtime quantitative PCR and the 2-ΔΔCT method. *Methods* **2001**, *25*, 402–408. [CrossRef] [PubMed]
44. Wu, T.; Zeng, N.; Li, W.; Wang, S.L.; Xu, F.S.; Shi, L. Genome-wide identification of the expansin gene family and differences in transcriptional responses to boron deficiency in *Brassica napus* L. *Plant Sci. J.* **2021**, *39*, 59–75.
45. Krishnamurthy, P.; Hong, J.K.; Kim, J.A.; Jeong, M.J.; Lee, Y.H.; Lee, S.I. Genome-wide analysis of the expansin gene superfamily reveals *Brassica rapa-specific* evolutionary dynamics upon whole genome triplication. *Mol. Genet. Genom.* **2015**, *290*, 521–530. [CrossRef]
46. Hao, X.; Li, X.Y.; La, G.X.; Dai, D.D.; Yang, T.G. Identification and bioinformatics analysis of cucumber expansin gene family in cucumber. *Fenzi Zhiwu Yuzhong Mol. Plant Breed.* **2015**, *13*, 2280–2289.
47. Ding, A.; Marowa, P.; Kong, Y. Genome-wide identification of the expansin gene family in tobacco (*Nicotiana tabacum*). *Mol. Genet. Genom.* **2016**, *291*, 1891–1907. [CrossRef]
48. Wang, R.X.; Yang, R.X.; Yin, P.; Liu, J.F.; Xu, J.C. Identification and characterization of the expansin gene in *Ginkgo biloba*. *Mol. Plant Breed.* **2021**, *19*, 1741–1749.

49. Shi, Y.; Xu, X.; Li, H.; Xu, Q.; Xu, J. Bioinformatics analysis of the expansin gene family in rice. *Hereditas* **2014**, *36*, 809–820. [CrossRef]
50. Lan, Y.C.; Huang, B.; Wei, J.; Jiang, S. Identification and bioinformatic analysis of the expansin gene family in *Physcomitrella patens*. *Guihaia* **2020**, *40*, 854–863.
51. Yang, R.X.; Liu, X.R.; Lan, B.L.; Wang, H.; Liu, X.; Xu, J.C. Genome identification and analysis of the expansin genes family in *Salix purpurea*. *Mol. Plant Breed.* **2021**, *19*, 2538–2549.
52. Hou, L.; Zhang, Z.; Dou, S.; Zhang, Y.; Pang, X.; Li, Y. Genome-wide identification, characterization, and expression analysis of the expansin gene family in Chinese jujube (*Ziziphus jujuba* Mill.). *Planta* **2019**, *249*, 815–829. [CrossRef]
53. Mao, Y. Preliminary Analysis of the Function of CpEXP2 gene of Plum Expansin. Master's Thesis, Southwest University, Chongqing, China, 2009.
54. Mayorga-Gómez, A.; Nambeesan, S.U. Temporal expression patterns of fruit-specific α- EXPANSINS during cell expansion in bell pepper (*Capsicum annuum* L.). *BMC Plant Biol.* **2020**, *20*, 241. [CrossRef]
55. Minami, A.; Yano, K.; Gamuyao, R.L.; Nagai, K.; Kuroha, T.; Ayano, M.; Nakamori, M.; Koike, M.; Kondo, Y.; Niimi, Y.; et al. Time-Course Transcriptomics Analysis Reveals Key Responses of Submerged Deepwater Rice to Flooding. *Plant Physiol.* **2018**, *176*, 3081–3102. [CrossRef]
56. Jiang, F.; Lopez, A.; Jeon, S.; de Freitas, S.T.; Yu, Q.; Wu, Z.; Labavitch, J.M.; Tian, S.; Powell, A.L.T.; Mitcham, E. Disassembly of the fruit cell wall by the ripening-associated polygalacturonase and expansin influences tomato cracking. *Hortic. Res.* **2019**, *6*, 17. [CrossRef]
57. Liu, Y.-L.; Chen, S.-Y.; Liu, G.-T.; Jia, X.-Y.; Haq, S.U.; Deng, Z.-J.; Luo, D.-X.; Li, R.; Gong, Z.-H. Morphological, physiochemical, and transcriptome analysis and CaEXP4 identification during pepper (*Capsicum annuum* L.) fruit cracking. *Sci. Hortic.* **2022**, *297*, 110982. [CrossRef]
58. Li, W.C.; Wu, J.Y.; Zhang, H.N.; Shi, S.Y.; Liu, L.Q.; Shu, B.; Liang, Q.Z.; Xie, J.H.; Wei, Y.Z. De Novo assembly and characterization of pericarp transcriptome and identification of candidate genes mediating fruit cracking in *Litchi chinensis* Sonn. *Int. J. Mol. Sci.* **2014**, *15*, 17667–17685. [CrossRef]
59. Kasai, S.; Hayama, H.; Kashimura, Y.; Kudo, S.; Osanai, Y. Relationship between fruit cracking and expression of the expansin gene MdEXPA3 in 'Fuji' apples (*Malus domestica* Borkh.). *Sci. Hortic.* **2008**, *116*, 194–198. [CrossRef]
60. Wang, Y.; Guo, L.; Zhao, X.; Zhao, Y.; Hao, Z.; Luo, H.; Yuan, Z. Advances in Mechanisms and Omics Pertaining to Fruit Cracking in Horticultural Plants. *Agronomy* **2021**, *11*, 1045. [CrossRef]
61. Zhu, Y.; Wu, N.; Song, W.; Yin, G.; Qin, Y.; Yan, Y.; Hu, Y. Soybean (*Glycine max*) expansin gene superfamily origins: Segmental and tandem duplication events followed by divergent selection among subfamilies. *BMC Plant Biol.* **2014**, *14*, 93. [CrossRef]
62. İncili, Ç.Y.; Arslan, B.; Çelik, E.N.Y.; Ulu, F.; Horuz, E.; Baloglu, M.C.; Çağlıyan, E.; Burcu, G.; Bayarslan, A.U.; Altunoglu, Y.C. Comparative bioinformatics analysis and abiotic stress responses of expansin proteins in Cucurbitaceae members: Watermelon and melon. *Protoplasma* **2023**, *260*, 509–527. [CrossRef]
63. Zhang, Q. Characterization and Functional Validation of Land Cotton Expansin Protein Gene Family. Master's Thesis, Northwest Agriculture and Forestry University, Xianyang, China, 2020.
64. Li, X.J.; Zao, H.L.; Lu, Y.C.; Zhao, Y.; Xiang, G.S.; Fan, W.; Yang, S.C. Identification and bioinformatic analysis of expansin family genes in *Panax notoginseng*. *Mol. Plant Breed.* **2021**, *19*, 6365–6375.
65. Zimmermann, R.; Sakai, H.; Hochholdinger, F. The *Gibberellic Acid Stimulated-Like* Gene Family in Maize and Its Role in Lateral Root Development. *Plant Physiol.* **2010**, *152*, 356–365. [CrossRef] [PubMed]
66. Yoo, S.; Gao, Z.; Cantini, C.; Loescher, W.H.; Nocker, S.V. Fruit ripening in sour cherry: Changes in expression of genes encoding expansins and other cell-wall-modifying enzymes. *J. Am. Soc. Hortic. Sci.* **2003**, *128*, 16–22. [CrossRef]
67. Wang, L.; Wang, Z.; Xu, Y.; Joo, S.-H.; Kim, S.-K.; Xue, Z.; Xu, Z.; Wang, Z.; Chong, K. *OsGSR1* is involved in crosstalk between gibberellins and brassinosteroids in rice. *Plant J.* **2009**, *57*, 498–510. [CrossRef]

horticulturae

MDPI

Article

Molecular Identification and Characterization of UDP-glycosyltransferase (UGT) Multigene Family in Pomegranate

Xueqing Zhao [1,2,*], Yingyi Feng [1,2], Ding Ke [1,2], Yingfen Teng [1,2], Ying Chen [3] and Renzeng Langjia [4]

1 Co-Innovation Center for Sustainable Forestry in Southern China, Nanjing Forestry University, Nanjing 210037, China; fengyingyi22@163.com (Y.F.); keding@njfu.edu.cn (D.K.); tyf3186521634@163.com (Y.T.)
2 College of Forestry, Nanjing Forestry University, Nanjing 210037, China
3 Zaozhuang Pomegranate Research Center, Zaozhuang 277300, China; dthy-0109@163.com
4 Forest Science Research Institute of Tibet, Lhasa 850000, China; 15728999555@163.com
* Correspondence: zhaoxq402@njfu.edu.cn

Abstract: Pomegranate (*Punica granatum* L.) is regarded as one of the functional fruits because of its large amounts of secondary metabolites. The glycosylation processes mediated by UDP-glycosyltransferases (UGTs) play a decisive role in regulating secondary metabolite availability. In this study, a genome-wide search identified 145 *UGT* genes in pomegranate, and further phylogenetic analysis defined 17 distinct groups: A to P and R. *PgUGT*s were dispersed unevenly across all eight chromosomes. Duplication events analysis revealed that both segmental and tandem duplications were the main mechanisms leading to gene family expansions. The comparison of exon–intron patterns identified 53 intron-less genes. A total of 24 types of cis-acting elements related to hormone, stress, and developmental responses were predicted in the promoter regions. Expression analysis of *PgUGT* genes using RNA-seq data and quantitative real-time PCR (qRT-PCR) verification suggested that *PgUGT* genes were expressed at specific stages of fruit development, and different *PgUGT* members likely played different roles in specific fruit developmental stages. In an attempt to identify the *UGT*s involved in the glycosylation of flavonoids, 44 *PgUGT*s were putatively determined, and 5 well-defined orthologous groups (OGs) were characterized by the regioselectivity of these enzymes. These results provide significant insight into the *UGT* multi-gene family in pomegranate, and will be helpful to further elucidate their roles involved in secondary and specialized metabolism in pomegranate.

Keywords: *Punica granatum*; UDP-glycosyltransferase; genome-wide identification; phylogenetic analysis; expression profile; flavonoid biosynthesis

Citation: Zhao, X.; Feng, Y.; Ke, D.; Teng, Y.; Chen, Y.; Langjia, R. Molecular Identification and Characterization of UDP-glycosyltransferase (UGT) Multigene Family in Pomegranate. *Horticulturae* **2023**, *9*, 540. https://doi.org/10.3390/horticulturae9050540

Academic Editor: Júlia Halász

Received: 10 April 2023
Revised: 24 April 2023
Accepted: 26 April 2023
Published: 28 April 2023

1. Introduction

Pomegranate (*Punica granatum* L.), native to Central Asia and the surrounding areas, is one of the members of the Lythraceae family [1]. Recent studies have proposed that pomegranate-derived secondary metabolites, especially polyphenolic compounds, are capable of exerting powerful antioxidant, anti-cancer, anti-inflammatory, antiparasitic, antihypertensive, and vascular-protective properties [2–4]. Consequently, pomegranate is well recognized as a 'super fruit' owing to its repository of bioactive substances, and thus, the plant acreages and fruit production of pomegranate have increased substantially over recent decades [5]. Given the potential applications of these bioactive compounds, it would be meaningful to comprehensively analyze the genes involved in bioactive compound biosynthesis.

Pomegranate accumulates a myriad of phenylpropanoid secondary metabolites, many of which are glycosylated in planta. Glycosylation, largely catalyzed by the glycosyltransferases (GTs), is a prominent modification reaction occurring in various biological processes.

In higher plants, glycosylation is usually regarded as the last step in the biosynthesis of secondary metabolites which affects the stability, solubility, and subsequent bioavailability of these metabolites [6]. GTs are a ubiquitous group of enzymes that can transfer sugar moieties from donor molecules to a wide range of acceptor substrates [7], thus generating a broad range of structurally diverse compounds. GTs account for approximately 1–2% of the gene products in the whole genome of an organism [8]. Additionally, GTs are one of the highly divergent multigene families due to their high degree of specificity and selectivity towards substrates. Until now, 115 GT families have been identified in the CAZy database (Carbohydrate Active enzymes Database, available online http://www.cazy.org), of which the GT1 family is the largest in plants and is commonly referred to as UDP-glycosyltransferases (UGTs) [9]. The majority of plant UGTs use UDP-glucose in the transfer reaction, while UDP-rhamnose, UDP-xylose, and UDP-galactose also exist [6]. Plant UGT sequences are characterized by a unique and conserved Plant Secondary Product Glycosyltransferase (PSPG) motif, a 44-amino-acid fragment at the C-terminal end of the protein [7,10], which is considered to be involved in the glycosylation of plant secondary metabolites or other natural products [11].

By taking advantage of the current wealth of omics-based resources, the comprehensive analyses of *UGT* families have been carried out, covering more than 40 plant species, and a large number of *UGT* family members have been found. In pomegranate, despite the fact that a wide range of secondary metabolites have been reported, including those glycosylated forms [12–15], only a few *UGT* genes have been identified and functionally characterized to be involved in secondary metabolite regulation. For instance, UGT95B2 preferentially glycosylates flavones/flavonols at more than one position in the molecule [16]. UGT84A23 and UGT84A24 exhibited β-glucogallin-forming activities [17]. UGT72BD1 used gallic acid as a substrate and produced a regiospecific product: gallic acid 4-O-glucoside [18]. Overall, the number of functionally characterized UGTs is still relatively low given the large abundance of UGTs in the genome of *P. granatum* L. Hence, in order to understand the biosynthesis pathway of secondary metabolites, it is necessary to systematically identify the *UGT* multigene family at the whole-genome level. The availability of the chromosomal-level genome of pomegranate presents an opportunity to explore the properties of *UGT* family genes in this versatile horticultural crop. In this study, we present a genome-wide analysis of *UGT* genes in pomegranate (*PgUGT*) based on the 'Tunisia' pomegranate genome data. All candidate *PgUGT* genes in the pomegranate genome were fully screened and a phylogenetic tree was constructed. Next, gene structural characteristics, including chromosome location, exon–intron structures, conserved motifs and cis-acting elements were analyzed. RNA-seq was carried out to investigate the expression profiles of *PgUGT* genes during fruit development phases. Finally, the phylogenetic trees were reconstructed and combined with those function-known UGTs from other plant species to explore UGT members acting towards flavonoid substrates. These results will contribute to future research elucidating the functions of *UGT*s in pomegranate.

2. Materials and Methods

2.1. Genome-Wide Identification of UGT Genes and Basic Physicochemical Properties of Proteins

The Hidden Markov Model (HMM) profile of the UDPGT domain (PF00201) was downloaded from Pfam website (http://pfam.xfam.org/, accessed on 6 July 2022). Then, the HMM model was built using the HMMER v3.0 software package (https://www.ebi.ac.uk/Tools/hmmer/, accessed on 7 July 2022) and was also searched against the pomegranate protein database (ASM765513V2) with E-values less than 1e−5. The candidate protein sequences of each UGT were further verified through the PFAM (http://pfam.xfam.org/, accessed on 8 July 2022), SMART (http://smart.embl-heidelberg.de/, accessed on 8 July 2022), and InterProScan (http://www.ebi.ac.uk/interpro/search/sequence/, accessed on 9 July 2022) databases to confirm the presence of the UDP-glycosyltransferase domain in order to remove the redundant sequences and isoforms.

Various physical and chemical parameters, including amino acid length (aa), molecular weight (MW), isoelectric point (pI), and instability index of all UGT proteins, were obtained using the online ExPASy program (http://web.expasy.org/protgaram, accessed on 11 July 2022) [19]. Transcript count for each *PgUGT* was obtained from the genome dataset. The subcellular localization of each PgUGT protein was predicted using the online CELLO V2.5 (http://cello.life.nctu.edu.tw, accessed on 12 July 2022).

2.2. Sequence Alignment and Phylogenetic Analysis

In order to classify PgUGTs based on a phylogenic tree, the selected amino acid sequences of PgUGTs were aligned with that from *Arabidopsis thaliana*, *Zea mays*, *Oryza sativa* and *Camellia sinensis* (Table S1) by MUSCLE in MEGA 11 (The Pennsylvania State University, University Park, PA, USA, 1993) with default settings [20]. A neighbor-joining (NJ) tree was constructed based on the JTT model and partial deletion using a 1000 bootstrap value. The phylogenetic tree was visualized and optimized by the online software Interactive Tree of Life program iTOL (https://itol.embl.de/, accessed on 15 July 2022) [21]. Grouping of PgUGTs was inferred by their phylogenetic position to the reference UGTs. To compare the phylogenetic groups of the UGTs in pomegranate and other species, we collected and summarized published records on the UGT families from other plant species.

2.3. Chromosomal Distribution of PgUGTs and Gene Duplications

The physical location of each *PgUGT* on the chromosome was retrieved based on pomegranate gene annotation files (GFF3 format) and visualized by MapChart 2.20 (Wageningen University & Research, Wageningen, The Netherlands, 2002) [22]. Segmental duplication events of the *PgUGT*s in the pomegranate genome were analyzed using TBtools software (South China Agricultural University, Guangzhou, Guangdong, China, 2020) [23]. The tandem duplications were defined according to the criteria published by Holub [24] and Gu et al. [25].

2.4. Exon–Intron Structures and Conserved Motifs

According to the general feature format file of *P. granatum*, the exon–intron structures of the *PgUGT*s were obtained and graphed with Tbtools software. The conserved motifs of the putative PgUGT proteins were predicted by using the on-line MEME procedure (http://meme-suite.org/, accessed on 2 August 2022) with maximum 15 motifs per sequence and annotated with InterPro database (https://www.ebi.ac.uk/interpro/, accessed on 4 August 2022). An image combined NJ phylogenetic tree, conserved motifs and gene exon–intron structure was drawn.

2.5. Promoter Cis-Acting Element Analysis

The 2000-bp sequences upstream of the start codons of *PgUGT* genes were extracted from pomegranate genome sequence by TBtools software. The presence of cis-acting elements was predicted by PlantCARE (http://bioinformatics.psb.ugent.be/webtools/plantcare/html/, accessed on 8 August 2022), and the results were visualized using TBtools.

2.6. Gene Expression Analysis Using RNA-seq

The fruits of four different developmental stages (10 June, 10 July, 10 August, and 10 September, designated as S1~S4, respectively) of the pomegranate 'Hongbaoshi' were used for transcriptomic analysis. For this, total RNA was firstly extracted according to the methods described by Yuan et al. [1], and the quality was assessed by electrophoresis and A260/A280. RNA-seq libraries were conducted and then sequenced on the Illumina HiSeq 2000 platform (5200 Illumina Way, San Diego, CA, USA, 2010). Three biological replicates for fruit development stages were prepared. The values of fragments per kilobase of per million mapped reads (FPKM) were used to calculate and evaluate transcript abundance. Genes with FPKM values <1.0 were defined to be minimally expressed and were removed from the data set. The FPKM values of *PgUGTs* were normalized with Log2, and heatmaps visualized by TBTools were presented to display the expression level of each *PgUGT*.

2.7. RNA Isolation, Reverse Transcription and Quantitative Real-Time PCR (qRT-PCR) Analysis

To validate the expression pattern of the selected genes, total RNA was extracted from fruit peels at the S1~S4 stages, respectively, by using RNA isolation system (Tiangen, Beijing, China), according to the manufacturer's instructions. The first-strand cDNAs were synthesized from the total RNA by using the RevertAid First Strand cDNA Synthesis Kit (Thermo Scientific, Waltham, MA, USA). qRT-PCR was performed with SuperReal PreMix Plus kit (Tiangen, Beijing, China) using the SYBR Green to detect gene expression. The primers used for qRT-PCR were listed in Table S2. The conditions for PCR conditions were performed following the previous reports [26]. The pomegranate *PgActin* (GenBank accession No. GU376750) was selected as an internal reference gene. Data from the individual runs were collated using the $2^{-\Delta\Delta CT}$ method [27]. All the reactions were performed using at least three replicates.

2.8. Functional Prediction of PgUGT Involved in Flavonoid Biosynthesis

In order to identify *PgUGTs* which participated in flavonoid biosynthetic pathway, 58 UGT proteins with known functions from other plant species were retrieved from Uniprot or NCBI database (Table S3). We constructed a NJ phylogenetic tree based on a collection of screened PgUGTs and 58 flavonoid UGT proteins to pre-screen flavonoid PgUGTs with parameters as described above in MEGA 11. The clustered PgUGTs were further filtered by removing those that do not use flavonoids as substrates when their functions were compared with annotation of *P. granatum* and *A. thaliana*. Subsequently, a phylogenetic tree based on the candidate flavonoid PgUGTs was reconstructed, and the filtered flavonoid UGTs were classified according to the method proposed by Yonekura-Sakakibara et al. [28].

3. Results

3.1. Identification of UGTs in Pomegranate

The comprehensive sequencing of the pomegranate genome greatly facilitated the identification of multi-gene families. In total, 180 candidate *PgUGT* genes were initially identified by HMM search. By subsequent verification of UDP-glycosyltransferase domain and removal of redundant sequences, a total of 145 putative *UGT* genes were screened and used for further analysis (Table S4). Most of the genes encoded proteins with the length in range of 400 to 500 amino acids, while only 24 were above 500 and 8 below 300 amino acids in size. Transcript counts for each gene were in a range of 1 to 4. The MW and pI ranged from 17.3 kDa to 60.9 kDa (average MW = 51.11 kDa) and from 4.62 to 9.89 (average pI = 5.82), respectively (Table S4). There were only 28 proteins with an instability index lower than 40, indicating that these UGT proteins were unstable. The predicted subcellular localization showed that 51, 28, 9, 2, and 2 of the encoded proteins were preferentially localized into cytoplasm, chloroplast, plasma membrane, mitochondria, and nucleus, respectively, while the rest were localized into any of these compartments (Table S4), revealing the various sub-cellular location of UGTs.

3.2. Phylogenetic Analysis of UGTs in Pomegranate

Accumulating phylogenetic analyses have demonstrated that plants UGTs could form 14 to 18 distinct groups depending on different plant species. In this study, we aligned all candidate 145 PgUGTs with 41 reference UGTs, including 32 from *Arabidopsis thaliana*, 4 from *Zea mays*, 3 from *Camellia sinensis*, and 2 from *Oryza sativa* to generate a phylogenetic tree for classification of PgUGTs. From Figure 1, we can see that 145 PgUGTs were clearly divided into 17 major phylogenetic groups, including 14 conserved that were defined as A-N according to the incorporated *Arabidopsis* UGTs in each group, and 3 additional new groups (O, P, R). Group Q did not contain any members of pomegranate.

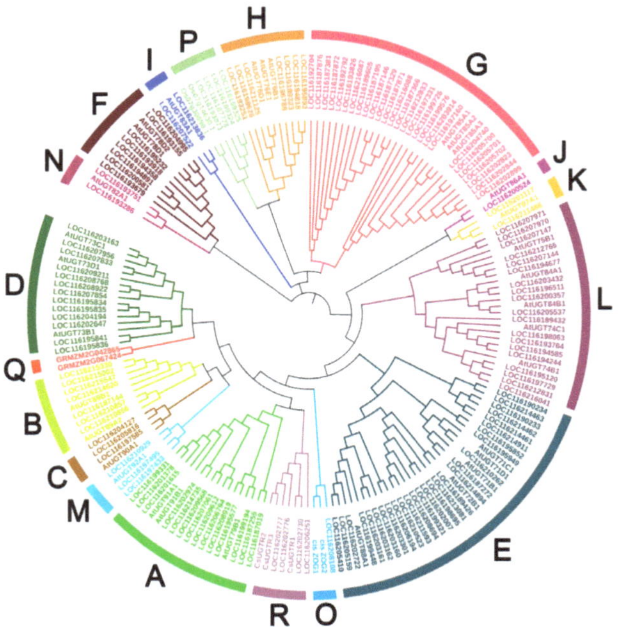

Figure 1. Phylogenetic analysis of UGT family genes in *P. granatum*. The phylogenetic tree was constructed by using the full-length sequences of 145 PgUGTs and 41 UGTs from *Arabidopsis thaliana*, *Zea mays*, *Camellia sinensis*, and *Oryza sativa*.

Next, we compared the number of *PgUGT*s in each phylogenetic group with that of other species available (Table 1). The results showed that the total number of *PgUGTs* was higher than that in white pear (*Pyrus bretschneideri*, 139); lower than apple (*Malus* × *domestica*, 237/241), Chinese bayberry (*Morella rubra*, 152), grape (*Vitis vinifera*, 228/181), and peach (*Prunus persica*, 168); and equal to pomelo (*Citrus grandis*). The members of the *PgUGT* family were unevenly distributed in groups A-P and R (Figure 1, Table 1). The most *PgUGT*s was observed in group E (28), followed by G (26), L (19), A (14), and D (13). There were 1, 4, and 4 members in the three new groups O, P, and R, respectively.

Table 1. Statistics of the number of *UGT*s in each phylogenetic groups from different plant species.

No.	Species	Total	A	B	C	D	E	F	G	H	I	J	K	L	M	N	O	P	Q	R	Reference
1	*Arabidopsis thaliana*	122	19	4	4	13	25	4	7	21	1	2	2	18	1	1	0	0	0	0	[29]
2	*Brassica napus*	251	17	10	10	36	48	2	14	35	2	3	6	61	4	3	0	0	0	0	[30]
3	*Brassica oleraca*	154	15	7	4	23	32	0	8	23	1	2	3	32	2	2	0	0	0	0	[30]
4	*Brassica rapa*	140	12	4	4	24	31	1	9	18	1	3	3	26	2	2	0	0	0	0	[30]
5	*Broussonetia papyrifera*	155	13	4	2	20	46	6	19	6	3	6	7	15	5	1	2	0	0	0	[31]
6	*Cajanus cajan*	120	2	2	1	36	33	0	9	0	5	0	0	12	2	0	6	12	0	0	[32]
7	*Camellia sinensis*	132	15	5	2	20	23	2	13	2	2	2	1	27	3	0	6	6	0	3	[33]
8	*Citrus grandis*	145	17	3	1	18	25	2	9	17	17	3	2	12	7	1	4	7	0	0	[34]
9	*Cucumis sativus*	85	10	1	2	12	13	0	11	5	0	2	1	17	2	1	3	5	0	0	[7]
10	*Epimedium pubescens*	339	105	9	23	24	49	5	15	17	11	6	6	49	2	1	15	0	2	0	[35]
11	*Gossypium arboreum*	146	20	8	3	21	28	7	13	7	7	4	1	13	4	1	1	8	0	0	[36]
12	*Gossypium arboreum*	143	19	7	1	21	30	5	9	8	7	4	1	17	3	1	1	8	0	1	[37]
13	*Gossypium barbadense*	220	26	12	2	34	37	10	14	13	14	5	2	29	5	1	2	10	0	4	[37]
14	*Gossypium hirsutum*	196	17	12	0	36	38	8	20	16	10	4	2	18	2	2	1	10	0	0	[36]
15	*Gossypium hirsutum*	220	27	14	2	29	41	12	14	15	11	5	2	27	4	2	2	9	0	4	[37]
16	*Gossypium raimondii*	152	15	9	2	20	30	8	5	8	7	4	1	15	4	1	1	12	0	0	[36]

Table 1. *Cont.*

No.	Species	Total	A	B	C	D	E	F	G	H	I	J	K	L	M	N	O	P	Q	R	Reference
17	*Gossypium raimondii*	149	13	9	1	21	29	8	5	8	7	4	1	20	3	1	1	12	0	6	[37]
18	*Glycine max*	182	25	3	1	43	36	1	15	3	18	3	2	19	4	1	5	3	0	0	[7]
19	*Glycine max*	149	5	1	2	38	46	6	16	2	4	0	0	18	5	0	6	0	0	0	[30]
20	*Glycine max*	208	21	3	0	46	52	8	16	3	17	7	0	19	5	1	6	4	0	0	[38]
21	*Linum usitatissimum*	137	16	5	6	21	22	1	19	6	9	4	5	19	3	1	0	0	0	0	[39]
22	*Lotus japonicus*	94	9	3	0	25	22	2	9	1	2	1	0	10	1	1	6	1	0	1	[38]
23	*Malus × domestica*	241	33	4	7	13	55	6	40	14	11	12	6	16	13	1	5	5	0	0	[7]
24	*Malus × domestica*	237	34	0	8	11	50	3	50	17	12	13	7	15	7	1	2	7	0	0	[40]
25	*Malus × domestica*	229	30	3	6	16	42	6	37	21	8	10	4	21	7	2	2	5	4	4	[41]
26	*Manihot esculenta*	121	14	4	3	18	18	3	12	7	11	2	3	22	3	1	0	0	0	0	[42]
27	*Medicago sativa*	409	0	4	2	100	62	6	0	13	134	0	10	43	9	0	10	14	0	2	[43]
28	*Medicago truncatula*	243	28	4	0	55	55	2	39	3	5	9	0	33	2	1	3	3	0	1	[38]
29	*Melilotus albus*	189	21	3	0	39	52	0	33	4	1	2	0	24	2	1	5	2	0	0	[44]
30	*Morella rubra*	152	8	8	5	9	33	4	26	11	0	8	8	24	4	0	4	0	0	0	[45]
31	*Nelumbo nucifera*	108	11	7	5	6	20	6	10	3	4	2	1	16	8	1	4	4	0	0	[46]
32	*Oryza sativa*	180	14	9	8	26	38	0	20	7	9	3	1	23	5	2	6	9	0	0	[7]
33	*Petunia hybrida*	129	27	1	2	12	20	2	8	3	5	6	1	10	5	1	22	4	0	0	[47]
34	*Phaseolus vulgaris*	168	19	3	2	33	33	5	18	3	15	3	0	17	4	1	6	5	0	1	[38]
35	*Populus trichocarpa*	178	12	2	6	14	49	0	42	5	5	6	2	23	6	1	3	2	0	0	[7]
36	*Prunus mume*	130	16	2	3	17	23	3	18	10	4	?	8	17	3	?	0	0	0	0	[48]
37	*Prunus persica*	168	10	2	4	19	29	4	34	9	5	7	7	18	14	1	1	4	0	0	[49]
38	*Punica granatum*	120	13	7	3	12	23	6	21	7	2	2	0	13	2	1	0	4	0	4	[50]
39	*Punica granatum*	145	14	7	3	13	28	8	27	7	2	1	2	19	3	2	1	4	0	4	This study
40	*Pyrus bretschneideri*	139	5	4	2	8	31	6	33	10	10	2	9	10	3	3	3	0	0	0	[51]
41	*Sorglum bicolor*	180	10	4	6	24	50	0	17	12	8	3	1	26	6	3	8	2	0	0	[7]
42	*Trifolium pratense*	121	11	3	0	29	39	1	13	3	1	2	0	12	1	0	2	3	0	1	[38]
43	*Triticum aestivum*	179	22	3	2	17	37	2	4	5	7	5	0	19	3	1	3	13	36	0	[52]
44	*Vitis vinifera*	228	25	4	6	9	45	8	29	7	13	7	2	33	5	1	3	0	0	0	[53]
45	*Vitis vinifera*	181	23	3	4	8	46	5	15	7	14	4	2	31	5	1	2	11	0	0	[7]
46	*Zea mays*	147	8	3	5	18	34	2	12	9	9	3	1	23	3	4	5	1	7	0	[54]

3.3. Chromosome Distribution and Gene Duplication

To summarize the genomic distribution of the *PgUGT* genes, the genetic mapping of *UGTs* on chromosomes was further investigated (Table S4, Figure 2). According to the genome annotation information retrieved from pomegranate genomic database, 143 *PgUGT*s were mapped on 8 specific chromosomes, and the remaining 2 *PgUGTs*, including *LOC116190233* and *LOC116190234*, could not be found on any specific pomegranate chromosome and were set on scaffolds. Chromosome 2 had the maximum number of 29 *PgUGT*s, whereas chromosome 6 contained the minimum 8 *PgUGT*s.

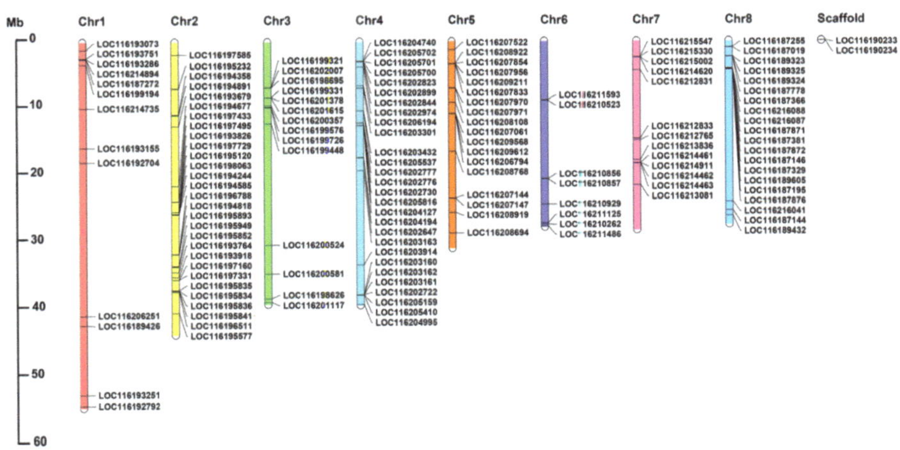

Figure 2. Chromosome distribution of *PgUGT* genes. The chromosome numbers are shown at the top of each chromosome.

The chromosome distribution of genes belonging to different groups is presented in Figure S1. We found that *PgUGTs* are distributed on chromosomes in clusters with sizes ranging from 2 to 6 genes per cluster, with a maximum 11 genes of the G group on chromosome 8.

To further understand how *PgUGT* genes were evolved, gene duplication events were further investigated in the pomegranate genome. Eventually, 23 segmental duplication events involving 32 genes were identified (Figure S2). In addition, 30 gene pairs were considered to originate from tandem duplication events (Table S5), and they were unequally distributed on all chromosomes except chromosome 1. These results suggested that both segmental and tandem duplication might play important roles for the generation of clusters of duplicated genes and for the expansion of the *PgUGT* family.

3.4. Conserved Motifs and Exon–Intron Organization of PgUGTs

To elucidate the structural features of *PgUGT* genes, the gene exon/intron structures and the protein motif structures were analyzed (Figure 3). In total, 15 MEME-predicted motifs were identified and subsequently annotated with the InterPro database. Motifs 1 and 2 were referred to the UDP-glucoronosyl and UDP-glucosyltransferase (UDPGT) domain, which are typical conserved domains found in all plant UGT proteins. The results showed that all PgUGTs contained motifs 1 and 2, indicating that the identification of PgUGT family members was reliable.

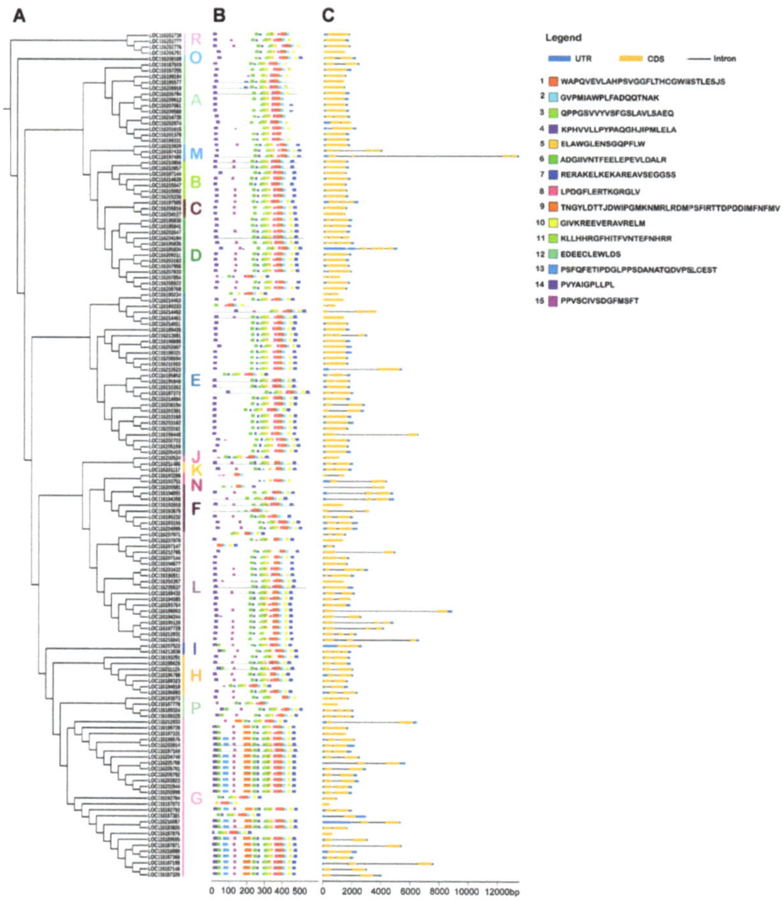

Figure 3. Conserved motifs and gene structure among *PgUGTs*. (**A**) The phylogenetic tree of 145 PgUGT proteins was constructed using the NJ method; (**B**) Conserved motifs in PgUGT proteins; (**C**) Exon–intron organization of *PgUGT* genes.

The number of motifs ranged from 4 to 15 in PgUGT sequences, and 19 PgUGT members contained all 15 motifs. In most PgUGT sequences, motif 4 was located at the N-terminal of the UGT sequence, and motif 7 was positioned in the C-terminal of the sequences. The results also demonstrated that the distribution of some motifs displayed group-specificity. For example, motif 13 was mainly present at the N-terminal of members in group G, while members of other groups did not contain motif 13. Similarly, motif 10 existed in group members at the C-terminal exclusive to B group. Motif 15 was not found in A, E, and O group members, but existed in most members of the D, F, H, G, and L groups. These differences in the distribution of these motifs might be related to the function differentiation of each group member. In general, the evolutionary relationship among PgUGT members was consistent with the types and locations of conserved motifs.

The characteristics of gene structure are an important basis for the analysis of the evolution and phylogeny of gene families. Exon–intron structure analysis indicated that the number of introns of 92 *PgUTGs* genes varied from 1 to 5, with 53 members lacking introns (Figure 3, Table S6). Out of the intron-containing *PgUGTs*, 58, 27, and 6 members had 1, 2, and 3 introns, respectively. Additionally, only one gene, *LOC116195834*, contained 5 introns. The characterization by fewer introns revealed a high conservation of gene structure in *PgUGT* gene family members.

In phylogenetic groups, the largest number of genes lacking introns was observed for group E, with 16 members, followed by 8 in A and 7 in B group. A total of 16 *PgUGTs* in group G contained 1 intron, followed by 9 in group L (Table S6). All members of the B and R groups had no introns. Generally, members within each group exhibited similar exon–intron organization style, which was consistent with the results obtained in the conserved motif structure. These results suggested that the *PgUGT* family members within groups were relatively conserved and diverged greatly among different groups.

3.5. Characterization of Cis-Acting Elements in the PgUGT Promoters

To understand the transcriptional regulation of *PgUGT* genes, the upstream promoter regions (2.0 kb in size) were used to predict potential cis-acting elements using the Plant-CARE database. A total of 24 types of cis-acting elements were observed in this study (Table 2, Figure S3). These elements were randomly distributed in the promoter regions of *PgUGTs* and were predicted to participate in hormone responses, stress responses, and developmental responses.

Among the cis-acting elements belonging to the hormone responses, abscisic acid-, auxin-, gibberellin-, MeJA-, and salicylic acid-responsive elements were observed in the promoters of *PgUGTs*, respectively. Stress-related response elements contained ARE, GC-motif, LTR, MBS, TC-rich repeats, and WUN-motif, which were involved in anaerobic induction, anoxic specific inducibility, low-temperature responsiveness, drought inducibility, defense and stress responsiveness, and wound responsiveness, respectively. This suggest that the expression of those genes containing the elements in promoters might be regulated by ambient pressure. Plant developmental elements contain cis-acting regulatory elements related to meristem expression (CAT-box), circadian control (circadian), endosperm expression (GCN4-motif), palisade mesophyll cells differentiation (HD-Zip 1), zein metabolism regulation (O2-site), flavonoid biosynthetic regulation (MYSI), seed-specific regulation (RY-element), and cell-cycle regulation (MSA), indicating that the *PgUGT* genes play vital roles in regulating physiological processes.

The number of cis-acting elements in genes was uneven (Tables 2 and S7). In particular, the most common motif was the ABRE elements associated with abscisic acid responsiveness, accounting for 30% of the hormone-responsive motifs, followed by the CGTCA motif, which was related to MeJA and accounted for 24% of the hormone-responsive motifs. Furthermore, there were 11 MYSI and 81 MBS belonging to MYB recognition and binding elements which were found in 10 and 63 *PgUGT* genes, revealing that the expression of these genes may be regulated by MYB transcription factors. Additionally, there were 92 and 13 *PgUGTs* containing LTR and WUN cis-acting elements, indicating that their transcription may be activated by low temperature and wound, respectively. Thus, the various cis-acting elements in the gene promoter region suggested that *PgUGT* genes play crucial roles in the complex hormone regulatory network and participate in diverse stress responses and secondary metabolite biosynthesis.

Table 2. Characteristics of cis-acting regulatory elements presented in the promoter regions of *PgUGT* genes.

Function	Promoter Name	Promoter Annotation	Total Number
Hormone response	ABRE	cis-acting element involved in abscisic acid responsiveness	393
	AuxRR-core	cis-acting regulatory element involved in auxin responsiveness	34
	CGTCA-motif	cis-acting regulatory element involved in MeJA responsiveness	312
	GARE-motif	gibberellin-responsive element	19
	P-box	gibberellin-responsive element	50
	TATC-box	cis-acting element involved in gibberellin responsiveness	30
	TCA-element	cis-acting element involved in salicylic acid responsiveness	70
	TGA-box	part of an auxin-responsive element	7
	TGA-element	auxin-responsive element	66
	TGACG-motif	cis-acting regulatory element involved in MeJA responsiveness	313
Stress response	ARE	cis-acting regulatory element essential for anaerobic induction	222
	GC-motif	enhancer-like element involved in anoxic-specific inducibility	52
	LTR	cis-acting element involved in low-temperature responsiveness	137
	MBS	MYB binding site involved in drought-inducibility	81
	TC-rich repeats	cis-acting element involved in defense and stress responsiveness	63
	WUN-motif	wound-responsive element	13
Developmental response	CAT-box	cis-acting regulatory element related to meristem expression	84
	circadian	cis-acting regulatory element involved in circadian control	34
	GCN4_motif	cis-regulatory element involved in endosperm expression	28
	HD-Zip 1	element involved in differentiation of palisade mesophyll cells	15
	O2-site	cis-acting regulatory element involved in zein metabolism regulation	73
	MBSI	MYB binding site involved in flavonoid biosynthetic gene regulation	11
	RY-element	cis-acting regulatory element involved in seed-specific regulation	12
	MSA-like	cis-acting element involved in cell cycle regulation	10

3.6. Temporal Expression Profiles of PgUGTs during Fruit Development

After filtering the low RNA-seq data, a total of 92 *PgUGT*s were further used to characterize the expression profiles in four fruit developmental stages. Next, the FPKM values were row-scaled, yielding normalized expression values for analysis. A hierarchical clustering analysis of their transcript levels indicated that these *PgUGT* genes could be divided into four expression patterns (Figure 4). For instance, the gene expression of cluster I and II exhibited low to medium levels, while cluster III and IV displayed a much higher level across fruit development. The 13 genes in cluster III maintained the highest level in four stages. Genes in Cluster IV were detected to display a fluctuated transcriptional abundance in S1~S4. These results suggested that the *PgUGT* genes in cluster III and IV may perform more vital glycosylation functions during pomegranate fruit development. Additionally, we found that the highly expressed *UGT*s were centered in groups D, E, G, L, and R (Figure S4).

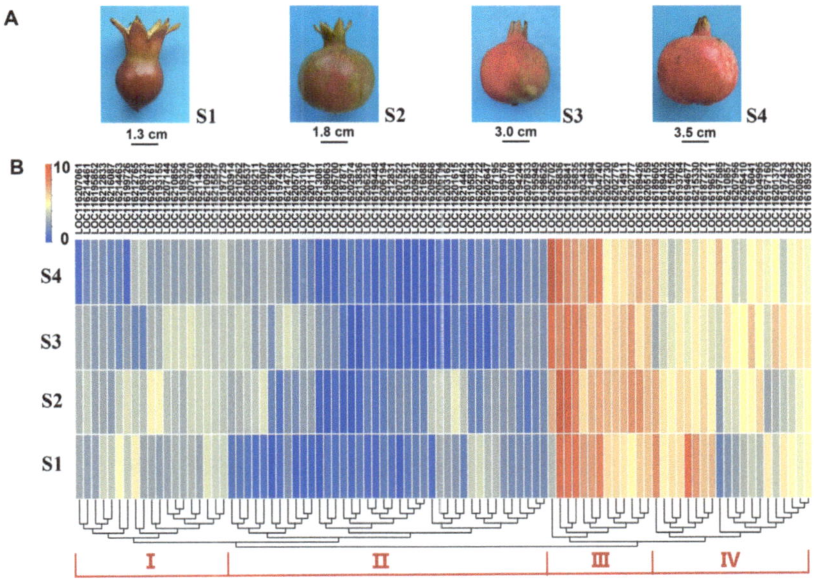

Figure 4. Expression profiles of *PgUGTs* across fruit developmental stages. (**A**) The fruits in four developmental stages. (**B**) The expression levels of *PgUGTs*. The scale represents the signal intensity of FPKM values. Red indicates a higher expression level, while blue indicates a lower expression level.

3.7. The Validation of PgUGT Expression with qRT-PCR

To confirm the reliability of the RNA-seq results, 12 representative *PgUGT* genes were selected and assayed using qRT-PCR (Figure 5). Most genes showed similar expression patterns with the FPKM values obtained by RNA-seq. The results of qRT-PCR in samples from S1 to S4 supported the reliability of the transcriptomic analysis described above.

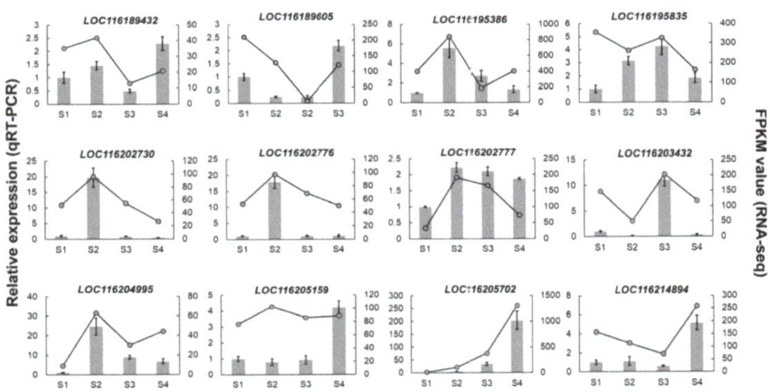

Figure 5. Validation of the expression of *PgUGTs* by qRT-PCR. Gray columns represent the expression levels tested by qRT-PCR; error bars indicate the standard deviation of qRT-PCR data (n = 3). For RNA-seq, each point is the mean of three biological replicates.

It can be seen that the selected *PgUGTs* exhibited different expression patterns during fruit development. Seven genes, including *LOC116189432*, *LOC116195386*, *LOC116202730*, *LOC116202776*, *LOC116202777*, *LOC116204995*, and *LOC116205159*, were highly expressed in the S2 stage, while there were two (*LOC116189605*, *LOC116195835*), one (*LOC116203432*), and two (*LOC116205702*, *LOC116214894*) genes which reached transcription summit in

S1, S3, and S4, respectively. These temporal gene expression patterns demonstrated that *PgUGTs* were more active in the S2 stage during fruit development.

3.8. Functional Prediction of PgUGT Involved in Flavonoid Biosynthesis

To determine putative *UGTs* for glycosylation reaction of flavonoids in pomegranate we generated an unrooted NJ phylogenetic tree based on the full-length proteins of PgUGTs and 58 UGTs which use mainly flavonoids as substrate acceptors. The preliminary results showed that 80 PgUGTs appeared to be involved in glycosylation of flavonoids by inferring their close phylogenetic relationships with known flavonoid UGT proteins in plants (Figure S5). The *PgUGTs* with flavonoid specificity fell into nearly all clustering groups except several minor groups, such as the J, K, M, N, and O groups. By subsequent functional screening, a total of putative 44 flavonoid PgUGTs were further used to reconstruct a NJ tree (Figure 6). The candidate *UGTs* for flavonoid biosynthesis were those distributed in A, B, C, D, E, F, L, and R groups. When these flavonoid UGTs were classified according to their regiospecificity, it can be seen that orthologous group 1 (OG1) occupied four PgUGT groups: B, C, D, and R. The other OGs each contained a single UGT group. Additionally, the high bootstrap values (>50%) on most nodes demonstrated the good reliability of the phylogenetic tree.

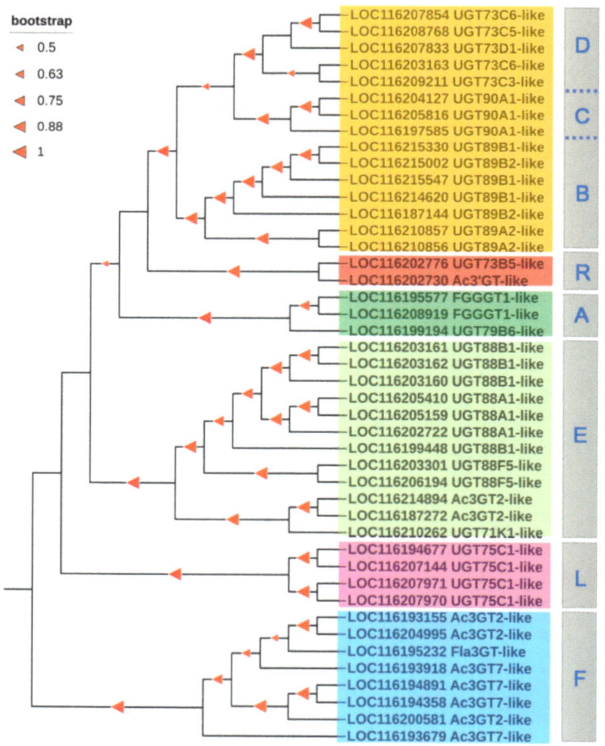

Figure 6. Phylogenetic tree of the flavonoid UGTs. A, B, C, D, E, F, L, and R are *PgUGT* groups as described in 3.2; OG, orthologous group; 3GT, 3-*O*-glycosyltransferase; 5GT, 5-*O*-glycosyltransferase; GGT, glycoside glycosyltransferase.

4. Discussion

Plant *UGT* is a large and functionally diverse family which plays an important role in the diversification of plant secondary metabolites. *PgUGTs* in 'Dabenzi' pomegranate have been recently reported [50]; however, the characterization of the *PgUGT* family in

'Tunisia', a noted soft-seeded cultivar, remains undefined. To deepen our understanding about the *UGT* family in pomegranate, we performed a genome-wide analysis on *PgUGTs*. Furthermore, the *UGTs* involved in flavonoid biosynthesis were identified and classified based on phylogenetic analysis, providing valuable information for further investigations into the catalytic functions of pomegranate UGTs.

4.1. The Classification of PgUGTs Based on Phylogenetic Tree

It has been reported that the ratio of *UGT* genes to total genes in vascular plants was in the range of 0.18% to 0.72%. In the present study, a total of 145 *UGT* genes were identified in the pomegranate 'Tunisia', accounting for approximately 0.4% of the total number of genes in the whole genome [55], lower than 0.5% in pomelo [34] and 0.6% in peach [49], but higher than maize (0.23%) [54] and soybean (0.26%) [30]. Recently, 120 *UGT* genes were screened in the 'Damenzi' pomegranate [50]. A possible reason for this discrepancy derived from the genome data of the two different cultivars 'Tunisia' and 'Dabenzi'.

Previous studies divided *UGTs* into A-N [56], O and P [7], Q [54], and R [33] groups successively. Based on phylogenetic analysis, we identified a total of 17 groups, including 14 groups (A-N) highly conserved in plants and 3 newly discovered ones: O, P, and R. However, this result was not consistent with that reported by Li et al. [50], who classified *PgUGTs* into 15 groups, lacking members in the K, O, and Q groups (Table 1). It was previously thought that group Q was specific to monocots, such as maize [54] and wheat [52]; however, the occurrence of Q members in *Malus* × *domestica* [41], *Epimedium pubescens* [35], and *Triticum aestivum* [52] suggested that group Q was not exclusive to monocots. Only one *UGT* in group R was found in *Camellia sinensis* [33] and other four species (*Lotus japonicus, Medicago truncatula, Phaseolus vulgaris,* and *Trifolium pratense*) [38], while there were 4, 4, and 6 *UGT* members belonging to the R group in *P. granatum*, *Malus* × *domestica* [41], and *Gossypium raimondii* [37], respectively, suggesting that group R may make an important contribution to the glycosylation of specific metabolites. Quantitatively, *PgUGTs* were concentrated in groups A, D, E, G, and L, indicating that members of these five groups expanded more rapidly than any other groups during plant evolution, as described by Caputi et al. [7]. In addition, group E contained the highest number of *UGT* members, consistent with views that group E expanded more than any other phylogenetic groups [7].

4.2. Segmental and Tandem Duplication Contribute to the Expansion of PgUGT Family

Gene duplication plays an important role in the occurrence of new gene functions and gene amplification. In plants, genomic duplications mainly arise from whole-genome duplication (WGD), segmental duplication, or tandem duplication [57]. Yuan et al. [1] reported that pomegranate underwent a paleotetraploidy event, resulting in at least two whole-genome duplications in the *P. granatum* genome. In this study, we identified segmental and tandem duplications between 23 and 30 *PgUGT* gene pairs, respectively, revealing that both tandem and segmental duplication events were the main driving forces for *PgUGT* expansion, which corresponded with previous work on the 'Dabenzi' pomegranate [50]. This phenomenon was also consistent with previous findings in *A. thaliana* [58], *Vitis vinifera* [53], and *Epimedium pubescens* [35]. However, other reports showed opposing observations that tandem duplication, rather than segmental duplication, was the major cause of *UGT* gene expansion [31,35,46]. In soybean, tandem duplication was not observed, and a series of segmental duplications caused *UGT* evolution [30]. In pear, segmental duplication was the dominate gene duplication event [51]. Together, these results demonstrated that the expansion of the *UGT* family driven by duplication events was species-specific.

4.3. PgUGT Transcription Analysis

Transcriptomic sequencing data can provide powerful complementary information to genomic analysis, guiding subsequent screening novel candidate genes for glycosylated secondary metabolite biosynthesis. The analysis results highlighted differential expression patterns, such as 33 *PgUGTs* (35.9%) showing high expression levels while 59 genes (64.1%)

showed low transcriptional levels. This result was inconsistent with the expression data of *Cier arietinum*, which showed 87.5% high-expression *CaUGT*s and 12.5% low-expression genes in developmental stages, indicating that the expression profiles were often species-, development-, and tissue-specific.

The highly expressed *PgUGT*s were centered in groups D, E, G, L, and R. It has been reported that members of these groups are involved in the glycosylation of a variety of polyphenols [7,56]. Therefore, the occurrence and high transcriptional levels of *UGT* members in D, E, G, and L groups may be associated with the diversity of phenolic compounds.

4.4. Identification of PgUGTs Involved in Flavonoid Biosynthesis

Flavonoids are abundant in vegetables, fruits, grains, and tea, and are known to have powerful antioxidant activity and provide a broad spectrum of health benefits [59]. In pomegranate, an increasing number of studies have described a large amount of flavonoids in fruits, flowers, leaves, seeds, and barks [60]. In plants, flavonoids are mostly glycosylated by UGTs with one or more sugar groups, leading to the diversity of flavonoids. Therefore, it is essential to identify UGT members with flavonoid substrate preferences.

In this study, a phylogenetic tree, constructed using stepwise-screened UGT sequences, identified 44 PgUGTs highly probably involved in flavonoid biosynthesis. Previously, the strategy of UGT specificity prediction has been used in *Vitis vinifera* [53], *Cicer arietinum* [61], *Epimedium pubescens* [35], and *Citrus sinensis* [62] by phylogenetic analysis with known UGT functions. According to the sugar acceptors and glycosidic linkages of characterized UGTs, Wilson et al. [63] placed UGTs with flavonoid acceptors into 11 groups, including A, B, C, D, E, F, G, H, L, Q, and R, while the A, B, D, E, F, and L groups were the most concentrated [7,33,35]. Our analysis confirmed the same distribution pattern of flavonoid PgUGTs among phylogenetic groups, suggesting the diversity and prosperity of glycosylation for flavonoid compounds in pomegranate.

It is generally accepted that sugar acceptor regiospecificity, rather than sugar donor specificity, is the basis for the clustering of flavonoid UGTs [64]. According to this criterion, flavonoid UGTs were categorized into unique clusters, including 3GT, 5GT, 7GT/3′GT, GGT, and CGT subfamilies [65,66]. Yonekura-Sakakibara et al. [28] classified the UGT families into 24 orthologous groups (OGs), while *A. thaliana* UGTs could be divided into ten OGs. In pomegranate, eight PgUGT members in the F group were categorized into OG23, which were regarded as flavonoid 3-*O*-glycosyltransferase belonging to the 3GT subfamily [28]. Four pomegranate UGT members in the L group were defined as UGT75C1-like proteins which may function as anthocyanin-5-O-glucosyltransferase as previously described [67]. Therefore, subfamily 5GT in group L clustered into OG14. There were three members belonging to A group, namely LOC116195577, LOC116208919, and LOC116199194; the former two were annotated as FGGT1-like (cyanidin 3-*O*-galactoside 2″-O-xylosyltransferase) and the last was UGT79B6-like. In Arabidopsis, UGT79B6 encodes a flavonoid GGT, also known as flavonoid 3-*O*-glucoside: 2″-O-glucosyltransferase [68]. Thus, flavonoid GGTs belonging to UGT79 group were divided into OG8.

The UGTs in both OG1 and OG7 showed functions as 7GT, 3′GT, GGT, and 3GT, demonstrating a broad plasticity in the position of glycosylation and formation of more than one glycoside product [28,65]. These UGT groups distributed into two OGs revealed that their function were established after divergence of OG1 and OG7 [69]. Until now, no function validation of UGT members in R group has been performed, though Cui et al. [33] inferred that UGTs belonging to R group in *Camellia sinensis* may be involved in the reaction of flavonoid glycosylation. Using phylogenetic analysis, members in R group were clustered closely with OG1, but distantly with OG7, indicating that the R group can be grouped into OG1 rather than OG7. Given that the genes in an OG diverged from a common ancestor, UGTs in R group may share a common origin with B, C, and D groups during plant lineage evolution.

This study tentatively identifies 44 *PgUGTs* involved in flavonoid biosynthesis on the basis of phylogenetic relationships. However, it is important to take into consideration that sometimes there were incongruences between the phylogenetic position and substrate specificities [7,69], which is to say that the same group may not use the same flavonoid as substrates, or distantly related UGTs glycosylate the same substrate, as reported in *Medicago truncatula* [70]. Thus, the selectivity for acceptors cannot only be inferred by their primary sequence information. Our results on flavonoid *PgUGTs* prediction should only be regarded as suggestive. It is essential to characterize the functions of glycosyltransferase enzyme combining with in vitro or in vivo assays.

5. Conclusions

In this study, a total of 145 *UGT* genes were identified from the genome of *Punica granatum* 'Tunisia', and the phylogenetic relationships, gene structure, gene duplications, and gene expression profiles of *PgUGT* were analyzed. In addition, 44 *PgUGT* members involved in flavonoid biosynthesis were tentatively identified. Their distribution in distinct OGs revealed the regiospecificity towards flavonoid substrates. Taken together, these data provide a useful basis for more precise annotation of each *PgUGT* and evaluation of those involved in glycosylation of a variety of secondary metabolites, especially flavonoid compounds. Further function confirmation of *PgUGTs* is required to be established by experimental evidence.

Supplementary Materials: The following supporting information can be downloaded at: https://www.mdpi.com/article/10.3390/horticulturae9050540/s1, Figure S1: The chromosome distribution of genes belonging to different groups; Figure S2: Segmental gene duplication events exhibited by *PgUGT* genes across eight pomegranate chromosomes; Figure S3: Cis-acting element analysis of the promoter regions of *PgUGT* genes; Figure S4: The expression profile of *PgUGT* members in different groups. Figure S5: Function preliminary prediction of PgUGTs involved in flavonoid biosynthesis; Table S1: Referenced UGTs used to construct phylogenetic tree; Table S2: The primers used in qRT-PCR; Table S3: Sequence information of 58 UGTs dataset; Table S4: Basic information of *UGT* genes identified in pomegranate; Table S5: The identified tandem duplication pairs of *PgUGT* genes; Table S6: Number of *PgUGTs* in each group according to intron amount; Table S7: The cis-acting elements of *PgUGTs*.

Author Contributions: Conceptualization, X.Z.; methodology, Y.F.; software, X.Z. and D.K.; validation, X.Z.; formal analysis, X.Z. and Y.T.; investigation, X.Z.; resources, Y.C. and R.L.; data curation, X.Z. and D.K.; writing—original draft preparation, X.Z.; writing—review and editing, X.Z., Y.F., D.K., Y.T.; Y.C. and R.L.; visualization, Y.F. and Y.T.; project administration, X.Z.; funding acquisition, X.Z. All authors have read and agreed to the published version of the manuscript.

Funding: This research was funded by National Natural Science Foundation of China (31901341) and the Priority Academic Program Development of Jiangsu High Education Institutions (PAPD).

Data Availability Statement: The RNA-seq data in this study were deposited in the NCBI Sequence Read Archive under the BioProject with the accession number PRJNA952822.

Acknowledgments: The authors are grateful to Zhaoxiang Hao of Zaozhuang Pomegranate Research Center for his help with the preparation of pomegranate samples.

Conflicts of Interest: The authors declare no conflict of interest.

References

1. Yuan, Z.; Fang, Y.; Zhang, T.; Fei, Z.; Han, F.; Liu, C.; Liu, M.; Xiao, W.; Zhang, W.; Wu, S.; et al. The pomegranate (*Punica granatum* L.) genome provides insights into fruit quality and ovule developmental biology. *Plant Biotechnol. J.* **2018**, *16*, 1363–1374. [PubMed]
2. Ge, S.S.; Duo, L.; Wang, J.; Yang, J.F.; Li, Z.Y.; Tu, Y. A unique understanding of traditional medicine of pomegranate, *Punica granatum* L. and its current research status. *J. Ethnopharmacol.* **2021**, *271*, 113877. [PubMed]
3. Ranjha, M.M.A.N.; Shafique, B.; Wang, L.; Irfan, S.; Safdar, M.N.; Murtaza, M.A.; Nadeem, M.; Mahmood, S.; Mueen-ud-Din, G.; Nadeem, H.R. A comprehensive review on phytochemistry, bioactivity and medicinal value of bioactive compounds of pomegranate (*Punica granatum*). *Adv. Tradit. Med.* **2023**, *23*, 37–57. [CrossRef]

4. Viuda-Martos, M.; Fernandez-Lopez, J.; Perez-Alvarez, J.A. Pomegranate and its many functional components as related to human health: A review. *Compr. Rev. Food Sci. Food Saf.* **2010**, *9*, 635–654.
5. Zhao, X.; Yuan, Z. Anthocyanins from pomegranate (*Punica granatum* L.) and their role in antioxidant capacities in vitro. *Chem Biodivers.* **2021**, *18*, e2100399. [PubMed]
6. Paquette, S.; Moller, B.L.; Bak, S. On the origin of family 1 plant glycosyltransferases. *Phytochemistry* **2003**, *62*, 399–413. [CrossRef]
7. Caputi, L.; Malnoy, M.; Goremykin, V.; Nikiforova, S.; Martens, S. A genome-wide phylogenetic reconstruction of family 1 UDP-glycosyltransferases revealed the expansion of the family during the adaptation of plants to life on land. *Plant J.* **2012**, *69*, 1030–1042. [CrossRef]
8. Albesa-Jove, D.; Guerin, M.E. The conformational plasticity of glycosyltransferases. *Curr. Opi. Struc. Biol.* **2016**, *40*, 23–32 [CrossRef]
9. Le Roy, J.; Huss, B.; Creach, A.; Hawkins, S.; Neutelings, G. Glycosylation is a major regulator of phenylpropanoid availability and biological activity in plants. *Front. Plant Sci.* **2016**, *7*, 735. [CrossRef]
10. Gachon, C.M.M.; Langlois-Meurinne, M.; Saindrenan, P. Plant secondary metabolism glycosyltransferases: The emerging functional analysis. *Trends Plant Sci.* **2005**, *10*, 542–549. [CrossRef]
11. Vogt, T.; Jones, P. Glycosyltransferases in plant natural product synthesis: Characterization of a supergene family. *Trends Plant Sci.* **2000**, *5*, 380–386. [CrossRef] [PubMed]
12. Fischer, U.A.; Carle, R.; Kammerer, D.R. Identification and quantification of phenolic compounds from pomegranate (*Punica granatum* L.) peel, mesocarp, aril and differently produced juices by HPLC-DAD-ESI/MSn. *Food Chem.* **2011**, *127*, 807–821.
13. Fourati, M.; Smaoui, S.; Ben Hlima, H.; Elhadef, K.; Ben Braiek, O.; Ennouri, K.; Mtibaa, A.C.; Mellouli, L. Bioactive compounds and pharmacological potential of pomegranate (*Punica granatum*) seeds—A review. *Plant Food Hum. Nutr.* **2020**, *75*, 477–486. [CrossRef] [PubMed]
14. Man, G.; Xu, L.; Wang, Y.; Liao, X.; Xu, Z. Profiling phenolic composition in pomegranate peel from nine selected cultivars using UHPLC-QTOF-MS and UPLC-QQQ-MS. *Front Nutr.* **2022**, *8*, 807447. [CrossRef] [PubMed]
15. Zeghad, N.; Abassi, E.A.; Belkhiri, A.; Demeyer, K.; Heyden, Y.V. Phenolic compounds profile from Algerian pomegranate fruit extract (*Punica granatum* L.) by UPLC-DAD-ESI-MS. *Chem. Afr.* **2022**, *5*, 1295–1303.
16. Wilson, A.E.; Wu, S.; Tian, L. PgUGT95B2 preferentially metabolizes flavones/flavonols and has evolved independently from flavone/flavonol UGTs identified in *Arabidopsis thaliana*. *Phytochemistry* **2019**, *157*, 184–193. [CrossRef]
17. Ono, N.N.; Qin, X.; Wilson, A.E.; Li, G.; Tian, L. Two UGT84 family glycosyltransferases catalyze a critical reaction of hydrolyzable tannin biosynthesis in pomegranate (*Punica granatum*). *PLoS ONE* **2016**, *11*, e0156319. [CrossRef]
18. Chang, L.; Wu, S.; Tian, L. Effective genome editing and identification of a regiospecific gallic acid 4-O-glycosyltransferase in pomegranate (*Punica granatum* L.). *Hortic. Res.* **2019**, *6*, 123. [CrossRef]
19. Artimo, P.; Jonnalagedda, M.; Arnold, K.; Baratin, D.; Csardi, G.; de Castro, E.; Duvaud, S.; Flegel, V.; Fortier, A.; Gasteiger, E.; et al. ExPASy: SIB bioinformatics resource portal. *Nucleic Acids Res.* **2012**, *40*, W597–W603. [CrossRef]
20. Tamura, K.; Stecher, G.; Kumar, S. MEGA11: Molecular evolutionary genetics analysis version 11. *Mol. Biol. Evol.* **2021**, *38*, 3022–3027. [CrossRef]
21. Letunic, I.; Bork, P. Interactive Tree of Life (iTOL) v5: An online tool for phylogenetic tree display and annotation. *Nucleic Acids Res.* **2021**, *49*, W293–W296. [CrossRef]
22. Voorrips, R.E. MapChart: Software for the graphical presentation of linkage maps and QTLs. *J. Hered.* **2002**, *93*, 77–78. [CrossRef] [PubMed]
23. Chen, C.; Chen, H.; Zhang, Y.; Thomas, H.R.; Xia, R. TBtools: An integrative toolkit developed for interactive analyses of big biological data. *Mol. Plant* **2020**, *13*, 1194–1202. [CrossRef] [PubMed]
24. Holub, E.B. The arms race is ancient history in Arabidopsis, the wildflower. *Nat. Rev. Genet.* **2001**, *2*, 516–527. [CrossRef] [PubMed]
25. Gu, Z.; Cavalcanti, A.; Chen, F.; Bouman, P.; Li, W. Extent of gene duplication in the genomes of *Drosophila*, nematode, and yeast. *Mol. Biol. Evol.* **2002**, *19*, 256–262. [CrossRef] [PubMed]
26. Zhao, X.; Yuan, Z.; Feng, L.; Fang, Y. Cloning and expression of anthocyanin biosynthetic genes in red and white pomegranate. *J. Plant Res.* **2015**, *128*, 687–696. [CrossRef]
27. Livak, K.J.; Schmittgen, T.D. Analysis of relative gene expression data using real-time quantitative PCR. *Methods* **2002**, *25*, 402–408. [CrossRef]
28. Yonekura-Sakakibara, K.; Hanada, K. An evolutionary view of functional diversity in family 1 glycosyltransferases. *Plant J.* **2011**, *66*, 182–193. [CrossRef]
29. Akere, A.; Chen, S.H.; Liu, X.; Chen, Y.; Dantu, S.C.; Pandini, A.; Bhowmik, D. Structure-based enzyme engineering improves donor-substrate regognition of *Arabidopsis thaliana* glycosyltransferases. *Bichem. J.* **2020**, *477*, 2791–2805. [CrossRef]
30. Rehman, H.M.; Nawaz, M.A.; Bao, L.; Shah, Z.H.; Lee, J.; Ahmad, M.Q.; Chung, G.; Yang, S.H. Genome-wide analysis of family-1 UDP-glycosyltransferases in soybean confirms their abundance and varied expression during seed development. *J. Plant Physiol.* **2016**, *206*, 87–97. [CrossRef]
31. Wang, F.; Su, Y.; Chen, N.; Shen, S. Genome-wide analysis of the UGT gene family and identification of flavonids in *Broussonetia papyrifera*. *Molecules* **2021**, *26*, 3449. [CrossRef]

32. Song, Z.; Niu, L.; Yang, Q.; Dong, B.; Wang, L.; Dong, M.; Fan, X.; Jian, Y.; Meng, D.; Fu, Y. Genome-wide identification and characterization of UGT family in pigeonpea (*Cajanus cajan*) and expression analysis in abiotic stress. *Trees* **2019**, *33*, 987–1002. [CrossRef]
33. Cui, L.; Yao, S.; Dai, X.; Yin, Q.; Liu, Y.; Jiang, X.; Wu, Y.; Qian, Y.; Pang, Y.; Gao, L.; et al. Identification of UDP-glycosyltransferases involved in the biosynthesis of astringent taste compounds in tea (*Camellia sinensis*). *J. Exp. Bot.* **2016**, *67*, 2285–2297. [CrossRef]
34. Wu, B.; Liu, X.; Xu, K.; Zhang, B. Genome-wide characterization, evolution and expression profiling of UDP-glycosyltransferase family in pomelo (*Citrus grandis*) fruit. *BMC Plant Biol.* **2020**, *20*, 459. [CrossRef] [PubMed]
35. Yao, Y.; Gu, J.; Luo, Y.; Wang, Y.; Pang, Y.; Shen, G.; Guo, B. Genome-wide analysis of UGT gene family identified key gene for the biosynthesis of bioactive flavonol glycosides in *Epimedium pubescens* Maxim. *Syn. Syst. Biotechnol.* **2022**, *7*, 1095–1107. [CrossRef] [PubMed]
36. Huang, J.; Pang, C.; Fan, S.; Song, M.; Yu, J.; Wei, H.; Ma, Q.; Li, L.; Zhang, C.; Yu, S. Genome-wide analysis of the family 1 glycosyltransferases in cotton. *Mol. Genet. Genom.* **2015**, *290*, 1805–1818. [CrossRef] [PubMed]
37. Sun, L.; Zhao, L.; Huang, H.; Zhang, Y.; Wang, J.; Lu, X.; Wang, S.; Wang, D.; Chen, X.; Chen, C.; et al. Genome-wide identifcation, evolution and function analysis of UGTs superfamily in cotton. *Front. Mol. Biosci.* **2022**, *9*, 965403. [CrossRef]
38. Krishnamurthy, P.; Tsukamota, C.; Ishimoto, M. Reconstruction of the evolutionary histories of UGT gene sumperfamily in *Legumes* clarifies the functional divergence of duplicates in specialized metabolism. *Int. J. Mol. Sci.* **2020**, *21*, 1855. [CrossRef]
39. Barvkar, V.T.; Pardeshi, V.C.; Kale, S.M.; Kadoo, N.Y.; Gupta, V.S. Phylogenomic analysis of UDP glycosyltransferase 1 multigene family in *Linum usitatissimum* identified genes with varied expression patterns. *BMC Genom.* **2012**, *13*, 175. [CrossRef]
40. Zhou, K.; Hu, L.; Li, P.; Gong, X.; Ma, F. Genome-wide identification of glycosyltransferases converting phloretin to phloridzin in *Malus* species. *Plant Sci.* **2017**, *265*, 131–145. [CrossRef]
41. Li, Y.; Li, P.; Zhang, L.; Shu, J.; Court, M.H.; Sun, Z.; Jiang, L.; Zheng, C.; Shu, H.; Ji, L.; et al. Genome-wide analysis of the apple family 1 glycosyltransferases identified a flavonoid-modifying UGT, MdUGT83L3, which is targeted by MdMYB88 and contributes to stress adaptation. *Plant Sci.* **2022**, *321*, 111314. [CrossRef] [PubMed]
42. Wu, C.; Dai, J.; Chen, Z.; Tie, W.; Yan, Y.; Yang, H.; Zeng, J.; Hu, W. Comprehensive analysis and expression profiles of cassava UDP-glycosyltransferases (UGT) family reveal their involvement in development and stress responses in cassava. *Genomics* **2021**, *113*, 3415–3429. [CrossRef]
43. Ao, B.; Han, Y.; Wang, S.; Wu, F.; Zhang, J. Genome-Wide analysis and profile of UDP-glycosyltransferases family in alfalfa (*Medicago sativa* L.) under drought stress. *Int. J. Mol. Sci.* **2022**, *23*, 7243. [CrossRef] [PubMed]
44. Duan, Z.; Yan, Q.; Wu, F.; Wang, Y.; Wang, S.; Zong, X.; Zhou, X.; Zhang, J. Genome-wide analysis of the UDP-glycosyltransferase family reveals its role in coumarin biosynthesis and abiotic stress in *Melilotus albus*. *Int. J. Mol. Sci.* **2021**, *22*, 10826. [CrossRef]
45. Ren, C.; Guo, Y.; Xie, L.; Zhao, Z.; Xing, Z.; Cao, Y.; Liu, Y.; Lin, J.; Grierson, D.; Zhang, B.; et al. Identification of UDP-rhamnosyltransferases and UDP-galactosyltransferase involved in flavonol glycosylation in *Morella rubra*. *Hortic. Res.* **2022**, *9*, uhac138. [CrossRef] [PubMed]
46. Li, H.; Yang, X.; Lu, M.; Chen, J.; Shi, T. Gene expression and evolution of family-1 UDPglycosyltransferases—Insights from an aquatic flowering plant (sacred lotus). *Aquat. Bot.* **2020**, *166*, 103270. [CrossRef]
47. Dong, L.; Tang, Z.; Yang, T.; Hao, F.; Deng, X. Genome-wide analysis of UGT genes in *Petunia* and identification of PhUGT51 involved in the regulation of salt resistance. *Plants* **2022**, *11*, 2434. [CrossRef]
48. Zhang, Z.; Zhuo, X.; Yan, X.; Zhang, Q. Comparative genomic and transcriptomic analyses of family-1 UDP glycosyltransferase in *Prunus Mume*. *Int. J. Mol. Sci.* **2018**, *19*, 3382. [CrossRef]
49. Wu, B.; Gao, L.; Gao, J.; Xu, Y.; Liu, H.; Cao, X.; Zhang, B.; Chen, K. Genome-wide identification, expression patterns, and functional analysis of UDP glycosyltransferase family in peach (*Prunus persica* L. Batsch). *Front. Plant Sci.* **2017**, *8*, 389. [CrossRef]
50. Li, G.; Li, J.; Qin, G.; Liu, C.; Liu, X.; Cao, Z.; Jia, B.; Zhang, H. Characterization and expression analysis of the UDP glycosyltransferase family in pomegranate (*Punica granatum* L.). *Horticulturae* **2023**, *9*, 119. [CrossRef]
51. Cheng, X.; Muhammad, A.; Li, G.; Zhang, J.; Cheng, J.; Qiu, J.; Jiang, T.; Jin, Q.; Cai, Y.; Lin, Y. Family-1 UDP glycosyltransferases in pear (*Pyrus bretschneideri*): Molecular identification, phylogenomic characterization and expression profiling during stone cell formation. *Mol. Biol. Rep.* **2019**, *46*, 2152–2175. [CrossRef] [PubMed]
52. He, Y.; Ahmad, D.; Zhang, X.; Zhang, Y.; Wu, L.; Jiang, P.; Ma, H. Genome-wide analysis of family-1 UDP glycosyltransferases (UGT) and identification of *UGT* genes for FHB resistance in wheat (*Triticum aestivum* L.). *BMC Plant Biol.* **2018**, *18*, 67. [CrossRef] [PubMed]
53. Wei, Y.; Mu, H.; Xu, G.; Wang, Y.; Li, S.; Wang, L. Genome-wide analysis and functional characterization of the UDP-glycosyltransferase family in grapes. *Horticulturae* **2021**, *7*, 204. [CrossRef]
54. Li, Y.; Li, P.; Wang, Y.; Dong, R.; Yu, H.; Hou, B. Genome-wide identification and phylogenetic analysis of family-1 UDP-glycosyltransferases in maize (*Zea mays*). *Planta* **2014**, *239*, 1256–1279. [CrossRef]
55. Luo, X.; Li, H.; Wu, Z.; Yao, W.; Cao, S. The pomegranate (*Punica granatum* L.) draft genome dissects genetic divergence between soft- and hard-seeded cultivars. *Plant Biotechnol. J.* **2019**, *18*, 955–968.
56. Ross, J.; Li, Y.; Lim, E.; Bowles, D.J. Higher plant glycosyltransferases. *Genome Biol.* **2001**, *2*, REVIEWS3004. [CrossRef]
57. Panchy, N.; Lehti-Shiu, M.; Shiu, S. Evolution of gene dulication in plants. *Plant Physiol.* **2016**, *171*, 1194–2316. [CrossRef]
58. Cannon, S.B.; Mitra, A.; Baumgarten, A.; Young, N.D.; May, G. The roles of segmental and tandem gene duplication in the evolution of large gene families in *Arabidopsis thaliana*. *BMC Plant Biol.* **2004**, *4*, 10. [CrossRef]

59. Panche, A.N.; Diwan, A.D.; Chandra, S.R. Flavonoids: An overview. *J. Nutr. Sci.* **2016**, *5*, e47. [CrossRef]
60. Zhao, X.; Shen, Y.; Yan, M.; Yuan, Z. Flavonoid profiles in peels and arils of pomegranate cutlivars. *J. Food Meas. Charact.* **2022**, *16*, 880–890. [CrossRef]
61. Sharma, R.; Rawat, V.; Suresh, C.G. Genome-wide identfication and tissue-specific expression analysis of UDP-glycosyltransferases genes confirm their abundance in *Cicer arietinum* (Chickea). *PLoS ONE* **2014**, *9*, e109715. [CrossRef] [PubMed]
62. Liu, X.; Lin, C.; Ma, X.; Tan, Y.; Wang, J.; Zeng, M. Functional characterization of a flavonoid glycosyltransferase in sweet orange (*Citrus sinensis*). *Front. Plant Sci.* **2018**, *9*, 166. [CrossRef] [PubMed]
63. Wilson, A.E.; Tian, L. Phylogenomic analysis of UDP-dependent glycosyltransferases provides insights into the evolutionary landscape of glycosylation in plant metabolism. *Plant J.* **2019**, *100*, 1273–1288. [CrossRef] [PubMed]
64. Noguchi, A.; Horikawa, M.; Fukui, Y.; Fukuchi-Mizutani, M.; Iuchi-Okada, A.; Ishiguro, M.; Kiso, Y.; Nakayama, T.; Ono, E. Local differentiation of sugar donor specificity of flavonoid glycosyltransferase in Lamiales. *Plant Cell* **2009**, *21*, 1556–1572. [CrossRef]
65. Yonekura-Sakakibara, K.; Saito, K. Function, structure, and evolution of flavonoid glycosyltransferases in Plants. In *Recent Advances in Polyphenol Research*; Romani, A., Lattanzio, V., Quideau, S., Eds.; John Wiley & Sons, Ltd.: Hoboken, NJ, USA, 2014; pp. 61–82.
66. Peng, M.; Shahzad, R.; Gul, A.; Subthain, H.; Shen, S.; Lei, L.; Zheng, Z.; Zhou, J.; Lu, D.; Wang, S.; et al. Differentially evolved glucosyltransferases determine natural variation of rice flavone accumulation and UV-tolerance. *Nat. Commun.* **2017**, *8*, 1875. [CrossRef]
67. Tohge, T.; Nishiyama, Y.; Hirai, M.Y.; Yano, M.; Nakajima, J.; Awazuhara, M.; Inoue, E.; Takahashi, H.; Goodenowe, D.B.; Kitajima, M.; et al. Functional genomics by integrated analysis of metabolome and transcriptome of Arabidopsis plants over-expressing an MYB transcription factor. *Plant J.* **2005**, *42*, 218–235. [CrossRef]
68. Yonekura-Sakakibara, K.; Nakabayashi, R.; Sugawara, S.; Tohge, T.; Ito, T.; Koyanagi, M.; Kitajima, M.; Takayama, H.; Saito, K. A flavonoid 3-O-glucoside:2″-O-glucosyltransferase responsible for terminal modification of pollen-specific flavonols in *Arabidopsis thaliana*. *Plant J.* **2014**, *79*, 769–782. [CrossRef]
69. Osmani, S.A.; Bak, S.; Møller, B.L. Substrate specificity of plant UDP-dependent glycosyltransferases predicted from crystal structures and homology modeling. *Phytochemistry* **2009**, *70*, 325–347. [CrossRef]
70. Modolo, L.V.; Blount, J.W.; Achnine, L.; Naoumkina, M.; Wang, X. A functional genomics approach to (iso)flavonoid glycosylation in the model legume *Medicago truncatula*. *Plant Mol. Biol.* **2007**, *64*, 499–518. [CrossRef]

horticulturae

Article

Unraveling the Pomegranate Genome: Comprehensive Analysis of R2R3-MYB Transcription Factors

Heming Suo [1], Xuan Zhang [1], Lei Hu [1], Huihui Ni [1], Renzeng Langjia [2], Fangyu Yuan [1], Maowen Zhang [1] and Shuiming Zhang [1,*]

[1] Department of Ornamental Horticulture, School of Horticulture, Anhui Agricultural University, Hefei 230036, China; shm1220073202@163.com (H.S.); xuan_xuan1211@163.com (X.Z.); l191563178@163.com (L.H.); nhh1477852456@163.com (H.N.); yuanfangyug@163.com (F.Y.)

[2] Forest Science Research Institute of Tibet, Lhasa 850000, China; rzenglangjia@163.com

* Correspondence: zhangshuiming@ahau.edu.cn

Abstract: R2R3-MYB TFs represent one of the most extensive gene families in plants and play a crucial role in regulating plant development, metabolite accumulation, and defense responses. Nevertheless, there has been no systematic investigation into the pomegranate R2R3-MYB family. In this study, 186 R2R3-MYB genes were identified from the pomegranate genome and grouped into 34 subgroups based on phylogenetic analysis. Gene structure analysis showed that the PgR2R3-MYB family in the same subgroup had a similar structure. Gene duplication event analysis revealed that the amplification of the PgMYB family was driven by Whole Genome Duplication (WGD) and dispersed duplication. In the upstream promoter sequence of the PgMYB gene, we identified a large number of plant hormones and environmental response elements. Using phylogenetic analysis and RNA-seq analysis, we identified three PgMYB TFs that may be involved in the regulation of lignin synthesis. Their expression patterns were verified by qPCR experiments. This study provides a solid foundation for further studies on the function of the R2R3-MYB gene and the molecular mechanism of lignin synthesis.

Keywords: lignin; genome-wide identification; phylogenetic analysis; gene expression

Citation: Suo, H.; Zhang, X.; Hu, L.; Ni, H.; Langjia, R.; Yuan, F.; Zhang, M.; Zhang, S. Unraveling the Pomegranate Genome: Comprehensive Analysis of R2R3-MYB Transcription Factors. *Horticulturae* **2023**, *9*, 779. https://doi.org/10.3390/horticulturae9070779

Academic Editor: Sherif M. Sherif

Received: 19 May 2023
Revised: 25 June 2023
Accepted: 6 July 2023
Published: 7 July 2023

1. Introduction

Pomegranate (*Punica granatum*) is an ancient cultivar that originated in Central Asia and spread throughout the world. There are two species of *Punica L.* The *P. granatum* is widely distributed throughout the world, while *Punica protopunica Balf. f.* is only found in certain regions. Pomegranates have been cultivated and evolved in China for more than 2000 years to form numerous varieties with rich genetic diversity [1,2]. Pomegranate is generally considered a healthy fruit [3]. Pomegranate seeds are rich in nutrients, including organic acids, sugars, minerals, and many other nutrients [4,5]. They are widely used in traditional medicine in the United States, Asia, Africa, and Europe to treat different types of diseases and have high edible and medicinal value [6]. Consumers prefer soft-seed pomegranate in market, which encourages investigators and breeders to progress and breed soft-seed cultivars [7].

The formation of pomegranate seed hardness is the process of continuous lignin accumulation in the endocarp cells and gradual thickening of the secondary cell wall, which is also the process of lignification of the endocarp cell wall [8]. Lignin, as the main component of the cell wall, interconnects with lignocellulose and hemicellulose to form a meshwork structure that forms the skeleton of plant cells [9]. Niu et al., pointed out that the seed coat cell wall thickness of hard-seeded pomegranate was significantly greater than that of soft-seeded pomegranate [10]. It has also been shown that pomegranate seed hardness is significantly and positively correlated with its lignin content [11,12].

Transcription factors (TFs) regulate gene expression by binding to distal and local cis elements of target genes. TFs are regulatory factors that play important roles in plant development, cell cycle, cell signaling and response to adversity. Typical TFs include transcriptional regulatory regions, nuclear localization signaling regions, oligomerization response regions, and DNA binding regions, and these functional regions determine the structure and characteristics of TFs [13]. Common TFs in plants are WRKY, ERF, bZIP, MYB, and other families, out of which the MYB family is one of the largest TFs families in plants. The MYB gene family is widely present in eukaryotic cells and is one of the largest in plants. The MYB protein consists of 1–4 highly conserved imperfect repeats (R1, R2, and R3). The MYB gene can be divided into four different types, namely 2R (R2R3-MYB), 3R (R1R2R3-MYB), 4R (R1R2R2R1/2-MYB), and 1R-MYB (MYB-related proteins), of which the R2R3-MYB gene accounts for the larger proportion [14]. R2R3-MYB is probably the TF that most directly regulates lignin biosynthesis and deposition, and it can regulate the expression of genes involved in the synthesis of phenylpropane substances, thereby affecting the content of lignin [15,16]. R2R3-MYB TFs play an important role in plant growth and development, such as regulating biological and abiotic stresses and affecting the synthesis of lignin and anthocyanins [17].

In recent years, research of the R2R3-MYB gene has made great progress, and members of this family have been identified in Arabidopsis [18], grape [19], kiwifruit [20], watermelon [21], Pitaya [22], Rhodiola [16], and apples [23]. According to previous studies, in cotton, GhMYB4 could impede lignin deposition in the cell wall by directly binding to and negatively regulating the expression of genes such as GhC4H-1/2, Gh4CL-4, GhCAD-3, and GhLac1, and the reduction of lignin content in GhMYB4 overexpressing cotton leads to enhanced cell wall permeability [24]. The Arabidopsis R2R3-MYB gene family was divided into 24 subgroups [18]. AtMYB58, AtMYB63, and AtMYB85 were TFs specific for lignin biosynthesis in Arabidopsis [25]. AtMYB46 and AtMYB83 were not only regulators of the lignin synthesis pathway but also effectively activated the entire process of secondary cell wall formation [26,27]. The overexpression of the OsMYB91 gene increased the ABA content in the plant and enhanced the ability of rice to resist drought stress [28]. MaMYB13 induced cold resistance by regulating the synthesis of VLCFAs and phenylpropane compounds in bananas [29].

The birth of new cultivars of fruit crops takes a long time, so the identification of related genes is an important work [3,30]. The purpose of this study was to search for lignin-related MYB family members and understand their expression patterns, so as to provide theoretical basis for breeding better soft-seed pomegranate cultivars. For this purpose, the pomegranate R2R3-MYB gene family was identified based on the pomegranate genome. We constructed phylogenetic trees of these genes, conducted intraspecific and interspecific collinearity analyses, and analyzed the structure and promoter cis-acting elements of the PgMYB family. In addition, an evolutionary tree was constructed using MYB members from other species known to regulate lignin synthesis to identify the MYB most likely to be involved in pomegranate lignin synthesis. Further, we selected soft-seed 'Tunisia' and hard-seed cultivars 'Hongyushizi' and 'Baiyushizi' (the lowest lignin content was found in seeds of Tunisia [31]) and drew an expression heatmap with their transcriptomic data, and the genes concerned were verified by qRT-PCR.

2. Materials and Methods

2.1. Plant Material

In this study, samples were collected from three selected cultivars: hard-seed pomegranates 'Hongyushizi' and 'Baiyushizi' and soft-seed pomegranate 'Tunisia'. They were nine-year-old cultivars planted in the experimental base of Anhui Agricultural University. Samples were collected 40, 80, and 120 days after anthesis, and three biological replicates were set for each sample. The seeds and arils were immediately extracted, placed in liquid nitrogen, and finally frozen in the laboratory at −80 °C.

2.2. Identification of R2R3-MYB Gene Family Members and Physicochemical Properties Analysis in Pomegranate

Some 126 Arabidopsis R2R2-MYB protein sequences were downloaded from TAIR (https://www.arabidopsis.org/, assessed on 22 January 2023). They were used as query sequences for blastp analysis of pomegranate genome and extracted sequences with E-value < 1×10^{-10} (other parameters default). To verify our results, the online tool Pfam was used to screen candidate PgMYB sequences to identify MYB domains. CD-search (https://www.ncbi.nlm.nih.gov/Structure/cdd/wrpsb.cgi, assessed on 23 January 2023) was used to view MYB protein domains. Sequences that did not contain the full domain of R2R3-MYB or incomplete reading regions were deleted. Ultimately, 186 members of the PgMYB gene family were identified. The ProtParam tool of Expasy (https://web.expasy.org/protparam, assessed on 25 January 2023) was used to analyze the physical and chemical properties. Online tool Euk-mPLoc 2.0 (http://www.csbio.sjtu.edu.cn/bioinf/euk-multi-2/, assessed on 25 January 2023) was used to predict protein subcellular localization.

2.3. Phylogenetic Analysis

To understand the evolutionary relationship of PgMYB, this study concentrated members of the MYB gene family of Arabidopsis and pomegranate and aligned these sequences using the ClustalW function (default parameters) of MGEA7. The study used the sequence alignment results to construct evolutionary trees in MEGA7 [32], utilizing the Neighbor Joining method. The bootstrap replication was set to 1000, and the parameters were default. Finally, in order to better display the results, the study imported the generated evolutionary tree into the online software ITOL (https://itol.embl.de/, assessed on 2 February 2023) for visualization [33]. In order to further screen PgMYB involved in lignin synthesis, we selected validated genes from chrysanthemum, flax, eucalyptus, populus, and Arabidopsis (CmMYB8 [34], EgMYB1 [35], LuMYB12, LuMYB113, LuMYB146 [36], PotMYB216 [37], and AtMYB4 [18]) to construct phylogenetic trees with alternative PgMYB genes.

2.4. Chromosomal Position and Collinearily Analysis

The genomes and annotation files of pomegranate, eucalyptus, and Arabidopsis were obtained from NCBI, and collinear analysis was performed using the MCScanX module of TBtools [38]. The TBtools service was used to perform chromosome localization and visualization.

2.5. Gene Structure and Protein Motif Analysis

The PgMYB protein sequences identified were used to construct phylogenetic trees. The intron, exon, and genomic localization information of the MYB family were all derived from NCBI (https://www.ncbi.nlm.nih.gov/, assessed on 5 February 2023). The online software MEME (http://meme-suite.org/tools/meme, assessed on 3 February 2023) was used to predict the motif of the pomegranate R2R3-MYB protein, and the number of predictions was set to 10 in the parameters, and the other parameters were set as the default parameters. TBtools was used for visualization.

2.6. Analysis of Promoter Cis-Acting Elements

The study obtained the 2000 bp upstream sequences of the PgMYB gene by the Gtf/GFF3 Sequences Extractor option in TBtools. The Plant CARE (http://bioinformatics.psb.ugent.be/webtools/plantcare/html/, assessed on 3 February 2023) service was used to perform promoter homeopathic element analysis. Finally, we used Excel to visualize the result.

2.7. RNA Extraction Library Construction, and Sequencing

Total RNA was extracted from samples using a Trizol reagent kit (Invitrogen, Carlsbad, CA, USA) according to the manufacturer's protocol. RNA quality was assessed on an

Agilent 2100 Bioanalyzer (Agilent Technologies, Palo Alto, CA, USA) and checked using RNase free agarose gel electrophoresis. After total RNA was extracted, we used Oligo(dT) beads to enrich mRNA, and prokaryotic mRNA was enriched by removing rRNA by a RiboZeroTM Magnetic Kit (Epicentre, Madison, WI, USA). Then, the enriched mRNA was fragmented into short fragments by using a fragmentation buffer and reversely transcribed into cDNA using a NEBNext Ultra RNA Library Prep Kit for Illumina (NEB #7530, New England Biolabs, Ipswich, MA, USA). The purified double-stranded cDNA fragments were end repaired, with a base added, and ligated to Illumina sequencing adapters. The ligation reaction was purified with the AMPure XP beads (1.0X). Ligated fragments were subjected to size selection by agarose gel electrophoresis and polymerase chain reaction (PCR) amplification. The resulting cDNA library was sequenced using Illumina Novaseq6000 by Gene Denovo Biotechnology Co. (Guangzhou, China).

2.8. Expression of R2R3-MYB Genes in Pomegranate

To obtain high quality clean reads, reads were filtered by fastp [39] (version 0.18.0). Short reads alignment tool Bowtie2 [40] (version 2.2.8) was used for mapping reads to the ribosome RNA (rRNA) database. The rRNA mapped reads then were removed. The remaining clean reads were further used in assembly and gene abundance calculation. Paired-end clean reads were mapped to the reference genome using HISAT2.2.4 [41]. The mapped reads of each sample were assembled by using StringTie v1.3.1 [42,43] in a reference-based approach. For each transcription region, a FPKM (fragment per kilobase of transcript per million mapped reads) value was calculated to quantify its expression abundance and variations, using RSEM software [44]. Finally, TBtools was used for visualization.

2.9. Real-Time Quantitative PCR (qRT-PCR)

Total RNA was extracted using the TRIzol® Plus RNA Purification Kit (Invitrogen, Carlsbad, CA, USA). The cDNA was synthesized using the SuperScript™ III First-Strand Synthesis SuperMix for qRT-PCR (Invitrogen, Carlsbad, United States) according to the manufacturer's protocol. Specific primers were designed by Beacon Designer 7.8 software (Table S1). qRT-PCR was performed with Power SYBR® Green PCR Master Mix (Roche, Basel, Switzerland). The reaction system is shown in Table S2. Three technical replicates were performed for each sample.

3. Results

3.1. Identification of R2R3-MYB Gene Family Members Physicochemical Properties of Pomegranate R2R3-MYB Proteins

In this study, 186 PgMYBs were identified from the pomegranate genome. To identify all members of the R2R3-MYB gene family in pomegranates, blastp was used to search, and a total of 206 candidate genes were obtained. This study removed 20 sequences that do not contain the full domain of R2R3-MYB or incomplete reading regions, and a total of 186 R2R3-MYB genes, called MYB1-MYB186, were obtained. The physicochemical properties of the 186 sequences were analyzed (Table S3), and the results showed that the amino acid length of the Pomegranate MYB gene family was between 94 (MYB100) and 1516 (MYB174). This is a large range, but most proteins do not differ greatly in length. The molecular weight of the proteins was 10.55 kDa–169.43 kDa, and the isoelectric point (pI) range was 4.81 (MYB96)–9.96 (MYB20), of which 100 proteins were acidic and pI was less than 7. Instability coefficients ranged from 39.19 (MYB103) to 73.97 (MYB2). The Aliphatic index was between 50.89 (MYB115) and 90.16 (MYB22). All PgMYBs were hydrophobic, and their average hydrophilicity was less than 0. Subcellular localization of 186 MYB sequences in pomegranate was predicted, with 184 structures localized on the nucleus and 19 on the cytoplasm. There were two localizations on the cytomembrane and one localization on the extracellular matrix.

3.2. Phylogenetic Analysis of the R2R3-MYB Gene Family

The Arabidopsis MYB family had been studied deeply, and many MYB functions had been verified. Therefore, researchers can infer the function of the same subgroup of genes from the phylogenetic tree. This study constructed phylogenetic trees with the R2R3-MYB gene families of *Arabidopsis thaliana* and pomegranate (Figure 1). In Dubos's study, the Arabidopsis R2R3-MYB gene family was divided into 25 subgroups [18]. The study referred to his classification result and divided the R2R3-MYB gene family of pomegranate into 34 subgroups, ensuring that each MYBs were classified. There are 31 subgroups of PgMYBs and AtMYBs, but a few members fail to cluster with AtMYBs. These branches that do not contain AtMYB are S26, S27, S28, and S32. The constructed phylogenetic tree not only facilitates the exploration of phylogenetic relationships within the PgMYB family but also enables identification of lignin-related PgMYBs, given that numerous AtMYB genes have been functionally characterized.

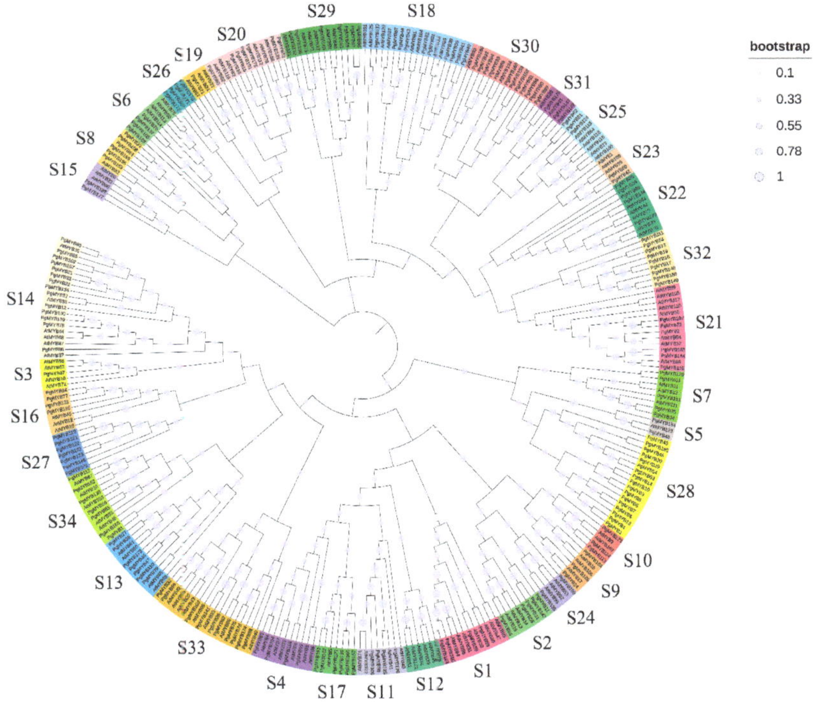

Figure 1. Phylogenetic relationships of R2R3-MYB gene family in pomegranate. The color blocks on the right indicate the different subgroups. The blue circle represents the bootstrap value: the larger the circle, the higher the bootstrap value.

Arabidopsis S3, S13, S21, and AtMYB46 and AtMYB83 have all been verified to be involved in lignin biosynthesis. Since the homologous genes usually have similar functions, the function of the corresponding orthologous genes in pomegranate can be inferred. Therefore, PgMYBs in the same subgroup as S3, S13, and S21 of Arabidopsis thaliana (PgMYB2, PgMYB37, PgMYB27, PgMYB28, PgMYB46, PgMYB47, PgMYB72, PgMYB79, PgMYB110, PgMYB107, PgMYB115, PgMYB164, PgMYB165, and PgMYB173) and homologous genes PgMYB66 and PgMYB83 of AtMYB46 and AtMYB83 may be involved in lignin biosynthesis. In order to further study the regulatory relationship of PgMYBs on lignin synthesis, the study selected validated genes from chrysanthemum, flax, eucalyptus, Populus, and Arabidopsis to construct phylogenetic trees with alternative PgMYB genes. The results showed that (Figure 2) nine of the PgMYBs (PgMYB37, PgMYB2,

PgMYB115, PgMYB165, PgMYB66, PgMYB83, PgMYB173, PgMYB110, and PgMYB79) were closely related, suggesting that these genes might have similar functions.

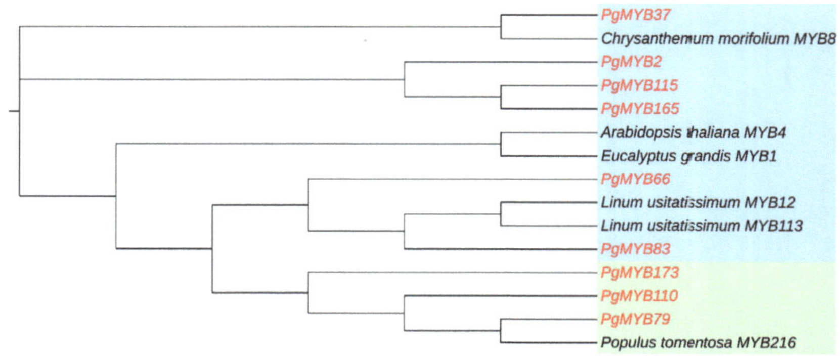

Figure 2. Phylogenetic analysis of Lignin-related PgMYB. The pomegranate MYB genes are highlighted in red.

3.3. Chromosomal Position and Collinearity Analysis

The pomegranate has a total of eight chromosomes, and 186 PGMYBs are unevenly distributed among them (Figure 3). In the figure, PgMYBs were concentrated in areas of the chromosome with high gene density. Chromosome 1 contains the most PgMYB genes (33 genes, 17.7%), followed by chromosome 4 (29 genes, 15.6%). Chromosome 6 contains the least amount of PgMYB genes (16 genes, 8.6%). Through collinearity analysis, 45 collinear gene pairs were found in PgMYB family.

To further investigate the potential evolutionary mechanisms of the R2R3-MYB subfamily, a comparative syntenic map was constructed based on pomegranate, *Arabidopsis*, and *Eucalyptus grandis* (Figure 4). This study found 138 orthologous gene pairs between pomegranate and *Arabidopsis thaliana* and 173 orthologous gene pairs between pomegranate and *Eucalyptus grandis*. This means that there is a closer evolutionary distance between eucalyptus and pomegranate than pomegranate and Arabidopsis. Further, MCScanX was used to detect duplication events in the pomegranate MYB gene family (Table S4). There are five duplication events in genes: WGD, dispersed, tandem, proximal, and singleton. This study found four of these duplication events in 145 PgMYBs (the duplication events of 41 genes remained undiscovered) except for singleton. Specifically, 53.10% (77) of PgMYBs were retained from WGD events. Secondly, 37.24% (56) of PgMYBs were from dispersed events. Only 6.20% (9) were from tandem events, and 2.06% (3) were from proximal events.

3.4. R2R3-MYB Gene Structures and Protein Domains Analysis in Pomegranate

To further elucidate the evolutionary relationship between pomegranate MYB genes, the phylogenetic tree of 186 PgMYB proteins was constructed (Figure 5a). In addition, the conservative motifs of the PgMYB proteins were predicted by the online software MEME (Figure 5b). This study found significant differences in the number of motifs for PgMYB TFs. The most complex genes were PgMYB21, PgMYB22, and PgMYB23, all containing nine motifs, while the simplest gene was PgMYB96, containing only Motif2. Motif1 (97.3%), Motif2 (96.8%), Motif3 (98.9%), and Motif4 (78.1%) are commonly found in PgMYB TFs. Motif5 was found in 40.6% of members, and Motif6 was found in 23% of members. Only a small subset of members contains Motif7 (5.3%), Motif8 (4.3%), Motif9 (5.3%), and Motif10 (4.3%). Members of the same subgroup in the evolutionary tree showed similar conserved motifs. For example, the members of the S19 subgroup all included Motif1, Motif2, Motif3, and Motif4, and the members of the S25 subgroup all included Motif1, Motif2, Motif3, and Motif8.

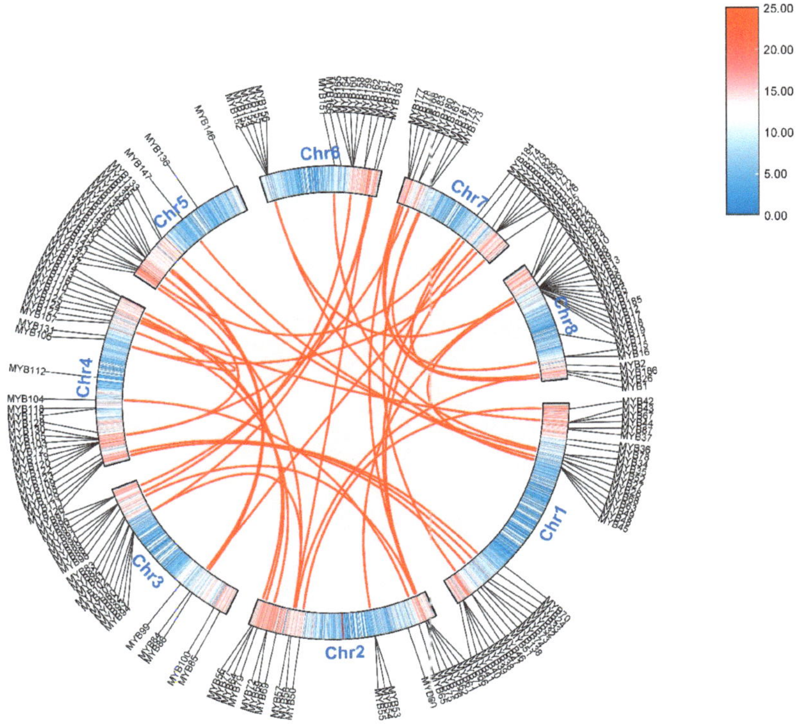

Figure 3. Chromosomal distribution of PgMYB genes. The color bands on the chromosomes indicate gene density, with red indicating high gene density and blue indicating low gene density. The red lines connect paralogous genes of PgMYB.

This study visualized the gene structure of the pomegranate R2R3-MYB family (Figure 5c). The phylogenetic development of the PgMYB gene family showed that the sequences of the same subgroup had similar exon and intron structures, but the overall results showed differences in the number of introns and exons in the pomegranate R2R3-MYB gene family. Even though there were significant differences in the length of MYB genes in pomegranate, there was still some regularity in their genetic structure. This study found that 147 members had three exons and 28 members had two or four exons. The gene structure containing two introns was common in the PgMYB family, where it was found in 150 members. PgMYB88, PgMYB89, PgMYB90, and PgMYB91 had 12 exons and 11 introns, the largest number of exons and introns of any member. In addition, PgMYB21 and PgMYB138 had eight exons and seven introns, PgMYB17, PgMYB18, and PgMYB19 had 11 exons and 10 introns.

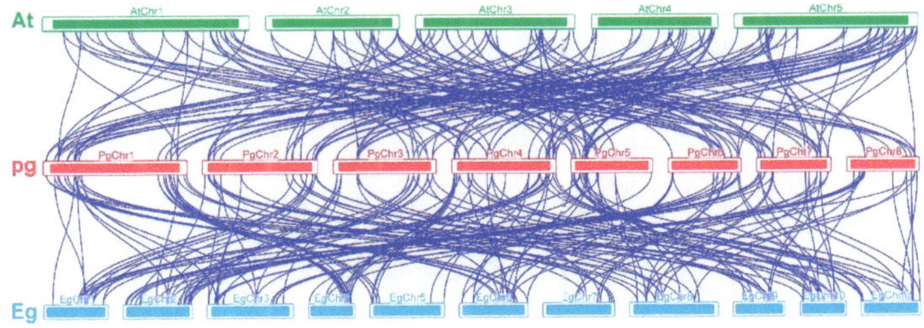

Figure 4. Genome-wide synteny analysis for R2R3-MYB genes among *Punica granatum*, *Eucalyptus grandis*, and *Arabidopsis thaliana*. The blue lines indicate ortholog gene pairs. Eg, *Eucalyptus grandis*; Pg, *Punica granatum*; At, *Arabidopsis thaliana*.

Figure 5. The gene structures and protein domains of PgMYB members, (**a**): phylogenetic tree of PgMYBs; (**b**): Protein motif of PgMYBs, they were named according to the E-value of the motif. The scale at the bottom indicates the sequence length. (**c**): Gene structures of PgMYBs, Green boxes indicated UTR, Yellow boxes indicated CDS, and gray lines indicated introns. UTR: untranslated region, CDS: Coding sequence.

3.5. Promoter Analysis of R2R3-MYB Genes in Pomegranate

To investigate the potential function of the pomegranate R2R3-MYB gene family, the study analyzed the 2000 bp sequence upstream of the PgMYB gene family CDS. A large number of cis-acting elements were detected (Figure 6). These elements were classified as promoter elements, plant growth elements, and environmental response elements. This study revealed the ubiquitous presence of ABRE, TGACE-motif, and CGTCA-motif in PgMYB promoter sequences, which played a crucial role in abscisic acid and methyl jasmonate responses. It indicates that MYB family generally had a hormone regulation effect. In addition, PgMYB exhibited a high prevalence of ARE, GC-motifs, and MBS elements which were implicated in low temperature, anaerobic, and drought stress responses. Suggesting the general responsiveness of PgMYB to abiotic stresses, it is likely that this transcription factor plays a crucial role in plant resistance against such environmental challenges.

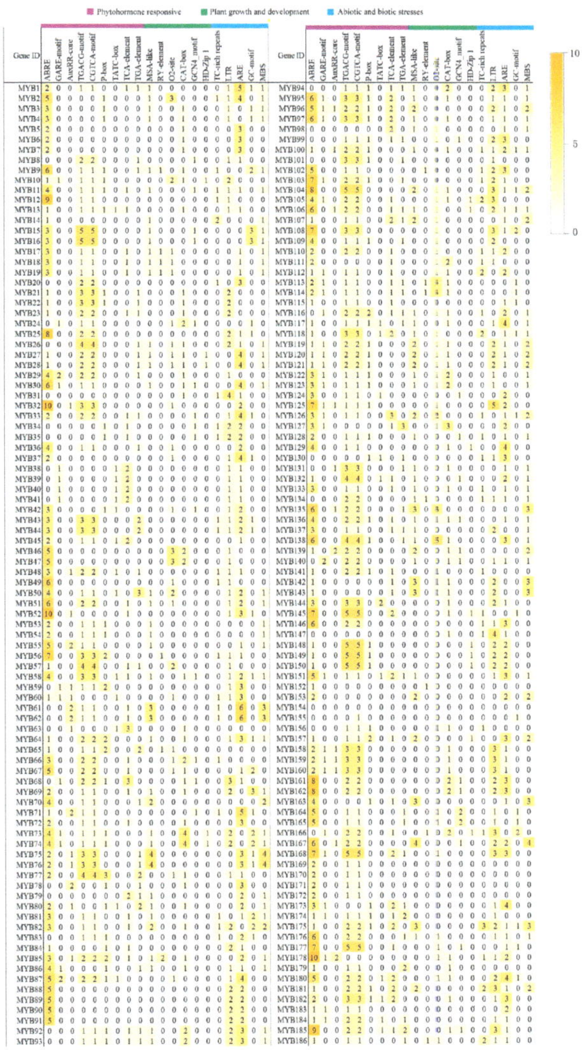

Figure 6. Number of cis-acting elements in the promoter region of PgMYB genes. The more cis-acting elements are predicted, the darker the colors appear in the diagram.

3.6. Expression of R2R3-MYB Genes

In phylogenetic analysis, a screening process was conducted to identify nine PgMYB genes that potentially participate in lignin biosynthesis. The expression information of these genes was obtained from RNA-seq data in this study (Table S5). The temporal (Figure 7a–c) and interspecific (Figure 7d–f) expression patterns of these genes were visualized using six heatmaps. Notably, PgMYB66 was conspicuously absent from the transcriptome data Given that the transcriptome data utilized in this study were obtained from pomegranate seeds, PgMYB66 may be expressed in other organs.

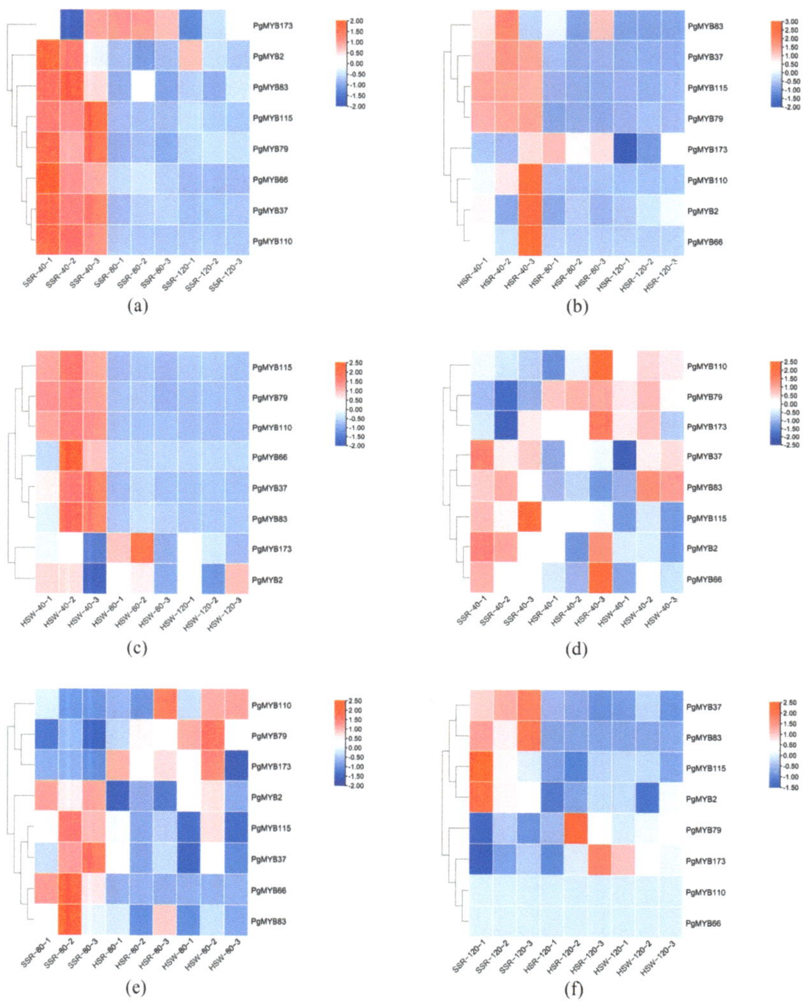

Figure 7. Expression patterns of PgMYB in pomegranate seeds. (**a**) Expression patterns of PgMYB at different developmental stages in Tunisia. (**b**) Expression patterns of PgMYB at different developmental stages in Hongyushizi. (**c**) Expression patterns of PgMYB at different developmental stages in Baiyushizi. (**d**) PgMYB expression patterns of different cultivars at 40 days after flowering. (**e**) Expression patterns of PgMYB in different cultivars at 80 days after flowering. (**f**) Expression patterns of PgMYB in different cultivars at 120 days after flowering. The scale on the left of the picture represents expression levels, with red indicating high expression and blue indicating low expression. SSR: Soft-seed cultivar "Tunisia"; HSW: hard-seed cultivar "Baiyushizi"; HSR: hard-seed cultivar "Hongyushizi".

Temporal expression analysis of the selected PgMYB genes (Figure 7a–c) indicated that their highest expression levels were observed in seeds 40 days after flowering. Notably, PgMYB173 exhibited the greatest level of expression among the three cultivars at 80 days after anthesis, while no significant differences were detected in the expression patterns of PgMYB2 during various stages of Baiyushizi (Figure 7c) seed development. By comparing the expression patterns of different cultivars simultaneously (Figure 7d–f), this study found similar expression patterns of these genes in different cultivars 40 (Figure 7d) days and 80 days (Figure 7e) after flowering. PgMYB110, PgMYB79, and PgMYB173 were highly expressed in Hongyushizi and Baiyushizi. PgMYB37, PgMYB83, PgMYB115, PgMYB2, and PgMYB66 were more highly expressed in Tunisia. PgMYB37, PgMYB83, PgMYB115, PgMYB2, and PgMYB66 were more highly expressed in Tunisia at 120 days after flowering (Figure 7f). PgMYB79 and PgMYB173 were highly expressed in Hongyushizi and Baiyushizi. PgMYB110 and PgMYB66 were expressed at low to no levels at 120 days after flowering. Notably, the expression of PgMYB2 and PgMYB115 in hard-seed cultivars (Hongyushizi and Baiyushizi) had been maintained at low levels, and PgMYB79 had been maintained at high levels throughout development.

3.7. The Validation of PgMYB Expression with qRT-PCR

The expressions of PgMYB2, PgMYB79, and PgMYB115 in Tunisia and Hongyushizi were detected by qRT-PCR (Figure 8). The results showed that the expression levels of all three were highest at 40 days after anthesis, decreased gradually with fruit development, and reached the lowest at 120 days after anthesis. Compared with the hard-seed cultivar 'Hongyushizi', PgMYB2 and PgMYB115 had higher expression in the soft-seed cultivar 'Tunisia', and PgMYB79 had higher expression in 'Hongyushizi'. The expression of all three genes exhibited a consistent temporal pattern, with the highest level observed at 40 days after flowering and a subsequent gradual decline. The expression pattern is similar to the heatmap.

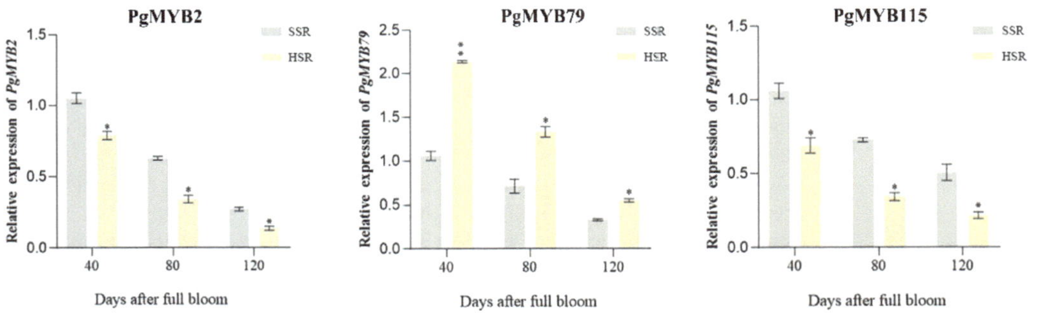

Figure 8. The validation of PgMYB Expression with qRT-PCR. The gray column represents 'Tunisia' (SSR) expression level tested by qRT-PCR; the yellow column represents 'Hongyushizi' (HSR) expression level; error bars indicate the standard deviation of qRT-PCR data. Asterisks indicate those values that are significantly different from the control (* $p < 0.05$, ** $p < 0.01$).

4. Discussion

Pomegranate has garnered significant attention from both consumers and researchers alike [45,46], owing to its exceptional nutritional, medicinal, antioxidant, and antibacterial properties [47,48]. However, the poor taste caused by the hardness of pomegranate seeds hinders further market expansion [3]. Lignin plays a pivotal role in the development of pomegranate seed hardness traits [12]. Current research on lignin biosynthesis in pomegranate has primarily focused on key enzymes that directly regulate lignin synthesis, with limited attention given to transcription factors. In this study, the pomegranate MYB transcription factor family was comprehensively analyzed for the first time. A total of 186 PgMYB genes were identified from the pomegranate genome, and their characteristics and structures were predicted and analyzed. Furthermore, MYB TFs potentially involved in lignin biosynthesis were identified through the utilization of phylogenetic tree and RNA-seq data.

Lignin is widely present in the plant body and is an important component of vascular plants' secondary cell walls, providing mechanical strength to resist damage from external factors [49]. MYB TFs are involved in lignin and cellulose biosynthesis and deposition and have important regulatory roles in secondary cell wall formation [50]. EgMYB2, the homologous gene of AtMYB46 in (*Eucalyptus grandis W. Mill ex Maiden*), binds to AC elements located on lignin synthetase promoters to regulate lignin synthesis [51]. Both maize ZmMYB167 [52] and poplar (*Populus tomentosa*) PtoMYB216 [53] have been shown to involve MYB TFs in lignin biosynthesis. This shows that the higher the homology between MYB class transcription factor members, the greater the similarity of their functions. Therefore, the functional characteristics of other MYB class transcription factor family members on the same branch can be predicted by the existing functional study of MYB class transcription factor family members in Arabidopsis.

In our study, PgMYB family was divided into 34 subgroups, which is consistent with the findings of Hou et al. [54]. The phylogenetic trees of pomegranate and Arabidopsis thaliana showed that most MYBs were in the same branch as Arabidopsis, indicating that pomegranate and Arabidopsis R2R3-MYB members had similar evolutionary origins, but S12 lacked pomegranate MYB members and S26, S27, S28, and S32 lacked Arabidopsis MYB members. Abundant evidence from previous studies has shown that the R2R3-MYB family is extensively involved in phenylpropane metabolism and regulates lignin synthesis [55,56]. Based on functional genes in *Arabidopsis thaliana*, S3, S13, and S21 were thought to be involved in the regulation of lignin synthesis. Based on homology, nine of these genes were found to be more likely involved in the regulation of lignin biosynthesis.

R2R3-MYB is a vast gene family with significant variations in membership properties and structure [57], which also holds true for pomegranate. In this study, the smallest PgMYB transcription factor comprises only 94 amino acids, while the largest one consists of 1516 amino acids. Pomegranate is a diploid plant species (2n = 16) [58], consisting of eight chromosomes per haploid set. The MYB gene has been identified on each chromosome. Notably, the quantity of R2R3-MYB genes in plant genomes does not necessarily correlate with genome size or ploidy level [57]. Collinear analysis showed that the MYB family had expanded extensively during evolution. There were 138 orthologous gene pairs between pomegranate and Arabidopsis and 173 orthologous gene pairs between pomegranate and Eucalyptus grandis. Due to the extensive replication of MYB TFs during evolution, new members are involved in specific functions [16]. In this study, large-scale WGD or segmental and dispersed duplication were detected in PgMYBs. Whole genome duplication is an important evolutionary origin in early plants [59], while dispersed duplication may have originated from transposable elements [60]. WGD and dispersed duplication are the main driving forces of MYB family amplification in pomegranate. The same phenomenon was found in the study of Li et al. [61].

In this study, PgMYB TFs belonging to the same subgroup exhibited identical motif structures. Similarly, genes in the same subgroup typically exhibit the same pattern of introns and exons, including the location and number of introns. These results suggest that the pattern of introns in pomegranates is highly conserved rather than random. In addition, the majority of PgMYB genes exhibit a typical splicing pattern consisting of three exons and two introns, which is consistent with observations in other plant species. We analyzed the cis-acting elements in the upstream sequence of PgMYB genes and found that they contain a large number of plant hormone response elements, which can be induced by light, drought, cold, and other environmental factors. In addition, we found MYB-binding sites upstream of certain PgMYB genes, indicating that these members might cooperate with other members to perform or enhance function.

Gene expression patterns provide important clues for gene function. We utilized RNA-seq data from nine candidate genes to produce expression heatmaps showing their expression patterns during fruit development in different cultivars. The results showed that the expression of PgMYB79 in hard-seed cultivars was higher than that in soft-seed cultivars at all stages, indicating that PgMYB79 was involved in the positive regulation of lignin synthesis. The expressions of PgMYB2 and PgMYB115 were higher in soft-seed cultivars, suggesting that PgMYB2 and PgMYB115 were involved in the negative regulation of lignin synthesis. Therefore, we employed qPCR to validate their expression profiles, which exhibited a general consistent pattern with the heatmaps.

5. Conclusions

In this study, 186 R2R3-MYB TF members were identified from the pomegranate genome, and a series of bioinformatics analyses were performed to reveal their characteristics. Furthermore, we screened PgMYB members that may participate in regulating lignin biosynthesis. This study provides valuable information for further studies on MYB and lignin synthesis.

Supplementary Materials: The following supporting information can be downloaded at: https://www.mdpi.com/article/10.3390/horticulturae9070779/s1, Table S1: Primer sequences and target genes for qPCR analysis; Table S2. qRT-PCR reaction system; Table S3: Basic information of the PgMYB gene family; Table S4: Duplication events of PgMYB; Table S5: Expression data of PgMYB.

Author Contributions: Conceptualization, X.Z.; Data curation, H.N. and L.H.; software, H.S. and M.Z.; formal analysis, X.Z. and H.S.; Writing—original draft, X.Z. and H.S.; Writing—review and editing, S.Z.; visualization, R.L., H.N. and F.Y.; supervision, S.Z.; funding acquisition, S.Z. All authors have read and agreed to the published version of the manuscript.

Funding: This research and the APC were funded by the Natural Science Foundation of Anhui Province, grant number 2008085MC100, and the Anhui Provincial Natural Science Research Project Fund, grant number KJ2019ZD19.

Data Availability Statement: The original transcriptome data used in this study have been submitted to the NCBI, and the data are stored in the SRA database. the accession number is "PRJNA914887".

Conflicts of Interest: The authors declare no conflict of interest.

References

1. Yuan, Z.H.; Fang, Y.M.; Zhang, T.K.; Fei, Z.J.; Han, F.M.; Liu, C.Y.; Liu, M.; Xiao, W.; Zhang, W.J.; Wu, S.; et al. The pomegranate (*Punica granatum* L.) genome provides insights into fruit quality and ovule developmental biology. *Plant Biotechnol. J.* **2018**, *16*, 1363–1374. [CrossRef] [PubMed]
2. da Silva, J.A.T.; Rana, T.S.; Narzary, D.; Verma, N.; Meshram, D.T.; Ranade, S.A. Pomegranate biology and biotechnology: A review. *Sci. Hortic.* **2013**, *160*, 85–107. [CrossRef]
3. Hu, L.; Zhang, X.; Ni, H.H.; Yuan, F.Y.; Zhang, S.M. Identification and Functional Analysis of CAD Gene Family in Pomegranate (*Punica granatum*). *Genes* **2023**, *14*, 26. [CrossRef] [PubMed]
4. Guo, L.H.; Ge, D.P.; Ren, Y.; Dong, J.M.; Zhao, X.Q.; Liu, X.Q.; Yuan, Z.H. The comparative analysis and identification of secondary metabolites between Tibet wild and cultivated pomegranates (*Punica granatum* L.) in China. *J. Integr. Agric.* **2022**, *21*, 736–750. [CrossRef]
5. Murthy, K.N.; Reddy, V.K.; Veigas, J.M.; Murthy, U.D. Study on wound healing activity of *Punica granatum* peel. *J. Med. Food* **2004**, *7*, 256–259. [CrossRef]
6. Alighourchi, H.; Barzegar, M.; Abbasi, S. Anthocyanins characterization of 15 Iranian pomegranate (*Punica granatum* L.) varieties and their variation after cold storage and pasteurization. *Eur. Food Res. Technol.* **2008**, *227*, 881–887. [CrossRef]
7. Zhang, X.H.; Zhao, Y.J.; Ren, Y.; Wang, Y.Y.; Yuan, Z.H. Fruit Breeding in Regard to Color and Seed Hardness: A Genomic View from Pomegranate. *Agronomy* **2020**, *10*, 991. [CrossRef]
8. Xue, H.; Cao, S.Y.; Li, H.X.; Zhang, J.; Niu, J.; Chen, L.N.; Zhang, F.H.; Zhao, D.G. De novo transcriptome assembly and quantification reveal differentially expressed genes between soft-seed and hard-seed pomegranate (*Punica granatum* L.). *PLoS ONE* **2017**, *12*, e0178809. [CrossRef]
9. Dalimov, D.N.; Dalimova, G.N.; Bhatt, M. Chemical composition and lignins of tomato and pomegranate seeds. *Chem. Nat. Compd.* **2003**, *39*, 37–40. [CrossRef]
10. Niu, J.; Cao, D.; Li, H.; Xue, H.; Chen, L.; Liu, B.; Cao, S. Quantitative proteomics of pomegranate varieties with contrasting seed hardness during seed development stages. *Tree Genet. Genomes* **2018**, *14*, 14. [CrossRef]

11. Cao, S.; Niu, J.; Cao, D.; Li, H.; Xue, H.; Chen, L.; Zhang, F.L.; Zhao, D. Comparative proteomics analysis of pomegranate seeds on fruit maturation period (*Punica granatum* L.). *J. Integr. Agric.* **2015**, *14*, 2558–2564. [CrossRef]
12. Qin, G.; Xu, C.; Ming, R.; Tang, H.; Guyot, R.; Kramer, E.M.; Hu, Y.; Yi, X.; Qi, Y.; Xu, X.; et al. The pomegranate (*Punica granatum* L.) genome and the genomics of punicalagin biosynthesis. *Plant J.* **2017**, *91*, 1108–1128. [CrossRef]
13. Du, H.; Yang, S.S.; Liang, Z.; Feng, B.R.; Liu, L.; Huang, Y.B.; Tang, Y.X. Genome-wide analysis of the MYB transcription factor superfamily in soybean. *BMC Plant Biol.* **2012**, *12*, 106. [CrossRef]
14. Millard, P.S.; Kragelund, B.B.; Burow, M. R2R3 MYB transcription factors-functions outside the DNA-binding domain. *Trends Plant Sci.* **2009**, *24*, 934–946. [CrossRef]
15. Jiang, C.H.; Gu, J.Y.; Chopra, S.; Gu, X.; Peterson, T. Ordered origin of the typical two- and three-repeat Myb genes. *Gene* **2004**, *326*, 13–22. [CrossRef]
16. Xu, B.J.; Chen, B.; Qi, X.L.; Liu, S.L.; Zhao, Y.B.; Tang, C.; Meng, X.L. Genome-wide Identification and Expression Analysis of RcMYB Genes in Rhodiola crenulate. *Front. Genet.* **2022**, *13*, 831611. [CrossRef]
17. Jacopo, M.; Pierdomenico, P.; Silvia, G. Novel R2R3 MYB transcription factors regulate anthocyanin synthesis in Aubergine tomato plants. *BMC Plant Biol.* **2023**, *23*, 148. [CrossRef]
18. Dubos, C.; Stracke, R.; Grotewold, E.; Weisshaar, B.; Martin, C.; Lepiniec, L. MYB transcription factors in *Arabidopsis*. *Trends Plant Sci.* **2010**, *15*, 573–581. [CrossRef]
19. Matus, J.T.; Aquea, F.; Arce-Johnson, P. Analysis of the grape *MYB R2R3* subfamily reveals expanded wine quality-related clades and conserved gene structure organization across *Vitis* and *Arabidopsis* genomes. *BMC Plant Biol.* **2008**, *8*, 83. [CrossRef]
20. Li, W.; Ding, Z.; Ruan, M.; Yu, X.; Peng, M.; Liu, Y. Kiwifruit R2R3-MYB transcription factors and contribution of the novel AcMYB75 to red kiwifruit anthocyanin biosynthesis. *Sci. Rep.* **2017**, *7*, 16861. [CrossRef]
21. Xu, Q.; He, J.; Dong, J.; Hou, X.; Zhang, X. Genomic survey and expression profiling of the MYB gene family in watermelon. *Hortic. Plant J.* **2018**, *4*, 1–15. [CrossRef]
22. Xie, F.F.; Hua, Q.Z.; Chen, C.B.; Zhang, Z.K.; Zhang, R.; Zhao, J.T.; Hu, G.B.; Chen, J.Y.; Qin, Y.H. Genome-Wide Characterization of R2R3-MYB Transcription Factors in Pitaya Reveals a R2R3-MYB Repressor HuMYB1 Involved in Fruit Ripening through Regulation of Betalain Biosynthesis by Repressing Betalain Biosynthesis-Related Genes. *Cells* **2021**, *10*, 1949. [CrossRef] [PubMed]
23. Cao, Z.H.; Zhang, S.Z.; Wang, R.K.; Zhang, R.F.; Hao, Y.J. Genome wide analysis of the apple MYB transcription factor family allows the identification of *MdoMYB121* gene confering abiotic stress tolerance in plants. *PLoS ONE* **2013**, *8*, e69955. [CrossRef] [PubMed]
24. Xiao, S.H.; Hu, Q.; Shen, J.L.; Liu, S.M.; Yang, Z.G.; Chen, K.; Klosterman, S.J.; Javornik, B.; Zhang, X.L.; Zhu, L.F. GhMYB4 downregulates lignin biosynthesis and enhances cotton resistance to Verticillium dahliae. *Plant Cell Rep.* **2021**, *40*, 735–751. [CrossRef]
25. Zhou, J.L.; Lee, C.H.; Zhong, R.Q.; Ye, Z.H. MYB58 and MYB63 Are Transcriptional Activators of the Lignin Biosynthetic Pathway during Secondary Cell Wall Formation in *Arabidopsis*. *Plant Cell* **2009**, *21*, 248–266. [CrossRef]
26. McCarthy, R.L.; Zhong, R.; Ye, Z.H. MYB83 is a direct target of SND1 and acts redundantly with MYB46 in the regulation of secondary cell wall biosynthesis in *Arabidopsis*. *Plant Cell Physiol.* **2009**, *50*, 1950–1964. [CrossRef]
27. Ko, J.H.; Jeon, H.W.; Kim, W.C.; Kim, J.Y.; Han, K.H. The MYB46/MYB83-mediated transcriptional regulatory programme is a gatekeeper of secondary wall biosynthesis. *Ann. Bot.* **2014**, *114*, 1099–1107. [CrossRef]
28. Zhu, N.; Cheng, S.F.; Liu, X.Y.; Du, H.; Dai, M.Q.; Yang, W.J.; Zhao, Y. The R2R3-type MYB gene OsMYB91 has a function in coordinating plant growth and salt stress tolerance in rice. *Plant Sci.* **2015**, *236*, 146–156. [CrossRef]
29. Li, Z.W.; Zhou, Y.J.; Liang, H.Z.; Li, Q.; Jiang, Y.M.; Duan, X.W.; Jiang, G.X. MaMYB13 is involved in response to chilling stress via activating expression of VLCFAs and phenylpropanoids biosynthesis-related genes in postharvest banana fruit. *Food Chem.* **2022**, *405*, 134957. [CrossRef]
30. Gomez-Casati, F.D.; Busi, V.M.; Barchiesi, J.; Peralta, A.D.; Hedin, N.; Bhadauria, V. Application of bioinformatics to plant breeding. *Curr. Issues Mol. Biol.* **2018**, *27*, 89–103. [CrossRef]
31. Gong, L.Y. Cloning and Expression Analysis of Genes Related with Lignin Synthesis in Pomegranate Seed. Ph.D. Thesis, Anhui Agricultural University, Hefei, China, 2014.
32. Kumar, S.; Stecher, G.; Tamura, K. MEGA7: Molecular Evolutionary Genetics Analysis Version 7.0 for Bigger Datasets. *Mol. Biol. Evol.* **2016**, *33*, 1870–1874. [CrossRef]
33. Letunic, I.; Bork, P. Interactive Tree of Life (iTOL) v5: An online tool for phylogenetic tree display and annotation. *Nucleic Acids Res.* **2021**, *49*, W293–W296. [CrossRef]
34. Zhu, L.; Guan, Y.X.; Zhang, Z.H.; Song, A.P.; Chen, S.M.; Jiang, J.F.; Chen, F.D. CmMYB8 encodes an R2R3 MYB transcription factor which represses lignin and flavonoid synthesis in chrysanthemum. *Plant Physiol. Biochem.* **2020**, *149*, 217–224. [CrossRef]
35. Soler, M.; Plasencia, A.; Larbat, R.; Pouzet, C.; Jauneau, A.; Rivas, S.; Pesquet, E.; Lapierre, C.; Truchet, I.; Grima-Pettenati, J. The Eucalyptus linker histone variant EgH1.3 cooperates with the transcription factor EgMYB1 to control lignin biosynthesis during wood formation. *New Phytol.* **2017**, *213*, 287–299. [CrossRef]
36. Tombuloglu, H. Genome-wide identification and expression analysis of R2R3, 3R-and 4R-MYB transcription factors during lignin biosynthesis in flax (*Linum usitatissimum*). *Genomics* **2020**, *112*, 782–795. [CrossRef]
37. Tian, Q.Y. Functional Characterization of PtoMYB216 and PtoMYB125 Involved in Transcriptional Regulation of Lignin Biosynthesis in Populus. Ph.D. Thesis, Southwest University, Chongqing, China, 2013.

38. Chen, C.J.; Chen, H.; Zhang, Y.; Thomas, H.R.; Frank, M.H.; He, Y.H.; Xia, R. TBtools: An Integrative Toolkit Developed for Interactive Analyses of Big Biological Data. *Mol. Plant.* **2020**, *13*, 1194–1202. [CrossRef]
39. Chen, S.F.; Zhou, Y.Q.; Chen, Y.R.; Gu, J. fastp: An ultra-fast all-in-one FASTQ preprocessor. *Bioinformatics* **2018**, *34*, 884–890. [CrossRef]
40. Langmead, B.; Salzberg, S.L. Fast gapped-read alignment with Bowtie 2. *Nat. Methods* **2012**, *9*, 357–359. [CrossRef]
41. Kim, D.; Langmead, B.; Salzberg, S.L. HISAT: A fast spliced aligner with low memory requirements. *Nat. Methods* **2015**, *12*, 357–360. [CrossRef]
42. Pertea, M.; Pertea, G.M.; Antonescu, C.M.; Chang, T.C.; Mendell, J.T.; Salzberg, S.L. StringTie enables improved reconstruction of a transcriptome from RNA-seq reads. *Nat. Biotechnol.* **2015**, *33*, 290–295. [CrossRef]
43. Pertea, M.; Kim, D.; Pertea, G.M.; Leek, J.T.; Salzberg, S.L. Transcript-level expression analysis of RNA-seq experiments with HISAT, StringTie and Ballgown. *Nat. Protoc.* **2016**, *11*, 1650–1667. [CrossRef] [PubMed]
44. Li, B.; Dewey, C.N. RSEM: Accurate transcript quantification from RNA-Seq data with or without a reference genome. *BMC Bioinform.* **2011**, *12*, 323. [CrossRef] [PubMed]
45. Al-Zoreky, N.S. Antimicrobial activity of pomegranate (*Punica granatum* L.) fruit peels. *Int. J. Food Microbiol.* **2009**, *134*, 244–248. [CrossRef] [PubMed]
46. Pantiora, P.D.; Balaouras, A.I.; Mina, I.K.; Freris, C.I.; Pappas, A.C.; Danezis, G.P.; Zoidis, E.; Georgiou, C.A. The Therapeutic Alliance between Pomegranate and Health Emphasizing on Anticancer Properties. *Antioxidants* **2023**, *12*, 187. [CrossRef] [PubMed]
47. Caruso, A.; Barbarossa, A.; Tassone, A.; Ceramella, J.; Carocci, A.; Catalano, A.; Basile, G.; Fasio, A.; Iacopetta, D.; Franchini, C.; et al. Pomegranate: Nutraceutical with promising benefits on human health. *Appl. Sci.* **2020**, *10*, 6915. [CrossRef]
48. Escarcega, G.H.; Sánchez-Chávez, E.; Álvarez, S.P.; Caballero, S.M.; Parra, J.M.S.; Flores-Córdova, M.A.; Salazar, N.A.S.; Barrios, D.L.O. Determination of antioxidant phenolic, nutritional quality and volatiles in pomegranates (*Punica granatum* L.) cultivated in Mexico. *Int. J. Food Prop.* **2020**, *23*, 979–991. [CrossRef]
49. Karlen, S.D.; Zhang, C.C.; Peck, M.L.; Smith, R.A.; Padmakshan, D.; Helmich, K.E.; Free, H.C.A.; Lee, S.; Smith, B.G.; Lu, F.C. Monolignol ferulate conjugates are naturally incorporated into plant lignins. *Sci. Adv.* **2016**, *2*, e1600393. [CrossRef]
50. Xiao, R.X.; Zhang, C.; Guo, X.R.; Li, H.; Lu, H. MYB Transcription Factors and Its Regulation in Secondary Cell Wall Formation and Lignin Biosynthesis during Xylem Development. *Int. J. Mol. Sci.* **2021**, *22*, 3560. [CrossRef]
51. Goicoechea, M.; Lacombe, E.; Legay, S.; Mihaljevic, S.; Rech, P.; Jauneau, A.; Lapierre, C.; Pollet, B.; Verhaegen, D.; Chaubet-Gigot, N.; et al. EgMYB2, a new transcriptional activator from Eucalyptus xylem, regulates secondary cell wall formation and lignin biosynthesis. *Plant J.* **2005**, *43*, 553–567. [CrossRef]
52. Bhatia, R.; Dalton, S.; Roberts, L.A.; Moron-Garcia, O.M.; Iacono, R.; Kosik, O.; Gallagher, J.A.; Bosch, M. Modified expression of ZmMYB167 in Brachypodium distachyon and Zea mays leads to increased cell wall lignin and phenolic content. *Sci. Rep.* **2019**, *9*, 8800. [CrossRef]
53. Tian, Q.Y.; Wang, X.Q.; Li, C.F.; Lu, W.X.; Yang, L.; Jiang, Y.Z.; Luo, K.M. Functional Characterization of the Poplar R2R3-MYB Transcription Factor PtoMYB216 Involved in the Regulation of Lignin Biosynthesis during Wood Formation. *PLoS ONE* **2013**, *8*, e76369. [CrossRef]
54. Hou, X.J.; Li, S.B.; Liu, S.R.; Hu, C.G.; Zhang, J.Z. Genome-Wide Classification and Evolutionary and Expression Analyses of Citrus MYB Transcription Factor Families in Sweet Orange. *PLoS ONE* **2014**, *9*, e112375. [CrossRef]
55. Ma, D.W.; Constabel, C.P. MYB Repressors as Regulators of Phenylpropanoid Metabolism in Plants. *Trends Plant Sci.* **2019**, *24*, 275–289. [CrossRef]
56. Liu, J.Y.; Osbourn, A.; Ma, P.D. MYB Transcription Factors as Regulators of Phenylpropanoid Metabolism in Plants. *Mol. Plant* **2015**, *8*, 689–708. [CrossRef]
57. Yang, J.H.; Zhang, B.H.; Gu, G.; Yuan, J.Z.; Shen, S.J.; Jin, L.; Lin, Z.Q.; Lin, J.F.; Xie, X.F. Genome-wide identification and expression analysis of the R2R3-MYB gene family in tobacco (*Nicotiana tabacum* L.). *BMC Genom.* **2022**, *23*, 432. [CrossRef]
58. Sowjanya, P.R.; Shilpa, P.; Patil, G.P.; Babu, D.K.; Sharma, J.; Sangnure, V.R.; Mundewadikar, D.M.; Natarajan, P.; Marathe, A.R.; Reddy, U.K. Reference quality genome sequence of Indian pomegranate cv. 'Bhagawa' (*Punica granatum* L.). *Front. Plant Sci.* **2022**, *13*, 947164. [CrossRef]
59. Bai, J.S.; Song, M.J.; Gao, J.; Li, G.T. Whole genome duplication and dispersed duplication characterize the evolution of the plant PINOID gene family across plant species. *Gene* **2022**, *829*, 146494. [CrossRef]
60. Kroon, M.; Lameijer, E.W.; Lakenberg, N.; Hehir-Kwa, J.Y.; Thung, D.T.; Slagboom, P.E.; Kok, J.N.; Ye, K. Detecting dispersed duplications in high-throughput sequencing data using a database-free approach. *Bioinformatics* **2016**, *32*, 505–510. [CrossRef]
61. Li, X.L.; Xue, C.; Li, J.M.; Qiao, X.; Li, L.T.; Yu, L.A.; Huang, Y.H.; Wu, J. Genome-Wide Identification, Evolution and Functional Divergence of MYB Transcription Factors in Chinese White Pear (*Pyrus bretschneideri*). *Plant Cell Physiol.* **2016**, *57*, 824–847. [CrossRef]

horticulturae

Article

Genome-Wide Identification of Laccase Gene Family from *Punica granatum* and Functional Analysis towards Potential Involvement in Lignin Biosynthesis

Jiangli Shi †, Jianan Yao †, Ruiran Tong, Sen Wang, Ming Li, Chunhui Song, Ran Wan, Jian Jiao and Xianbo Zheng *

College of Horticulture, Henan Agricultural University, Zhengzhou 450002, China; a18839327647@163.com (R.T.); 13619895268@163.com (S.W.); liya1702@163.com (M.L.); songchunhui060305@henau.edu.cn (C.S.); wanxayl@henau.edu.cn (R.W.); jiaojian@henau.edu.cn (J.J.)

* Correspondence: xbzheng@henau.edu.cn

† These authors contributed equally to this work.

Abstract: Laccase (LAC) is the key enzyme responsible for lignin biosynthesis. Here, 57 *PgLACs* from pomegranate were identified and distributed on eight chromosomes and one unplaced scaffold. They were divided into six groups containing three typical Cu-oxidase domains. Totally, 51 *cis*-acting elements in the promoter region of *PgLACs* are involved in response to ABA, GA, light, stress, etc., indicating diverse functions of *PgLACs*. The expression profiles of 13 *PgLACs* during the seed development stage showed that most *PgLACs* expressed at a higher level earlier than at the later seed development stage in two pomegranate cultivars except *PgLAC4*. Also, *PgLAC1/6/7/16* expressed at a significantly higher level in soft-seed 'Tunisia'; on the contrary, *PgLAC37* and *PgLAC50* with a significantly higher expression in hard-seed 'Taishanhong'. Combined with their distinguishing *cis*-acting elements, it was concluded that *PgLAC1/6/7* may respond to GA via TATC-box and GARE-motif, and *PgLAC16* repressed the promotor activity of embryo mid-maturation genes via RY-element so as to contribute to softer seed formation, whereas *PgLAC37/50* may participate in seed formation and accelerate seed maturity via ABRE and G-box elements. Collectively, the dramatic gene expressions of *PgLAC1/6/7/16/37/50* will provide valuable information to explore the formation of soft- and hard-seed in pomegranate.

Keywords: pomegranate; lignin biosynthesis; seed hardness; laccase; gene expression

Citation: Shi, J.; Yao, J.; Tong, R.; Wang, S.; Li, M.; Song, C.; Wan, R.; Jiao, J.; Zheng, X. Genome-Wide Identification of Laccase Gene Family from *Punica granatum* and Functional Analysis towards Potential Involvement in Lignin Biosynthesis. *Horticulturae* **2023**, *9*, 918. https://doi.org/10.3390/horticulturae9080918

Academic Editor: Davide Neri

Received: 25 July 2023
Revised: 8 August 2023
Accepted: 9 August 2023
Published: 11 August 2023

1. Introduction

Pomegranate (*Punica granatum* L.) is a commercial fruit tree with an old cultivation history and belongs to *Lytharceae* family [1]. Nowadays, pomegranate is grown commercially in many counties, such as Iran, China, Spain, and Turkey. Pomegranate fruit is very popular with consumers worldwide due to its delicious taste and benefits to human health. The increasing research demonstrates that pomegranate fruit, as well as its extracts, contain abundant bioactive components, e.g., flavonoids, anthocyanins, organic acids, ellagitannins, phenolic acids, and possess unique antihelminthic, antimicrobial, and antioxidant effects [2–6]. Generally, the edible part of pomegranate fruit is juicy pulp, named arils, by which seeds are surrounded. According to seed hardness, pomegranate cultivars are divided into four groups: soft seed, semi-soft seed, semihard seed, and hard seed [7]. Seeds in soft-seed pomegranate cultivars are easily swallowed and edible. However, the disadvantage is spitting seeds for hard-seed pomegranate cultivars, which seriously affects the taste and consumer appreciation. As we know, lignin is a principal structural component of cell walls in higher plants, and the pomegranate seed formation is closely related to lignin biosynthesis and metabolism [8].

Lignin significantly influences the physical properties and enhances the strength and hardness of cells in plants [9]. Laccase (LAC) is a copper-containing polyphenol oxidase, which is a key enzyme and is broadly present in plants, bacteria, insects, and fungi [10]. In recent years, LACs have been identified in rice [10], *Arabidopsis thaliana* [11], poplar (*Populus trichocarpa*) [12], and some horticultural plants, such as peach (*Prunus persica*) [13], citrus (*Citrus sinensis*) [14], pear (*Pyrus bretschneideri*) [15], tea plant (*Camellia sinensis*) [16], and eggplant (*Solanum melongena*) [17]. Although the LAC gene family from different species is divided into five to eight groups with quite different amino acid sequences, their catalytic sites are relatively conserved [18]. LAC can catalyze the oxidation of various aromatic and non-aromatic substrates via three catalytic sites with four Cu ions, and thereby LAC family possesses multiple biological functions [18,19]. *AtLAC15* expressed specifically in seed coat cell walls of *Arabidopsis*, and oxidative polymerization of epicatechin and soluble PAs led to seed coat browning of *Arabidopsis* [11,20]. The up-regulation of *DkLAC2* increased proanthocyanidin accumulation in persimmon fruit by short tandem target mimic STTM-miR397 [21]. On the other hand, plant LACs are involved in lignin biosynthesis. Eight *AtLACs* expressed at a high level in the inflorescence stems, leading to deposited lignin. Also, *AtLAC4*, *AtLAC11*, and *AtLAC17* were strongly expressed in stems and promoted the constitutive lignification of *Arabidopsis* stem [20,22,23]. The *LACCASE 5* mutant decreased 10% of Klason lignin content and modified the ratio of the syringyl to guaiacyl units [24]. *AtLAC4* regulated by MiR397b may result in plant biomass production with less lignin in flowering plants [25]. The gene expression of *ChLac8* from *Cleome hassleriana* in the *Arabidopsis caffeic acid o-methyltransferase* mutant led to the C-lignin formation in the stems [26]. In the aspect of fruit trees, *PpLAC20* and *PpLAC30* were identified as candidate genes involved in peach lignin biosynthesis [13]. Six *PbLACs* were likely associated with lignin synthesis and stone cell formation in pear fruit, and *PbLAC1* significantly increased lignin deposition and thickened cell walls in transgenic *Arabidopsis* [15]. These recent progresses further stressed the importance of laccases in lignin biosynthesis; thus, more genetic evidence from other species will contribute to illuminating the LAC function in lignin polymerization.

Seed hardness is not only an important index of fruit quality but also directly decides consumers' preferences. To clarify the function of the laccase family in lignin metabolism and seed formation in pomegranate fruit, *PgLAC* family members were first explored in pomegranate in the present study. The bioinformatic analysis of *PgLAC* family members, including sequence characteristics, exon–intron structures, and conserved motifs, was performed. Moreover, the specific expression patterns of the *PgLAC* members were elucidated during the four seed development stages of soft- and hard-seed pomegranates. The results will provide the key candidate genes for seed formation in pomegranate, contributing to breeding the new germplasm.

2. Materials and Methods

2.1. Plant Material

P. granatum cv. 'Tunisia' and 'Taishanhong' plants were grown in the Fruit Tree Experimental Station, Henan Agricultural University, Zhengzhou, Henan, China. The fruits were collected at 30 d, 45 d, 70 d, and 120 d after full flowering and divided into two groups after removing arils. One group quickly evaluated seeds' seed hardness and lignin content, and the other was stored at −80 °C for qRT-PCR analysis.

2.2. Determination of Seed Hardness and Lignin Content

Seed hardness was measured with GY-4 Digital Fruit Hardness Analyzer (Xandpi Instrument Co., Ltd., Leqing, China). The average value of 20 seeds was calculated from three technique replicates and expressed as Kg/cm^2. The seeds were dried at 80 °C till reaching the constant weight, ground to powder, and sieved with 0.425 mm aperture. Then, 2 mg samples were evaluated for lignin content using MZS-1-G Kit (Comin Biotechnology Co., Ltd., Suzhou, China), expressed as mg/g.

2.3. Identification and Physicochemical Properties of PgLAC Family Members

Pomegranate genome data (ASM765513v2) was downloaded from the website (https://www.ncbi.nlm.nih.gov/assembly/GCF_007655135.1, accessed on 12 March 2022). The sequences of 17 AtLAC proteins were obtained from Uniprot website (https://www.uniprot.org/, accessed on 12 March 2022). *PgLACs* sequences from 'Tunisia' were obtained by Blast Wrapper (E-value < 1×10^{-5}) in TBtools software with *AtLACs* sequences and were matched with three Cu-oxidase domains (PFAM00394, PFAM07731, and PFAM07732) on the NCBI CDD (Conserved Domain Database; https://www.ncbi.nlm.nih.gov/Structure/bwrpsb/bwrpsb.cgi, accessed on 18 March 2022). Subsequently, the Genbank Accession Numbers of *PgLACs* were obtained on NCBI BLAST alignment (https://blast.ncbi.nlm.nih.gov/Blast.cgi, accessed on 18 March 2022). The protein isoelectric point (pI) and molecular weight (MW) were accessed using online Expasy Protparam (https://web.expasy.org/compute_pi/, accessed on 25 March 2022). The protein isoelectric point (pI) is calculated using pK values of amino acids, and molecular weight (MW) is calculated by the addition of average isotopic masses of amino acids in the protein and the average isotopic mass of one water molecule.

2.4. Bioinformation Analysis of PgLAC Family Members

The subcellular localization was predicted on the WoLF PSORT (https://wolfpsort.hgc.jp/, accessed on 16 July 2023). The phylogenetic tree was constructed using Clustal W method of MAGE 7.0 software (Mega Limited, Auckland, New Zealand) and optimized on the online website Interactive Tree of Life (http://iTOL.embl.de, accessed on 19 May 2023). The amino acid sequence alignment was performed with neighbor-joining (NJ), and the parameters were set as maximum composite likelihood, complete deletion, and bootstrap 1000 of MAGE 7.0 software. PgLAC proteins were clustered based on the published LAC proteins from other plant species (details in Table S1). Conserved motifs of PgLAC proteins were analyzed using the MEME online software (https://meme-suite.org/meme/tools/meme, accessed on 18 May 2023). Exon-intron structures, chromosomal locations, and gene duplication of *PgLAC* genes were visualized using Gene Structure Shower, Gene Location Visualize, One-Step MCScanX, and Advanced Circos of TBtools software.

2.5. Analysis of Cis-Acting Elements and Protein Interaction Networks

The promotor sequences were obtained from 2000-bp upstream sequences from the start codon of *PgLAC* genes and predicted *cis*-acting elements on PlantCARE (http://bioinformatics.psb.ugent.be/webtools/plantcare/html/, accessed on 18 May 2023), and illustrated with TBtools software. To explore gene co-expression patterns, the protein interaction networks were drawn on the website String (http://cn.string-db.org, accessed on 20 May 2023), and Arabidopsis thaliana was chosen as the species parameter.

2.6. RNA Extraction and Quantitative RT-PCR (qRT-PCR) Analysis of PgLAC Family Members

The total RNA of seeds was extracted using a Quick RNA Isolation Kit (0416-50-GK, Huayueyang, Beijing, China), and the cDNA was synthesized using HiScript III RT SuperMix for qPCR (+gDNA wiper) (Vazyme, Nanjing, China). qRT-PCR was run on ABI 7500 PCR instrument (Applied Biosystems, Foster, CA, USA) using ChamQ Universal SYBR qPCR Master Mix (Vazyme, Nanjing, China). The PCR reaction was performed at 95 °C for 5 min, 40 cycles of 95 °C for 10 s, and 60 °C for 30 s. The relative expression level was calculated by $2^{-\Delta\Delta CT}$ method [27]. The *PgActin* (XM_031530994.1) was used as an internal reference. All the primers were designed on the website Primer-Blast of NCBI (https://www.ncbi.nlm.nih.gov/tools/primer-blast/, accessed on 6 August 2023) and listed in (Table S2). Statistical analysis was carried out with SPSS Statistics v. 20 (IBM, Chicago, IL, USA) with a significant difference of $p < 0.05$ and $p < 0.01$. The gene expression level was drawn using GraphPad Prism 8 software (San Diego, CA, USA).

3. Results

3.1. Comparison of Seed Hardness and Lignin Content during Seed Development Stage

Lignin content was an important index to evaluate the seed hardness in pomegranate. Figure 1a presented seed phenotypic characteristics during four seed development stages of the two pomegranate cultivars. From Figure 1b,c, it was found that seed hardness and lignin content both increased steadily in 'Taishanhong' and 'Tunisia' seeds as seed development and were highly significantly lower in 'Tunisia' seeds than in 'Taishanhong' ones ($p < 0.01$).

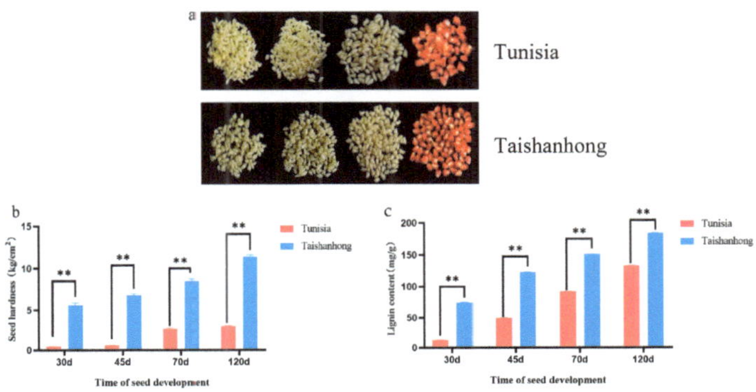

Figure 1. Seed appearance (**a**), seed hardness (**b**), and lignin content (**c**) during four seed development stages from 'Tunisia' and 'Taishanhong'. ** indicated significant differences between groups at $p < 0.01$, respectively.

3.2. Identification of PgLAC Gene Family Members

In the present study, 57 *LAC* genes were identified from pomegranate, distributed on eight chromosomes and Unplaced Scaffold, and named *PgLAC1–PgLAC57* according to their chromosome positions of the pomegranate genome (Figure 2). It was found that 57 *PgLAC* genes were mainly distributed Chr 3 (*PgLAC9-19*), Chr 4 (*PgLAC20-36*), and Chr5 (*PgLAC37-49*), containing 41 *PgLAC* genes, while Chr 6, Chr7, and Unplaced Scaffold only contained one *PgLAC* gene, respectively.

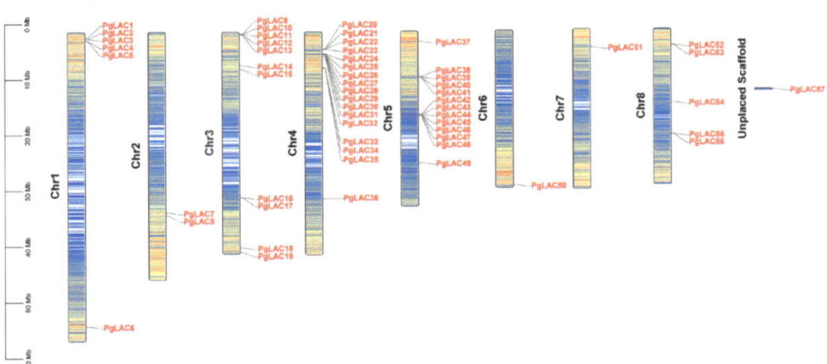

Figure 2. Chromosome distribution of the PgLAC genes. Blue and red represented from less to more of gene density on the chromosome.

Gene duplication was performed in the pomegranate genome to further understand gene evolvement. The results displayed that eight *PgLACs* gene pairs were considered to originate from segmental duplication events across six chromosomes except Chromosome 4 and Chromosome 7, including *PpLAC2/39*, *PgLAC4/7*, *PpLAC7/38*, *PpLAC7/50*, *PpLAC8/19*, *PpLAC8/53*, *PpLAC18/37*, and *PpLAC38/50* (Figure S2 and Table S3). Also, 13 gene pairs were considered as tandem duplication events on Chromosome 1, Chromosome 3, Chromosome 4, and Chromosome 5 (Table S4). The results suggested that both segmental and tandem duplication occurred in the *PgLAC* gene family, which was closely related to the expansion of the PgLAC family.

3.3. Bioinformatic Characteristics of PgLAC Gene Family Members

The amino acids encoded by the *PgLAC* genes ranged from 397 aa (*PgLAC10*) to 614 aa (*PgLAC36*), and only two *PgLAC* genes (*PgLAC10* and *PgLAC52*) were lower than 500 aa (Table 1). Their MWs ranged from 44.53 to 68.98 kDa, and the theoretical isoelectric points (pI) were from 4.51 to 9.90 (Table 1). Moreover, the subcellular localization displayed that 21 PgLAC proteins were located on chloroplast; secondly, vacuolar membrane and cytosol each with 11 PgLAC proteins, 5 on endoplasmic reticulum, 2 on extracellular, and the least was nucleus and mitochondrion each with one PgLAC protein.

To explore the evolutionary relationships of PgLACs, the phylogenetic tree of the amino acid sequences of 57 PgLACs was constructed with LAC proteins from other plant species (including 17 LACs from *Arabidopsis thaliana*, 53 from *Populus trichocarpa*, 27 from *Citrus reticulata* Blanco, and 48 from *Solanum melongena*) (Figure 3). The results demonstrated that 202 LAC proteins were divided into six groups. Group I had 37 LAC proteins, and 33 LACs belonged to Group II. Moreover, Group VI contained the maximum LAC proteins, up to 77 LACs, while only 12 LAC proteins were clustered into Group III. Group IV and V had 25 and 18 LACs, respectively. Also, the maximum LAC proteins from pomegranate were distributed in Group VI, reaching 34, whereas Group III had only one PgLAC51. Considering that *AtLAC4*, *AtLAC11*, and *AtLAC17* were responsible for lignin polymerization [23], it was concerned that *PgLACs* (*PgLAC4*, *5*, *6*, *7*, *15*, *38*, and *50*) were clustered into Group I with *AtLAC4* and *AtLAC11*, furthermore, *PgLACs* (*PgLAC1*, *16*, *17*, *18*, *32*, and *37*) along with *AtLAC17* belonged to Group II (Figure 3), which indicated that they had similar function.

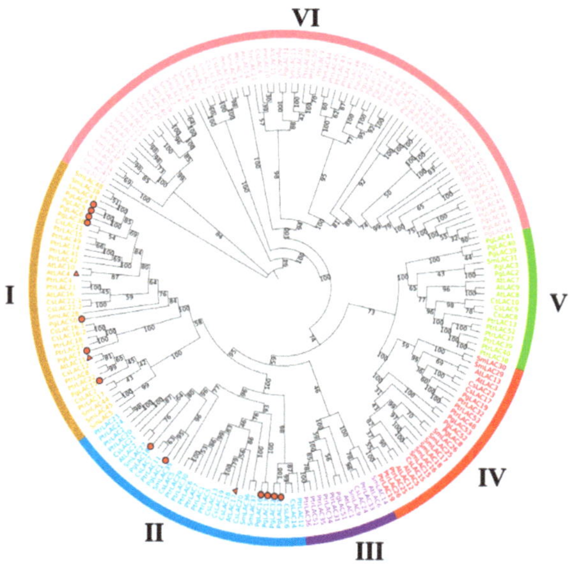

Figure 3. The phylogenetic tree of pomegranate, *Arabidopsis thaliana*, *Populus trichocarpa*, *Citrus reticulata* Blanco, and *Solanum melongena*. PgLACs from *P. granatum*, AtLACs from *A. thaliana*, PtrLACs from *P. trichocarpa*, CsLACs from *C. reticulata*, and SmLACs from *S. melongena*.

Table 1. Bioinformation analysis and physic–chemical properties of *PgLACs*.

Gene Name	Gene ID	Protein ID	Gene Length (bp)	Chromosome Position	Amino Acid (aa)	pI	MW (kDa)	Subcellular Localization
PgLAC1	LOC116200081	XP_031386621.1	1811	Chr1	573	9.59	63.46	Chloroplast
PgLAC2	LOC116192628	XP_031377084.1	1822	Chr1	568	8.22	62.61	Vacuolar membrane
PgLAC3	LOC116193231	XP_031377897.1	1811	Chr1	572	8.84	62.79	Vacuolar membrane
PgLAC4	LOC116193760	XP_031378369.1	1796	Chr1	560	9.48	61.62	Chloroplast
PgLAC5	LOC116193760	XP_031378376.1	1793	Chr1	559	9.48	61.49	Chloroplast
PgLAC6	LOC116192848	XP_031377387.1	1763	Chr1	559	9.47	61.53	Chloroplast
PgLAC7	LOC116195123	XP_031379957.1	1885	Chr2	556	7.06	61.22	Vacuolar membrane
PgLAC8	LOC116197663	XP_031383717.1	1805	Chr2	571	8.59	62.92	Chloroplast
PgLAC9	LOC116199437	XP_031385642.1	1890	Chr3	595	5.17	66.76	Endoplasmic reticulum
PgLAC10	LOC116199438	XP_031385643.1	1706	Chr3	397	5.44	44.53	Cytoskeleton
PgLAC11	LOC116199436	XP_031385641.1	1896	Chr3	601	5.05	67.30	Cytosol
PgLAC12	LOC116199364	XP_031385547.1	1806	Chr3	601	5.04	67.60	Chloroplast
PgLAC13	LOC116199435	XP_031385640.1	1909	Chr3	602	5.67	67.59	Vacuolar membrane
PgLAC14	LOC116202020	XP_031389389.1	1877	Chr3	590	5.08	66.49	Chloroplast
PgLAC15	LOC116199704	XP_031386025.1	1784	Chr3	564	7.70	62.49	Vacuolar membrane
PgLAC16	LOC116198560	XP_031384589.1	1870	Chr3	591	9.38	65.15	Endoplasmic reticulum
PgLAC17	LOC116198765	XP_031384857.1	1876	Chr3	591	9.38	65.12	Endoplasmic reticulum
PgLAC18	LOC116199682	XP_031386005.1	1817	Chr3	581	9.48	64.04	Chloroplast
PgLAC19	LOC116201632	XP_031388782.1	1837	Chr3	584	9.10	64.85	Mitochondrion
PgLAC20	LOC116205486	XP_031393972.1	1668	Chr4	550	4.94	61.26	Cytosol
PgLAC21	LOC116205486	XP_031393974.1	1838	Chr4	547	4.98	60.95	Cytosol
PgLAC22	LOC116205486	XP_031393971.1	1846	Chr4	595	5.10	66.46	Cytosol
PgLAC23	LOC116205995	XP_031394575.1	1887	Chr4	595	5.26	67.21	Cytosol
PgLAC24	LOC116202230	XP_031389549.1	1837	Chr4	575	5.01	63.50	Vacuolar membrane
PgLAC25	LOC116202230	XP_031389550.1	1747	Chr4	541	4.83	59.61	Cytosol
PgLAC26	LOC116202996	XP_031390491.1	1707	Chr4	568	6.16	62.63	Vacuolar membrane
PgLAC27	LOC116203824	XP_031391629.1	1790	Chr4	569	6.72	63.34	Extracellular
PgLAC28	LOC116204024	XP_031391913.1	1807	Chr4	569	5.92	62.36	Cytosol
PgLAC29	LOC116206410	XP_031395142.1	1620	Chr4	519	4.51	56.86	Cytosol
PgLAC30	LOC116206410	XP_031395141.1	1765	Chr4	570	4.66	62.78	Vacuolar membrane
PgLAC31	LOC116206444	XP_031395180.1	1789	Chr4	564	8.89	62.93	Peroxisome
PgLAC32	LOC116205510	XP_031393999.1	1881	Chr4	591	9.78	65.91	Chloroplast
PgLAC33	LOC116206011	XP_031394594.1	1809	Chr4	565	5.21	63.18	Chloroplast
PgLAC34	LOC116206011	XP_031394592.1	2074	Chr4	599	5.21	66.79	Endoplasmic reticulum
PgLAC35	LOC116206011	XP_031394593.1	2074	Chr4	599	5.21	66.79	Endoplasmic reticulum
PgLAC36	LOC116206216	XP_031394887.1	1885	Chr4	614	5.04	68.98	Chloroplast
PgLAC37	LOC116206933	XP_031395630.1	1852	Chr5	587	9.90	65.32	Chloroplast
PgLAC38	LOC116209331	XP_031398793.1	1873	Chr5	563	9.31	61.69	Vacuolar membrane
PgLAC39	LOC116207137	XP_031395863.1	1816	Chr5	571	8.76	63.02	Chloroplast
PgLAC40	LOC116209617	XP_031399176.1	1813	Chr5	558	9.12	62.73	Chloroplast
PgLAC41	LOC116209616	XP_031399174.1	1853	Chr5	558	9.4	62.76	Chloroplast
PgLAC42	LOC116209508	XP_031399023.1	1874	Chr5	595	5.08	67.17	Chloroplast
PgLAC43	LOC116209509	XP_031399024.1	1876	Chr5	595	4.94	67.03	Vacuolar membrane
PgLAC44	LOC116209332	XP_031398797.1	1784	Chr5	512	4.77	57.58	Cytosol
PgLAC45	LOC116209332	XP_031398799.1	1812	Chr5	506	5.24	57.14	Chloroplast
PgLAC46	LOC116209332	XP_031398796.1	1739	Chr5	531	4.77	60.24	Cytoskeleton
PgLAC47	LOC116209332	XP_031398795.1	1910	Chr5	588	5.1	66.57	Cytosol
PgLAC48	LOC116209332	XP_031398794.1	1925	Chr5	593	4.86	67.00	Peroxisome
PgLAC49	LOC116209566	XP_031399095.1	1812	Chr5	585	4.87	65.04	Vacuolar membrane
PgLAC50	LOC116210060	XP_031399722.1	1857	Chr6	578	8.6	63.75	Chloroplast
PgLAC51	LOC116213035	XP_031403701.1	1851	Chr7	593	7.29	65.69	Chloroplast
PgLAC52	LOC116189420	XP_031374936.1	1708	Chr8	481	6.41	52.59	Nucleus
PgLAC53	LOC116189420	XP_031374934.1	1808	Chr8	576	6.78	63.14	Chloroplast
PgLAC54	LOC116187280	XP_031371794.1	1855	Chr8	607	5.5	68.71	Chloroplast
PgLAC55	LOC116189029	XP_031374373.1	1596	Chr8	603	5.05	67.58	Extracellular
PgLAC56	LOC116189029	XP_031374374.1	1910	Chr8	543	4.91	61.46	Cytosol
PgLAC57	LOC116189768	XP_031375356.1	1762	Unplaced Scaffold	564	8.54	63.07	Peroxisome

Subsequently, the amino acid sequences of the above-mentioned 13 PgLACs were aligned with the LACs of other species. The results showed that the PgLAC proteins had higher similarity with other LAC proteins and contained three Cu oxidase domains, namely, Cu-oxidase, Cu-oxidase_2, and Cu-oxidase_3 (Figure 4).

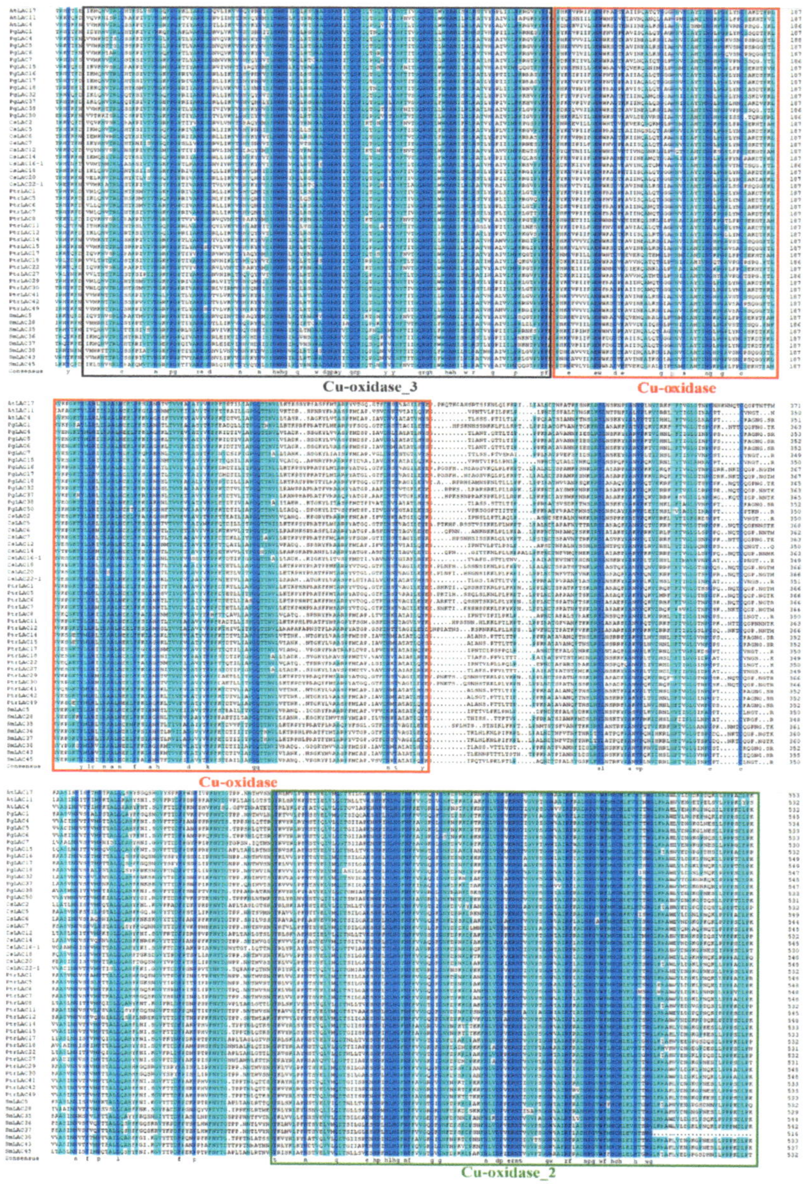

Figure 4. Alignment analysis of PgLAC proteins with other LAC proteins. PgLACs from *P. granatum*, AtLACs from *A. thaliana*, PtrLACs from *P. trichocarpa*, CsLACs from *C. reticulata*, and SmLACs from *S. melongena*.

In addition, the LAC family in other species was investigated and listed (Table 2). The LACs were clustered into five or six subfamilies, except eight subfamilies from *S. miltiorrhiza* and *Solanum melongena*. Group I and V from most species contained more LACs. However, *P. granatum* and *S. melongena* had the most members in Group VI and VIII, respectively. A total of 93 LACs were identified from *Glycine max*, ranking first. Secondly, 65 LACs were obtained from *S. miltiorrhiza*. Among four fruit tree plants, *P. granatum* had 57 LACs, *P. persica* for 48, *P. bretschneideri* for 41, and *C. sinensis* for 27. At present, the number of LACs from *A. thaliana* is less, only 17. Based on the phylogenetic relationships, the 57 PgLACs can be divided into six subfamilies (I–VI), and Group I contained 7 members, 6 members in Group II each contained, 1 member in Group III, 4 members in Group IV, 5 members in Group V, and 34 members in Group VI.

Table 2. LACs characteristics from different plant species.

Species	I	II	III	IV	V	VI	VII	VIII	Total	Reference
Arabidopsis thaliana	2	4	4	3	3	1	0	0	17	[11]
Camellia sinensis	7	6	4	12	12	2	0	0	43	[16]
Citrus sinensis	9	7	3	1	6	0	1	0	27	[14]
Glycine max	15	13	8	8	49	0	0	0	93	[28]
Oryza sativa	7	2	5	5	11	0	0	0	30	[10]
Panicum virgatum	9	4	8	9	19	0	0	0	49	[29]
Populus trichocarpa	12	12	6	12	6	5	0	0	53	[12]
Prunus persica	14	4	1	12	17	0	0	0	48	[13]
Punica granatum	7	6	1	4	5	34	0	0	57	This study
Pyrus bretschneideri	10	10	2	11	1	7	0	0	41	[15]
Salvia miltiorrhiza	9	3	2	5	10	32	1	3	65	[30]
Solanum melongena	4	7	8	4	1	1	3	20	48	[17]
Sorghum bicolor	4	3	4	8	8	0	0	0	27	[31]

3.4. Motif Distribution and Exon/Intron Analysis of PgLAC Family Members

The MEME result showed that 10 conserved motifs were presented in *PgLACs* (Figure 5). The length of the 10 motifs was 21–50 aa, and the motif sequences were provided in Figure S1, which encoded multicopper oxidase and belonged to typical plant laccases. Among them, motif1, motif5, motif8 encoded multicopper Cu-oxidase_3; motif3 and motif7 encoded multicopper Cu-oxidase; motif2, motif4, motif6, and motif9 encoded multicopper Cu-oxidase_2. As shown in Figure 5, 50 *PgLACs* all contained the 10 motifs, and most *PgLACs* ended with the order of motif9, motif6, motif4, and motif2 except *PgLAC45*, suggesting that *PgLAC* gene members possessed relatively conserved sequences. Additionally, among 57 members of *PgLAC* family, *PgLAC26*, *PgLAC33*, *PgLAC34*, and *PgLAC35* all contained one more motif 2; motif 5 did not occur in *PgLAC10*, *PgLAC29*, *PgLAC44*, *PgLAC46*, *PgLAC52*, and *PgLAC56*; *PgLAC10* and *PgLAC46* also missed motif 1 and motif 8. Therefore, the motif distribution displayed the specificity of the gene structure of *PgLACs*, perhaps resulting in different functions. To better understand the structural characteristics of *PgLAC* genes, their exon–intron structures were explored. As shown in Figure 5, 57 *PgLACs* exhibited diverse intron/exon patterns, and the number of exons ranged from 4 to 10. Among *PgLACs*, 23 and 20 *PgLACs* contained 7 and 6 exons, respectively.

3.5. Analysis of Cis-Acting Elements in PgLACs Promotors

The *cis*-acting elements were obtained from the 2000-bp upstream sequence of *PgLACs*, so as to investigate the possible function of *PgLACs*. As shown in Figure 6, in *PgLACs* promotors were observed 51 *cis*-acting elements involving hormone response, stress response, and development response. Among hormone-responsive elements, the abscisic acid responsiveness element (ABRE) was the most common and appeared on the upstream sequences of 48 *PgLACs*, reaching a maximum of 10 on the upstream sequences of *PgLAC12*. The second one was MeJA responsiveness elements (CGTCA-motif and TGACG-motif), which existed on the upstream sequences of 42 *PgLACs* with 1–3, furtherly, 5 *PgLACs* (*PgLAC10*, *PgLAC11*, *PgLAC20*, *PgLAC21*, and *PgLAC49*) contained 4 or 5 (Figure 6). Also, other important hormone-responsive elements were discovered, such as salicylic acid responsiveness element (TCA-element), auxin responsiveness elements (TGA-element, AuxRR-core, and AuxRE), and gibberellin responsiveness elements (GARE-motif, P-box, and TATC-box). Thereby, abscisic acid and methyl jasmonate may greatly participate in modulating the expression of *PgLACs*. Among the *cis*-acting elements involved in stress response, the light-responsive elements were the most abundant in 53 *PgLACs* promoters and G-box in 51 *PgLACs* promotors. In particular, the upstream sequences of *PgLAC1* and *PgLAC53* contained 10 Box 4, respectively, and the upstream sequences of *PgLAC50* had 10 G-box. Meanwhile, the low-temperature responsiveness element (LTR), MBSI (involved in flavonoid biosynthesis), drought stress responsiveness element (MBS), defense and stress responsiveness element (TC-rich repeat), etc., were discovered. Regarding plant development, eight *cis*-acting elements were detected on the upstream sequences of *PgLACs*. O_2-site involved in zein metabolism regulation was the most common, existing in 22 *PgLACs* promotors, while the least for HD-Zip 1, only 1 in *PgLAC50* promotor (Figure 6).

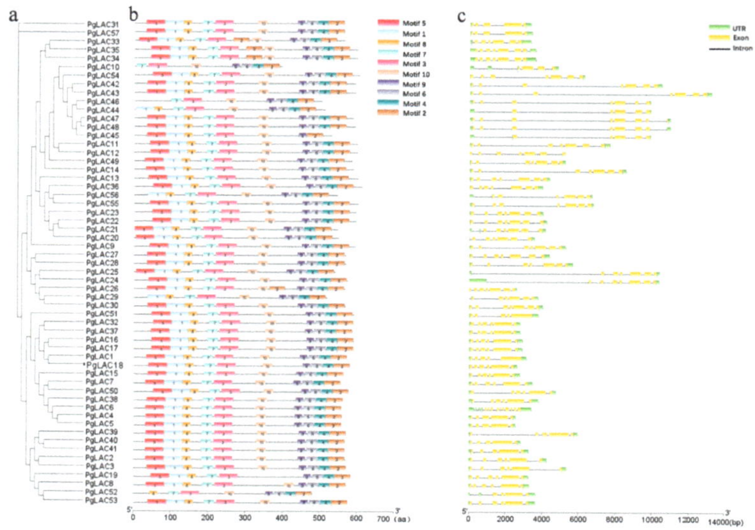

Figure 5. Motif structure and exon/intron analysis of *LAC* family members from pomegranate. (**a**) The phylogenetic tree of 57 pomegranate laccase proteins. (**b**) motif distribution. (**c**) exon/intron analysis. The scales at the bottom of the image indicated motif length (amino acid numbers) and the estimated exon/intron length (bp). Yellow box indicated exons, while black lines indicated introns.

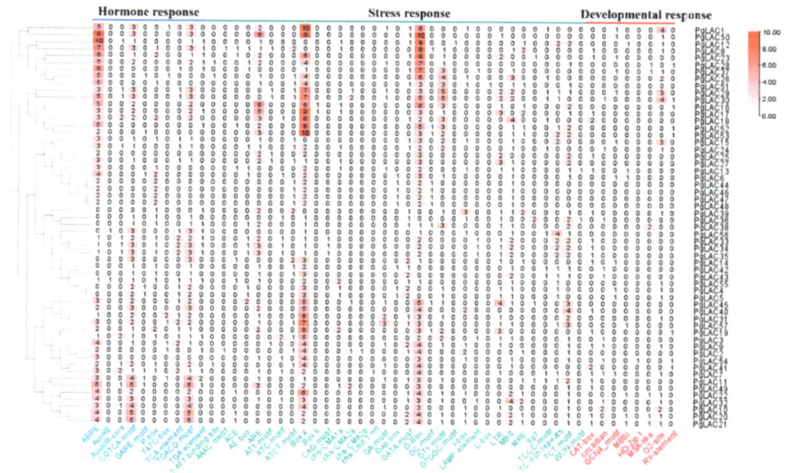

Figure 6. *Cis*-acting elements analysis on the promoter of *PgLACs*. The numbers represented the number of *cis*-acting elements in each *PgLAC* promotor. The blue represents the hormone response, the green the stress response, and the red for the developmental response.

3.6. Analysis of Protein Interaction Networks

The co-expression of 57 PgLAC proteins was predicted using the String protein interaction database; *A. thaliana* was the model species. The stronger reaction between the two proteins, the thicker the linkage line. As Figure 7a shown, PgLAC proteins were identical to AtLAC1, AtLAC3, AtLAC5, IRX12 (AtLAC4), AtLAC7, AtLAC11, AtLAC14, TT10 (AtLAC15), and AtLAC17, respectively. Also, IRX12, AtLAC11, and AtLAC17 had high homology and co-expressed. PgLAC4, PgLAC5, PgLAC6, PgLAC7, PgLAC38, and PgLAC50 were identical to IRX12 and co-expressed with fasciclin-like arabinogalactan-protein FLA11 which was related to secondary cell-wall cellulose synthesis, and galacturonosyltransferase GAUT12 which involved in pectin assembly (Figure 7b). PgLAC1, PgLAC16, PgLAC17, PgLAC18, PgLAC32, and PgLAC37 were identical to AtLAC17, and co-expressed with IRX12, chitinase-like protein CTL2, cellulose synthase A catalytic subunit 4 (CESA4), and cellulose synthase A catalytic subunit 8 (IRX1) (Figure 7c). PgLAC15 was identical to AtLAC11 and co-expressed with DMP2, which is involved in membrane remodeling and xylem cysteine peptidase XCP1 (Figure 7d).

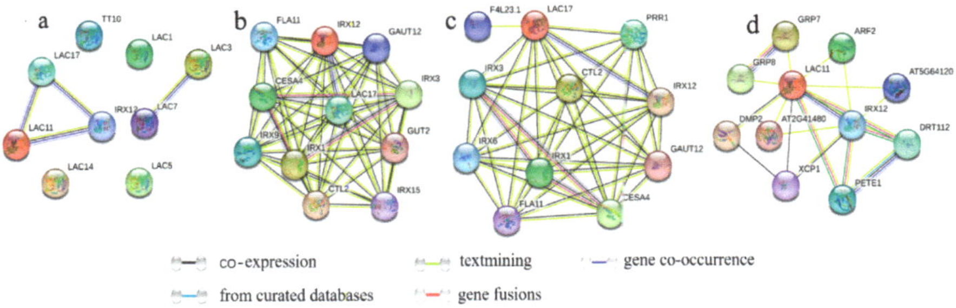

Figure 7. Analysis of interaction networks of PgLAC proteins (**a**); PgLAC4, PgLAC5, PgLAC6, PgLAC7, PgLAC38 and PgLAC50 (**b**); PgLAC1, PgLAC16, PgLAC17, PgLAC18, PgLAC32 and PgLAC37 proteins (**c**); PgLAC15 protein (**d**). CESA4: cellulose synthase A catalytic subunit 4; CTL2: chitinase-like protein 2; GAUT12: galacturonosyltransferase 12; IRX1: cellulose synthase A catalytic subunit 8; IRX15: irregular xylem protein; IRX3: cellulose synthase A catalytic subunit 7; XCP1: xylem cysteine peptidase 1; DMP2: transmembrane protein.

3.7. Expression Profiles of PgLACs during the Seed Development

To clarify the potential function of *PgLACs*, the expression level of 13 *PgLACs* (*PgLAC1, PgLAC4, PgLAC5, PgLAC6, PgLAC7, PgLAC15, PgLAC16, PgLAC17, PgLAC18, PgLAC32, PgLAC37, PgLAC38*, and *PgLAC50*) was assessed during four seed development stages of soft/hard-seed pomegranate using RT-PCR method. The *PgLACs* exhibited different expression levels during seed development; importantly, a significant difference in expression levels was found between soft- and hard-seed pomegranates (Figure 8). And, the expression of most *PgLACs* was higher during the earlier stage of seed development (30–70 d after full flowering) than at the later stage of seed development (120 d after full flowering) between the two pomegranate cultivars except *PgLAC4*. Moreover, 5 *PgLACs* (*PgLAC1/7/32/38/50*) had not a significant difference in the gene expression level at 120 d after full flowering, which demonstrated that the formation of seed hardness depended on the earlier stage of seed development. Meanwhile, during the earlier seed development stage, *PgLAC1* and *PgLAC7* expressed at a significantly higher level in soft-seed 'Tunisia' than in hard-seed 'Taishanhong'; on the contrary, *PgLAC50* displayed a significantly higher expression in 'Taishanhong' than in 'Tunisia' ($p < 0.01$). Similarly, during the whole seed development, *PgLAC6* and *PgLAC16* expressed at a significantly higher level in 'Tunisia' than in 'Taishanhong', while the expression of *PgLAC37* always kept a significantly higher level in 'Taishanhong' than in 'Tunisia' pomegranate ($p < 0.01$) (Figure 8). Therefore, it was inferred that the soft-seed development might be a close relationship with *PgLAC1, PgLAC6,*

PgLAC7, and *PgLAC16*; correspondingly, *PgLAC37* and *PgLAC50* may greatly participate in hard-seed development in pomegranate. In addition, the peak of gene expression at 30 d, 45 d, and 70 d after full flowering each appeared 4 *PgLACs* gene in 'Tunisia', while in 'Taishanhong', 7 *PgLACs* at 30 d after full flowering, and 3 *PgLACs* at 70 d and 2 *PgLACs* at 120 d after full flowering. Collectively, Combined with Figure 1, the earlier stage of seed development was the key to the formation of seed hardness and lignin accumulation for hard-seed pomegranate.

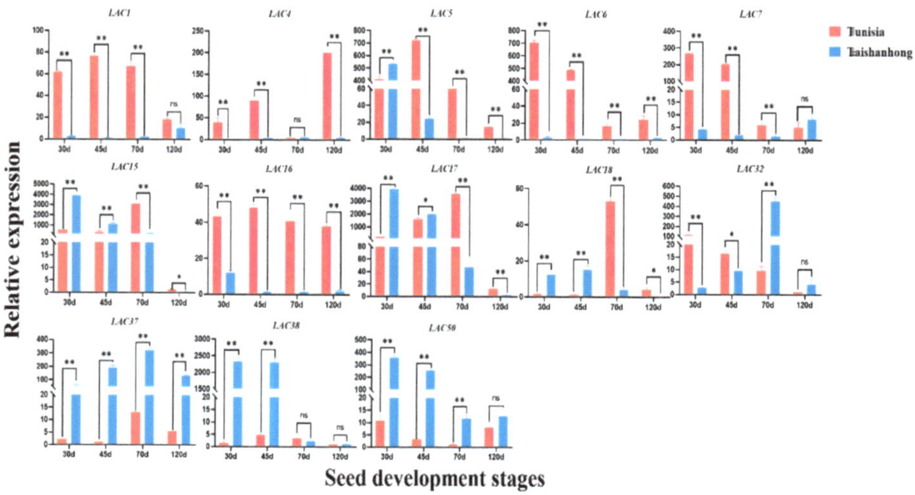

Figure 8. Gene expression profiles of *PgLACs* during seed development stage. * and ** indicated significant differences between the two cultivars at $p < 0.05$ and $p < 0.01$, respectively.

4. Discussion

In general, hard-seed pomegranate cultivars are more resistant to cold stress, whereas soft-seed ones are more popular with consumers due to easily swallowed seeds [32,33]. Seed hardness has become the first preference for more customers. In China, 'Taishanhong' (hard-seed) and 'Tunisia' (soft-seed) are widely cultivated pomegranate cultivars, which play an important role in promoting the pomegranate industry. In the present study, it was found that seed hardness and lignin content both increased steadily in 'Taishanhong' and 'Tunisia' seeds as seed development, with a significantly lower level in 'Tunisia' seeds than in 'Taishanhong' ones. Furthermore, lignin content at 45 d after full flowering increased by 7.7% than 30 d after full flowering, whereas that at 120 d after full flowering increased by 3.3% than 70 d after full flowering; thereby, more lignin deposition may be conducted at the earlier stage of seed formation. Similarly, Niu et al. also proved that the soft-seeded variety contained lower lignin at the early fruit developmental stage [34]. To explore the formation of seed hardness is essential to breed new soft-seed pomegranate germplasm, furtherly, other fruit trees.

Plant laccase enzymes belong to copper-containing polyphenol oxidase and polymerize monolignols into lignin, participating in lignin biosynthesis and various abiotic/biotic stresses [14,35]. Laccases have a multiple-gene family, ranging from 17 (*Arabidopsis*) to 94 members (soybean) [11,28]. In the current study, 57 laccase genes were identified from the *P. granatum* genome, and 41 of 57 *PgLACs* distributed Chr 3, Chr 4. and Chr5. The putative sequences of amino acid ranged from 397 aa (*PgLAC10*) to 614 aa (*PgLAC36*), and the large difference also presented 367–1559 aa in peach, 133–595 aa in citrus, and 485–1136 aa in pear [13–15]. Most PgLACs belong to alkaline pI, similar to PbLACs [15]. Most PgLAC proteins were located on chloroplast with 21 vacuolar membranes and cytosol, each with 11 by the predicted subcellular localization. They also contained the three typical conserved Cu-oxidase domains and had high homology with *A. thaliana*, *P. trichocarpa*,

C. reticulata, and *S. melongena*. The exon number of 57 *PgLACs* varied from 4 to 10, and for other horticultural plants, 3–6 exons for citrus [14], 4–12 exons for pear [15], and 1–13 exons for eggplant [17]. The comparison of exons among species indicated diverse functions in pomegranate laccases.

The proteins interaction networks of 57 PgLAC showed PgLAC1, PgLAC16, PgLAC17, PgLAC18, PgLAC32, and PgLAC37 were co-expressed with CTL2, IRX1, and IRX12 involved in cell wall biosynthesis, further supporting the involvement of the PgLACs in lignin biosynthesis. A previous study reported that the three *Arabidopsis LACs* (*AtLAC4*, *AtLAC17*, and *AtLAC11*) were responsible for lignin polymerization [23]. The phylogenetic tree was constructed and showed that 57 *PgLACs* were divided into six groups, seven *PgLACs* (*PgLAC4*, *5*, *6*, *7*, *15*, *38*, and *50*) in Group I with *AtLAC4* and *AtLAC11*, and six *PgLACs* (*PgLAC1*, *16*, *17*, *18*, *32*, and *37*) in Group II along with *AtLAC17*. The results implied the 13 *PgLACs* may possess similar functions to the *Arabidopsis LACs*. Therefore, the expression profiles of the 13 *PgLACs* were investigated during the four seed development stages. We provided the evidence that most *PgLACs* expressed at a higher level during the earlier stage of seed development than at the later stage of seed development in the two pomegranate cultivars except *PgLAC4*, which was in line with higher lignin content at the earlier seed development stage. However, the distinct difference in the gene expression levels of individual *PgLAC* existed between cultivars, higher *PgLAC1*, *PgLAC6*, *PgLAC7*, and *PgLAC16* in the soft-seed cultivar, whereas higher *PgLAC37* and *PgLAC50* in the hard-seed pomegranate cultivar.

G-box widely presents on the promoter of many plant genes with a palindromic DNA motif of CACGTG [35]. Previous reports found that G-box participated in the co-expression system for nuclear photosynthetic genes and influenced organ differentiation [36]. In the current study, the gene expression of *PgLAC37* and *PgLAC50* during the first 70 d after full flowering displayed higher with a significant difference in 'Taishanhong' than in 'Tunisia'; further, the more G-box elements presented in *PgLAC37* with 7 G-box and *PgLAC50* with 10 ones, suggesting that the two genes were the key candidate gene for the formation of the hard seed in pomegranate. Similarly, more ABRE elements involved in abscisic acid (ABA) responsiveness were also observed in the *PgLAC37* (6 ABRE) and *PgLAC50* (9 ABRE). ABA is involved in secondary cell-wall formation [37], and the late-wood formation of *Pinus radiata* and *Pinus sylvestris* is correlated with an increase in ABA concentration [38,39]. Taken together, the G-box and ABRE elements in the promotors of *PgLAC37* and *PgLAC50* may be an essential reason for modulating the higher expression of *PgLAC37* and *PgLAC50* during the earlier seed development stage of 'Taishanhong' pomegranate, thus, the two genes greatly participated in seed formation and accelerated seed maturity.

Gibberellin (GA) is a primarily growth-regulating phytohormone and regulates diverse biological processes. Previous studies revealed that GA induced berry seedless and regulated flower development, berry set, expansion, and ripening in grapes [40–42]. TATC-box and GARE-motif are known to both respond to GA. Interestingly, we found TATC-box in only *PgLAC1* promotor and GARE-motif in the promotors of *PgLAC6* and *PgLAC7*, as well as *PgLAC4* and *PgLAC5* among 13 *PgLACs*. Collectively, higher expression of *PgLAC1/4/6/7* in 'Tunisia' may be induced by GA and then produce softer seeds. The five *PgLACs* played an indispensable role in the formation of softer seeds. RY-element is found predominantly in seed-specific promoters [43] and mediates repression of embryo mid-maturation genes involved in the accumulation of storage compounds [44]. In the current study, RY-element was observed only in the *PgLAC16* promotor, which suggests that the dramatically higher expression of *PgLAC16* may inhibit the accumulation of storage compound during the seed development stage, thus hindering the formation of seed hardness in soft-seed pomegranate. In conclusion, *PgLAC1/4/6/7/16* with higher expression in 'Tunisia' will greatly contribute to exploring the soft seed formation, which is potential candidate gene for breeding soft-seed pomegranate with GA application.

5. Conclusions

Laccase is the key enzyme on the lignin biosynthesis pathway, closely correlated with seed hardness. The LAC family was first identified from the pomegranate genome. A total of 57 *PgLACs* were divided into six groups containing typical Cu-oxidase domains. Exon-intron structure and motif analysis predicted that the *PgLACs* had diverse functions in lignin biosynthesis. Combined with *cis*-acting elements and the gene expression patterns, *PgLAC37* and *PgLAC50* were the key candidate genes for the formation of hard seed in pomegranate, attributed to more G-box and ABRE elements in their promotors, which regulated the expression of *PgLAC37* and *PgLAC50*, participated in seed formation, accelerated seed maturity, finally, produced harder seed. In soft-seed pomegranate, higher expression of *PgLAC1/4/6/7* may contribute to soft-seed formation responsive to GA via GARE motif and TATC-box. And the *PgLAC16* promotor containing RY-element may regulate soft seed development by reducing the accumulation of storage compounds in seeds. Collectively, the results of our study will provide important gene information and a new perspective for breeding hard- and soft-seed pomegranate cultivars.

Supplementary Materials: The following are available online at https://www.mdpi.com/article/10.3390/horticulturae9080918/s1, Figure S1: The sequence information for each motif (Motif 1-Motif 10); Figure S2: Segmental duplication of the *PgLAC* genes in pomegranate; Table S1: LACs information in the phylogenetic tree; Table S2: The primers used in qRT-PCR; Table S3: Segmental duplication pairs of the *PgLAC* genes within the pomegranate genome; Table S4: The identified tandem duplication pairs of *PgLAC* genes.

Author Contributions: Conceptualization, J.S. and J.Y.; formal analysis, R.T. and S.W.; visualization, J.Y., R.T., S.W. and M.L.; investigation, J.Y., R.T. and R.W.; resources, C.S. and J.J.; writing—original draft, J.S. and J.Y.; writing—review and editing, J.S., J.Y. and X.Z. All authors have read and agreed to the published version of the manuscript.

Funding: This research was funded by the Special Fund for Henan Agriculture Research System (Grant No. HARS-22-09-Z2).

Data Availability Statement: The data presented in this study are available.

Conflicts of Interest: The authors declare no conflict of interest.

References

1. Yuan, Z.; Fang, Y.; Zhang, T.; Fei, Z.; Han, F.; Liu, C.; Min, L.; Xiao, W.; Zhang, W.; Wu, S.; et al. The pomegranate (*Punica granatum* L.) genome provides insights into fruit quality and ovule developmental biology. *Plant Biotechnol. J.* **2017**, *7*, 1363–1374. [CrossRef] [PubMed]
2. Bishayeea, A.; Mandalb, A.; Bhattacharyyac, P.; Bhatiad, D. Pomegranate exerts chemoprevention of experimentally induced mammary tumorigenesis by suppression of cell proliferation and induction of apoptosis. *Nutr. Cancer* **2016**, *68*, 120–130. [CrossRef] [PubMed]
3. Karimi, M.; Sadeghi, R.; Kokini, J. Pomegranate as a promising opportunity in medicine and nanotechnology. *Trends Food Sci. Technol.* **2017**, *69*, 59–73. [CrossRef]
4. Kandylis, P.; Kokkinomagoulos, E. Food applications and potential health benefits of pomegranate and its derivatives. *Foods* **2020**, *9*, 122. [CrossRef]
5. Ambigaipalan, P.; de Camargo, A.C.; Shahidi, F. Phenolic compounds of pomegranate byproducts (outer skin. mesocarp. divider membrane) and their antioxidant activities. *J. Agric. Food Chem.* **2016**, *64*, 6584–6604. [CrossRef]
6. Morittu, V.M.; Mastellone, V.; Tundis, R.; Loizzo, M.R.; Tudisco, R.; Figoli, A.; Cassano, A.; Musco, N.; Britti, D.; Infascelli, F.; et al. Antioxidant, biochemical, and in-life effects of *Punica granatum* L. natural juice vs. clarified juice by polyvinylidene fluoride membrane. *Foods* **2020**, *9*, 242. [CrossRef]
7. Zarei, A.; Zamani, Z.; Fatahi, R.; Mousavi, A.S.; Seyed, A. A mechanical method of determining seed-hardness in pomegranate. *J. Crop Improv.* **2013**, *27*, 444–459. [CrossRef]
8. Xue, H.; Cao, S.; Li, H.; Zhang, J.; Niu, J.; Chen, L.; Zhang, F.; Zhao, D. De novo transcriptome assembly and quantification reveal differentially expressed genes between soft-seed and hard-seed pomegranate (*Punica granatum* L.). *PLoS ONE* **2017**, *12*, e0178809. [CrossRef]
9. Zarei, A.; Zamani, Z.; Fatahi, R.; Mousavi, A.; Salami, S.A.; Avila, C.; Canovas, F.M. Differential expression of cell wall related genes in the seeds of soft-and hard-seeded pomegranate genotypes. *Sci. Hortic.* **2016**, *205*, 7–16. [CrossRef]

10. Liu, Q.; Luo, L.; Wang, X.; Shen, Z.; Zheng, L. Comprehensive analysis of rice laccase gene (*OsLAC*) family and ectopic expression of *OsLAC10* enhances tolerance to copper stress in *Arabidopsis*. *Int. J. Mol. Sci* **2017**, *18*, 209. [CrossRef]
11. Turlapati, P.V.; Kim, K.W.; Davin, L.B.; Lewis, N.G. The laccase multigene family in *Arabidopsis thaliana*: Towards addressing the mystery of their gene function(s). *Planta* **2011**, *233*, 439–470. [CrossRef] [PubMed]
12. Liao, B.; Wang, C.; Li, X.; Man, Y.; Ruan, H.; Zhao, Y. Genome-wide analysis of the *Populus trichocarpa* laccase gene family and functional identification of *PtrLAC23*. *Front. Plant Sci.* **2023**, *13*, 1063813. [CrossRef] [PubMed]
13. Qui, K.; Zhou, H.; Pan, H.; Sheng, Y.; Yu, H.; Xie, Q.; Chen, H.; Cai, Y.; Zhang, J.; He, J. Genome-wide identification and functional analysis of the peach (*P. persica*) laccase gene family reveal members potentially involved in endocarp lignification. *Trees* **2022**, *36*, 1477–1496. [CrossRef]
14. Xu, X.; Zhou, Y.; Wang, B.; Ding, L.; Wang, Y.; Luo, L.; Zhang, Y.; Kong, W. Genome-wide identification and characterization of laccase gene family in *Citrus sinensis*. *Gene* **2018**, *689*, 114–123. [CrossRef] [PubMed]
15. Cheng, X.; Li, G.; Ma, C.; Abdullah, M.; Zhang, J.; Zhao, H.; Qing, J.; Cai, Y.; Lin, Y. Comprehensive genome-wide analysis of the pear (*Pyrus bretschneideri*) laccase gene (*PbLAC*) family and functional identification of *PbLAC1* involved in lignin biosynthesis. *PLoS ONE* **2019**, *14*, e0210892. [CrossRef] [PubMed]
16. Yu, Y.; Xing, Y.; Liu, F.; Zhang, X.; Li, X.; Zhang, J.; Sun, X. The laccase gene family mediate multi-perspective trade-offs during tea plant (*Camellia sinensis*) development and defense processes. *Int. J. Mol. Sci.* **2021**, *22*, 12554. [CrossRef]
17. Wan, F.; Zhang, L.; Tan, M.; Wang, X.; Wang, G.; Qi, M.; Liu, B.; Gao, J.; Pan, Y.; Wang, Y. Genome-wide identification and characterization of laccase family members in eggplant (*Solanum melongena* L.). *PeerJ* **2022**, *10*, e12922. [CrossRef]
18. Janusz, G.; Pawlik, A.; Świderska-Burek, U.; Polak, J.; Sulej, J.; Jarosz-Wilkołazka, A.; Paszczyński, A. Laccase properties, physiological functions, and evolution. *Int. J. Mol. Sci.* **2020**, *21*, 966. [CrossRef]
19. Liu, M.; Dong, H.; Wang, M.; Liu, Q. Evolutionary divergence of function and expression of laccase genes in plants. *J. Genet.* **2020**, *99*, 23. [CrossRef]
20. Pourcel, L.; Routaboul, J.M.; Kerhoas, L.; Caboche, M.; Lepiniec, L.; Debeaujon, I. *TRANSPARENT TESTA10* encodes a laccase-like enzyme involved in oxidative polymerization of flavonoids in *Arabidopsis* seed coat. *Plant Cell* **2005**, *17*, 2966–2980. [CrossRef]
21. Zaman, F.; Zhang, M.; Liu, Y.; Wang, Z.; Xu, L.; Guo, D.; Luo, Z.; Zhang, Q. DkmiR397 regulates proanthocyanidin biosynthesis via negative modulating *DkLAC2* in Chinese PCNA Persimmon. *Int. J. Mol. Sci.* **2022**, *23*, 3200. [CrossRef]
22. Berthet, S.; Demont-Caulet, N.; Pollet, B.; Bidzinski, P.; Cézard, L.; Bris, P.L.; Borrega, N.; Hervé, J.; Blondet, E.; Balzergue, S.; et al. Disruption of *LACCASE4* and 17 results in tissue-specific alterations to lignification of *Arabidopsis thaliana* stems. *Plant Cell* **2011**, *23*, 1124–1137. [CrossRef] [PubMed]
23. Zhao, Q.; Nakashima, J.; Chen, F.; Yin, Y.; Fu, C.; Yun, J.; Shao, H.; Wang, X.; Wang, Z.; Dixon, R.A.; et al. *LACCASE* is necessary and nonredundant with *PEROXIDASE* for lignin polymerization during vascular development in *Arabidopsis*. *Plant Cell* **2013**, *25*, 3976–3987. [CrossRef] [PubMed]
24. Wang, Y.; Bouchabke-Coussa, O.; Lebris, P.; Antelme, S.; Soulhat, C.; Gineau, E.; Dalmais, M.; Bendahmane, A.; Morin, H.; Mouille, G.; et al. LACCASE5 is required for lignification of the *Brachypodium distachyon* culm. *Plant Physiol.* **2015**, *168*, 192–204. [CrossRef] [PubMed]
25. Wang, C.; Zhang, S.; Yu, Y.; Luo, Y.; Liu, Q.; Ju, C.; Zhang, Y.; Qu, L.; Lucas, W.J.; Wang, X.; et al. MiR397b regulates both lignin content and seed number in *Arabidopsis* via modulating a laccase involved in lignin biosynthesis. *Plant Biotechnol. J.* **2014**, *12*, 1132–1142. [CrossRef]
26. Wang, X.; Zhuo, C.; Xiao, X.; Wang, X.; Docampo-Palacios, M.; Chen, F.; Dixon, R.A. Substrate specificity of LACCASE8 facilitates polymerization of caffeyl alcohol for c-lignin biosynthesis in the seed coat of cleome hassleriana. *Plant Cell* **2020**, *32*, 3825–3845. [CrossRef] [PubMed]
27. Livak, K.J.; Schmittgen, T.D. Analysis of relative gene expression data using real-time quantitative PCR and the $2^{-\triangle\triangle CT}$ method. *Methods* **2001**, *25*, 402–408. [CrossRef]
28. Wang, Q.; Li, G.; Zheng, K.; Zhu, X.; Ma, J.; Wang, D.; Tang, K.; Feng, X.; Leng, J.; Yu, H.; et al. The soybean laccase gene family: Evolution and possible roles in plant defense and stem strength selection. *Genes* **2019**, *10*, 701. [CrossRef]
29. Li, R.; Zhao, Y.; Sun, Z.; Wu, Z.; Wang, H.; Fu, C.; Zhao, H.; He, F. Genome-Wide identification of switchgrass laccases involved in lignin biosynthesis and heavy-metal responses. *Int. J. Mol. Sci.* **2022**, *23*, 6530. [CrossRef]
30. Li, C.; Li, D.; Zhou, H.; Li, J.; Lu, S. Analysis of the laccase gene family and miR397-/miR408-mediated posttranscriptional regulation in *Salvia miltiorrhiza*. *PeerJ* **2019**, *7*, e7605. [CrossRef]
31. Wang, J.; Feng, J.; Jia, W.; Fan, P.; Bao, H.; Li, S.; Li, Y. Genome-Wide identification of sorghum bicolor laccases reveals potential targets for lignin modification. *Front. Plant Sci.* **2017**, *8*, 714. [CrossRef] [PubMed]
32. Shi, J.; Gao, H.; Wang, S.; Wu, W.; Tong, R.; Wang, S.; Li, M.; Jian, Z.; Wan, R.; Hu, Q.; et al. Exogenous arginine treatment maintains the appearance and nutraceutical properties of hard-and soft-seed pomegranates in cold storage. *Front. Nutr.* **2022**, *9*, 828946. [CrossRef] [PubMed]
33. Luo, X.; Li, H.; Wu, Z.; Yao, W.; Zhao, P.; Cao, D.; Yu, H.; Li, K.; Poudel, K.; Zhao, D.; et al. The pomegranate (*Punica granatum* L.) draft genome dissects genetic divergence between soft-and hard-seeded cultivars. *Plant Biotechnol. J.* **2019**, *18*, 955–968. [CrossRef] [PubMed]
34. Niu, J.; Cao, D.; Li, H.; Xue, H.; Chen, L.; Liu, B.; Cao, S. Quantitative proteomics of pomegranate varieties with contrasting seed hardness during seed development stages. *Tree Genet. Genomes* **2018**, *14*, 14. [CrossRef]

35. Giuliano, G.; Pichersky, E.; Malik, V.S.; Timko, M.P.; Scolnik, P.A.; Cashmore, A.R. An evolutionarily conserved protein binding sequence upstream of a plant light-regulated gene. *Proc. Natl. Acad. Sci. USA* **1988**, *85*, 7089–7093. [CrossRef]
36. Kobayashi, K.; Obayashi, T.; Masuda, T. Role of the G-box element in regulation of chlorophyll biosynthesis in *Arabidopsis* roots *Plant Signal. Behav.* **2012**, *7*, 922–926. [CrossRef]
37. Liu, C.; Yu, H.; Rao, X.; Li, L.; Dixon, R.A. Abscisic acid regulates secondary cell-wall formation and lignin deposition in *Arabidopsis thaliana* through phosphorylation of NST1. *Proc. Natl. Acad. Sci. USA* **2021**, *118*, 23. [CrossRef]
38. Jenkins, P.A.; Shepherd, K.R. Seasonal changes in levels of indole-acetic acid and abscisic acid in stem tissues of *Pinus radiata* *N. Z. J. Bot.* **1974**, *4*, 511–519.
39. Wodzicki, T.J.; Wodzicki, A.B. Seasonal abscisic acid accumulation in stem cambial region of *Pinus silvestris*, and its contribution to the hypothesis of a late-wood control system in conifers. *Physiol. Plant.* **1980**, *48*, 443–447. [CrossRef]
40. Cui, M.; Wang, W.; Guo, F.; Fan, X.; Guan, L.; Zheng, T.; Zhu, X.; Jia, H.; Fang, J.; Wang, C.; et al. Characterization and temporal-spatial expression analysis of *LEC1* gene in the development of seedless berries in grape induced by gibberellin. *Plant Growth Regul.* **2020**, *90*, 585–596. [CrossRef]
41. Cheng, C.; Xu, X.; Singer, S.D.; Li, J.; Zhang, H.; Gao, M.; Wang, L.; Song, J.; Wang, X. Effect of GA3 treatment on seed development and seed related gene expression in grape. *PLoS ONE* **2013**, *8*, e80044. [CrossRef] [PubMed]
42. Wang, P.; Xuan, X.; Su, Z.; Wang, W.; Abdelrahman, M.; Jiu, S.; Zhang, X.; Liu, Z.; Wang, X.; Wang, C.; et al. Identification of miRNAs-mediated seed and stone-hardening regulatory networks and their signal pathway of GA-induced seedless berries in grapevine (*V. vinifera* L.). *BMC Plant Biol.* **2021**, *21*, 442. [CrossRef] [PubMed]
43. Lelievre, J.M.; Oliveira, L.O.; Nielsen, N.C. 5′-CATGCAT-3′ elements modulate the expression of glycinin genes. *Plant Physiol.* **1992**, *98*, 387–391. [CrossRef] [PubMed]
44. Guerriero, G.; Martin, N.; Golovko, A.; Sundström, J.F.; Rask, L.; Ezcurra, I. The RY/Sph element mediates transcriptional repression of maturation genes from late maturation to early seedling growth. *New Phytol.* **2009**, *184*, 552–565. [CrossRef]

horticulturae

Article

Effects of 1–MCP Treatment on Postharvest Fruit of Five Pomegranate Varieties during Low-Temperature Storage

Ran Wan [1,†], Jinhui Song [1,†], Zhenyang Lv [1], Xingcheng Qi [1], Zhiliang Feng [1], Zhenfeng Yang [1], Xinyue Cao [1], Jiangli Shi [1], Zaihai Jian [1], Ruiran Tong [1], Qingxia Hu [1,*] and Yanhui Chen [1,2,*]

1 College of Horticulture, Henan Agricultural University, Zhengzhou 450002, China; wanxayl@henau.edu.cn (R.W.); sjh20120511@163.com (J.S.); 17637630353@163.com (Z.L.); 15294747067@163.com (X.Q.); 18336374852@163.com (Z.F.); yzf15038898023@163.com (Z.Y.); cxy15638978077@163.com (X.C.); sjli30@henau.edu.cn (J.S.); jianzaihai@henau.edu.cn (Z.J.); a18839327647@163.com (R.T.)

2 Henan Key Laboratory of Fruit and Cucurbit Biology, College of Horticulture, Henan Agricultural University, Zhengzhou 450002, China

* Correspondence: hortdept@henau.edu.cn (Q.H.); chenyanhui188@henau.edu.cn (Y.C.)

† These authors contributed equally to this work.

Abstract: Pomegranate fruit production and consumption are restricted by appropriate postharvest handling practices. 1–MCP (1–methylcyclopropene) is a natural preservative of fruits and vegetables; however, its effects on the storage of different pomegranate varieties have not been extensively investigated. Herein, the effects of 1.0 μL L^{-1} 1–MCP on postharvest pomegranate fruit of three soft-seed 'Mollar', 'Malisi', and 'Tunisan soft seed' and two semi-soft-seed 'Moyuruanzi' and 'Dongyan' were investigated over 90 d (days) under low-temperature storage at 4 ± 0.5 °C with a relative humidity of 85–90%. Several indexes of exterior and interior quality were recorded, the sensory quality was evaluated, and the respiration and ethylene production were also determined. The results showed that peel browning was generally more severe in the soft-seed varieties than in the semi-soft-seed varieties. Significantly lighter peel browning presented in the three soft-seed fruits from 45 d after the 1–MCP treatment, with 35%, 19%, and 28% less than those controls at 90 d, correspondingly. However, 1–MCP only significantly decreased peel browning in the semi soft-seed fruits at 60 days. A prominent decrease in weight loss was recorded in all five varieties, with 'Malisi' showing the largest and 'Dongyan' the smallest difference between the 1–MCP and control treatments. Through the results of color, physiological, and chemical changes, as well as sensory properties, better color and total acceptance were found with higher titratable acids and vitamin C but with decreased anthocyanins in most fruits treated with 1–MCP. In contrast to the control, remarkable suppression of ethylene production peaks in all whole fruits and periodical increase in respiration rates in the soft-seed whole fruits were activated at 30–60 d after storage by the 1–MCP treatment, roughly when peel browning occurred and began increasing. Overall, our findings provided a crucial foundation for extending the application of 1–MCP in postharvest preservation of pomegranates.

Keywords: postharvest pomegranate; 1–MCP; peel browning; fruit quality; weight loss; respiration; ethylene production

Citation: Wan, R.; Song, J.; Lv, Z.; Qi, X.; Feng, Z.; Yang, Z.; Cao, X.; Shi, J.; Jian, Z.; Tong, R.; et al. Effects of 1–MCP Treatment on Postharvest Fruit of Five Pomegranate Varieties during Low-Temperature Storage. *Horticulturae* **2023**, *9*, 1031. https://doi.org/10.3390/horticulturae9091031

Academic Editors: Tong Chen and Hongliang Zhu

Received: 19 June 2023
Revised: 23 August 2023
Accepted: 29 August 2023
Published: 13 September 2023

1. Introduction

Pomegranates are native to the Middle East and widely cultivated worldwide [1]. Pomegranates (*Punica granatum* L.) belonging to *Punica* of *Lythraceae* are well-liked for their distinctive flavors and biologically active ingredients [2,3]. In China, the pomegranate harvest season typically lasts from August to October. Different protocols for large storing of pomegranates have been developed in different countries [4–6]. However, they are still highly perishable commodities along the postharvest value chain, from harvest to consumption, because of peel browning, weight loss, color and flavor deterioration, chilling injury, and quality loss, which reduce the storability and affect the consumer acceptance of the fruits, leading to direct financial loss [5–7].

Low-temperature storage is widely and effectively used for the postharvest preservation of pomegranates according to Liu et al. (cv. soft-seed 'Tunisian' and hard-seed 'Jingpitian' and 'Lishanhong') [8], Belay et al. and Lufu et al. (cv. hard-seed 'Wonderful') [7,9], Shi et al. (cv. soft-seed 'Tunisian' and hard-seed 'Yudazi') [10], and Caleb et al. (cv. 'Acco' and 'Herskawitz') [4]. Pomegranate fruit can be categorized into three categories based on aril hardness, namely, soft-seed, semi-soft-seed, and hard-seed types. Pomegranate fruits with hard seeds, such as 'Yudazi', have a longer storage life of 90–100 days at room temperature. The 'Tunisan soft seed', 'Mollar', and 'Malisi' varieties of soft-seed pomegranates, which were recently imported from Tunisia and Israel into China [11–14], have a storage life of only 30 days at room temperature. Little is known regarding the postharvest storage of the soft-seed varieties mentioned above and the Chinese native varieties, semi-soft-seed 'Moyuruanzi' and 'Dongyan'. Studying the quality variations of the five varieties during low-temperature storage was one of the aims of this study.

1–Methylcyclopropen (1–MCP) is a highly effective, non-toxic, and chemically stable preservative widely used to extend postharvest storage time and prevent the decay of horticultural products [15]. Studies have shown that low temperature combined with 1–MCP treatment can effectively delay the postharvest ripening and senescence process of most climacteric fruits, such as bananas, apples, and peaches, with the effects of promoting freshness and extending shelf-life [15–18]. However, 1–MCP treatment also positively affects postharvest storage and preservation of non-climacteric fruits [19–23]. The treatment of 1 mL L^{-1} 1–MCP effectively delayed the pedicel browning of fresh fruits of different grape cultivars [19]. With the ability to inhibit the increase in the respiration intensity of winter jujubes during low-temperature storage, 1–MCP treatment can ensure excellent quality by maintaining the preferable pericarp color and reducing weight loss and moldy rate [20]. The treatment of 300 nL L^{-1} 1–MCP can significantly decrease the lychee peel browning of 'Mauritius' and 'McLean's Red' under MAP packaging, while maintaining the vibrant color of the peel [21].

The non-climacteric nature of pomegranate fruit during development and ripening has been proven [6]. Gamrasni et al. [24] found that the pomegranate 'Wonderful' fruit maintained a good flavor compared to the control after the 900 nL L^{-1} 1–MCP treatment and storage at 7 °C for 120 days. Other researchers found that 1.0 μL L^{-1} 1–MCP effectively promoted 'Tunisia', 'Wonderful', 'Tunisian soft-seed', and 'Dahongpao' pomegranate fruit quality by reducing peel browning rates and preserving flavors during low-temperature storage [25–28]. Although non-climacteric fruit exhibit a declined low respiration and low ethylene production during maturation and ripening, ethylene does participate in the regulation of maturation and some physiological changes [29]. According to Valdenegro et al. [5], 1–MCP as a typical ethylene antagonist did not significantly inhibit the respiration rate and ethylene release of 'Wonderful' during low-temperature storage at 2 °C. Differently, other researchers observed 1.0 μL L^{-1} 1–MCP significantly reduced the respiration intensity of 'Tunisia' and 'Tunisian soft-seed' during low-temperature storage at 4 °C [25–27]. It has been found that the effects of a 1–MCP application on respiration, ethylene production, and fruit quality may depend on the variety, tissue, and storage conditions [5,23,30]. 1–MCP is a natural preservative of fruits and vegetables; however, its effects on the storage of different pomegranate varieties have not been investigated extensively. Therefore, this work also aimed to explore the effect of the 1–MCP treatment on postharvest fruit of the five different pomegranate varieties mentioned above during a low-temperature storage. Our work will contribute to the practical application of 1–MCP in postharvest preservation of pomegranates.

2. Materials and Methods

2.1. Plant Materials

Pomegranate fruit of soft-seed varieties, 'Tunisan soft seed', 'Malisi', and 'Mollar', as well as semi-soft-seed varieties, 'Moyuruanzi' and 'Dongyan', from a 8-year-old commercial vineyard (Liugou Village, Gaocun Town, Xingyang City, Henan, China) under natural field conditions were harvested at commercial maturity by experienced staff (typical peel and aril color and juice soluble solids higher than >14%). At harvest, the undamaged and healthy fruits were representatively selected for uniformity in color, size (about 0.3–0.4 kg/fruit), and appearance without sunburn, cracks, bruises, cuts, decay, or disease. The harvest time of 'Tunisan soft seed', 'Malisi', 'Mollar', and 'Moyuruanzi' were on 5 October, and it was on 27 October 2021 for 'Dongyan', a late-mature landrace. Three-hundred pomegranate fruit of each variety were transported into the laboratory within 3 h after harvest and then were sorted, cleaned, and arranged into three replications.

2.2. 1–MCP Treatment

A total of ninety harvested pomegranate fruit of each variety for one replication were randomly and equally put in fifteen flat polyethylene bags with the specification of 1200 mm long, 800 mm wide, and 0.05 mm thick. They were either left untreated or treated with AnsiP-S stickers (1–methylcyclopropene, Lytone Enterprise, Inc., Taipei, Taiwan, China) and then sealed for 24 h at 4 ± 0.5 °C with a relative humidity of 85–90%. The treatment method of Guo et al., 2019 [27] was used, and the effective concentration of 1–MCP was 1.0 μL L^{-1}. This concentration of 1–MCP has been verified with good effect on preservation of several different pomegranate varieties such as 'Tunisian', 'Wonderful', and 'Dahongpao' [25–28]. Thereafter, all bags were converted to a semi-closed state and transferred to the low-temperature storage at 4 ± 0.5 °C for up to 90 days (d). Six pomegranate fruit for each treatment for one replication were taken out from the low-temperature storage for data analysis at 0, 15, 30, 45, 60, and 90 d after harvest. Fruit quality, ethylene production, and respiration rate were assessed in whole fruit and arils. Arils were artificially and quickly separated and transferred to measurement or making juices. Decreasing the effects of the higher room temperature than arils themselves from the low-temperature surroundings should be considered as much as possible. Aril juices were prepared by squeezing the arils through a double-layer gauze and then were frozen at −20 °C for further analysis. The analyses were performed for all samples at the same time.

2.3. Exterior Quality Index

According to Kashash et al. [30], pomegranate peel browning, the principal non-pathogenic disorder occurring on the fruit surface, was divided into five levels based on the browning areas: level 0 (no browning symptoms and 0 browning area), level 1 (browning areas between 1 and 25%), level 2 (browning areas between 26 and 50%), level 3 (browning areas between 51 and 75%), and level 4 (browning areas > 75%). Then, the browning index was calculated according to the formula described by Zhang et al. [28]: the browning index = $\sum$ (browning level × the number of fruits of that level)/(5 × total number of fruits). A higher browning index represented peel browning more severe. From all arils of each sample, 100 arils were randomly selected and weighted to determine the hundred-aril mass with values being presented as g. The L*, a*, and b* values of aril color were measured using a HP-C210 visible light colorimeter (Hanpu Photoelectric Technology Co., Ltd., Shanghai, China) to obtain the brightness values (L*), the chromatic values [$C^* = (a^{*2} + b^{*2})^{1/2}$], and the hue angle values [$H^* = \tan^{-1}(b^*/a^*)$].

2.4. Interior Quality Index

The total soluble solids of aril juice were measured by a WTY handhold refractometer (Chengdu Qingyang Huarui Optical Instrument Factory, Chengdu, China), with values being presented as Brix degrees (%). The titratable acid contents were tested using the acid–base neutralization titration method by titrating 5 mL of juice to reach the endpoint of pH 8.2 with 0.1 M NaOH and recording the titration volume. The resulting data were expressed as citric acid percentage. The ratios of soluble solid and titratable acid were then calculated. The methods described above were all based on Gao et al., 2022 [31], and each measurement was repeated at least three times.

The vitamin C contents of aril juice were detected by a 2,6-dichlorophenol indophenol (DCIP) titration method described by Gao et al., 2022 [31]. About 0.5 g of fresh arils, ground with liquid nitrogen, was mixed with 50 mL of 2% (m/v) oxalicacid solution. Then, 10 mL of the solution was transferred to a triangular flask (50 mL), and the DCIP solution that had been calibrated was immediately performed for sample solution titration. The terminal point was recorded with a reddish appearance at 15 s fadeless. Vitamin C contents of each sample were determined by the consumed volume of the DCIP solution. Total anthocyanin contents were determined according to the pH differential method described by Shi et al., 2022 [32]. Absorbance was measured in a UV-VIS spectrophotometer (L9, INESA, Inc., Shanghai, China) at 510 and 700 nm in buffers at pH 1.0 and 4.5, respectively. The results were expressed as milligrams of cyanidin-3-glucoside per L of pomegranate juice. Each detection above was performed three times.

2.5. Ethylene Production

After the postharvest treatments (control, 1–MCP), ethylene production of whole fruit and arils were analyzed at different periods after 0, 15, 30, 45, 60, and 90 days during storage using gas chromatography (GC; GC-2010 PLUS, Shimadzu, Kyoto, Japan) with a flame ionization detector [26]. Two whole fruits were randomly selected from each sample and placed into a 2 L plastic box to rest for 3 h at room temperature. The plastic boxes were cubic and had a sealed lid fitted with a rubber stopper. Then, 1000 μL of gas was taken with a micro-sampler from the cubic plastic box and injected into a GC Packed Column (GDX-502, Shimadzu, Kyoto, Japan) for analysis. Nitrogen was used as the carrier gas at the flow rate of 40 mL min^{-1}, with the injection port set to 100 °C and the detector (FID, Lanzhou Institute of Chemical Physics, Lanzhou, China) to 150 °C. A 10 μL L^{-1} ethylene standard was used for equipment calibration. Ethylene production [μL h^{-1} kg^{-1} (FW)] = $c \times V \times m^{-1} \times t^{-1}$, where c is the ethylene content as determined by gas chromatography; V is the sealed container's space volume—sample volume (L); m is the sample mass (kg); t is the standing time (h). The arils of each sample were weighted 0.2–0.3 kg to analyze the respective ethylene production according to this method. Each experiment was repeated at least three times.

2.6. Respiration Rate

Two whole fruit from each sample were taken out from low temperature and kept still for 3 h at room temperature before being placed into a 2 L cubic plastic box with a sealed lid fitted with a rubber stopper for another 3 h at room temperature. Finally, carbon dioxide concentration was measured using the portable carbon dioxide analyzer (F-950, FELIX Company, Camas, WA, USA), and respiration rates were calculated according to Guo et al. [27]. Respiration rate [mg h^{-1} kg^{-1} (FW)] = $c \times V \times 44 \times 273 \times (m \times t \times 22.4 \times 293)^{-1}$, where c is the carbon dioxide concentration; V is the sealed container's space volume—sample volume (L); m is the sample mass (kg); t is the storage time (h); 44: the molar mass value of carbon dioxide, 44 kg/mol; 273: the thermodynamic temperature at 0 °C in standard condition, 273 k; 22.4: the volume of 1 mmol gas in standard condition, 22.4 mL/mmol; 293: the thermodynamic temperature at 20 °C in standard condition, 293 k. According to the method mentioned above, the arils of each sample were weighted 0.2–0.3 kg to analyze the corresponding respiration rates. Each experiment was repeated at least three times.

2.7. Sensory Evaluation

After 90 days of low-temperature storage, six fruit from each sample were randomly selected. Descriptive sensory analyses were performed by 10 panelists, developed by Osondu et al. [33], which were then thoroughly evaluated through four aspects. First, fruit pericarp and aril were scored according to several color grades: those with superior color received 10–9 points; those with good color received 8–7 points; those with acceptable but limited marketability received 6–5 points; and the rest received under 5 points. Second, fruit were scored according to their level of flavor: if they had a typical flavor, they received 9–10 points, if good, they received 7–8 points; if moderately acceptable, they received 6–5 points, if acceptably but limitedly marketable, they received 3–4 points, and the rest received 1–3 points. Third, fruit were scored according to the level of odor grades: fruit with no odor received 9–10 points; fruit with a little odor was given 7–8 points; fruit with some odor received 6–5 points; if obvious but tolerable odor, they received 3–4 points, and the rest received 1–3 points. Finally, overall acceptance would be directly scored through comprehensive sense: fruit with 10–9 points were liked; fruit with 8–7 points were moderately liked; fruit with 6–5 points were not liked nor disliked; fruit with 4–3 points were moderately disliked; and fruit with 2–1 points were especially disliked. The final score used to assess the sensory qualities of each sample was the average value of the aspects mentioned above.

2.8. Statistical Analysis

One-way analysis of variance was used to analyze all the data using Statistical Package for Social Sciences version (IBM. SPSS Inc., Chicago, IL, USA), with three replications of each experiment. Duncan's multiple comparison was applied at a $p = 0.05$ probability level to evaluate significant differences among data points and between the control and 1–MCP treatments. GraphPad Prism 9 (GraphPad Software Inc., San Diego, CA, USA) was used to draw the figures, and Adobe Photoshop CC 2017 (Adobe Systems Inc., San Jose, CA, USA) was used to combine them.

3. Results

3.1. Effects of 1–MCP on Exterior Quality

3.1.1. Browning Index

The investigated five pomegranate varieties except 'Dongyan' experienced obviously fewer browning lesions in fruit treated with 1–MCP compared to the untreated control, especially 90 d after low-temperature storage. When 'Dongyan' and 'Moyuruanzi' fruit were treated with 1–MCP, there were fewer browning lesions than in fruit from other types or the control (Figure 1). The Browning index of the five varieties increased with storage time. Rather, the browning in the peels of the treated or untreated soft-seed fruit of 'Mollar' and 'Malisi' occurred and then rapidly developed from 45 d; before that, there was no visible browning. In 'Tunisian soft seed', peel browning occurred early at 30 d with the control, when none was found with the 1–MCP treatment. The beginning time point of peel browning in the two semi-soft-seed fruit was at 60 d (Figure 2a). With the 1–MCP treatment, 'Mollar', 'Malisi', and 'Tunisan soft seed' reduced the peel browning index from 0 at harvest to 0.20, 0.28, and 0.20 after 90 days of storage, which were significantly 35%, 19%, and 28% less than those controls (0.31, 0.34, and 0.28), respectively (Figure 2a). On the contrary, peel browning presented from 60 d after storage both in 'Moyuruanzi' and 'Dongyan' semi-soft-seed fruit. At this point, their peel browning indexes with the 1–MCP treatment (0.044 and 0.022, respectively) were significantly lower than those with the control (0.089 and 0.033, respectively). However, they exhibited no significant difference between the control and 1–MCP treatments at 90 days (around 0.20) (Figure 2a). Additionally, the results showed that compared to semi-soft-seed fruit, these soft-seed fruits had a considerably greater peel browning (Figures 1 and 2).

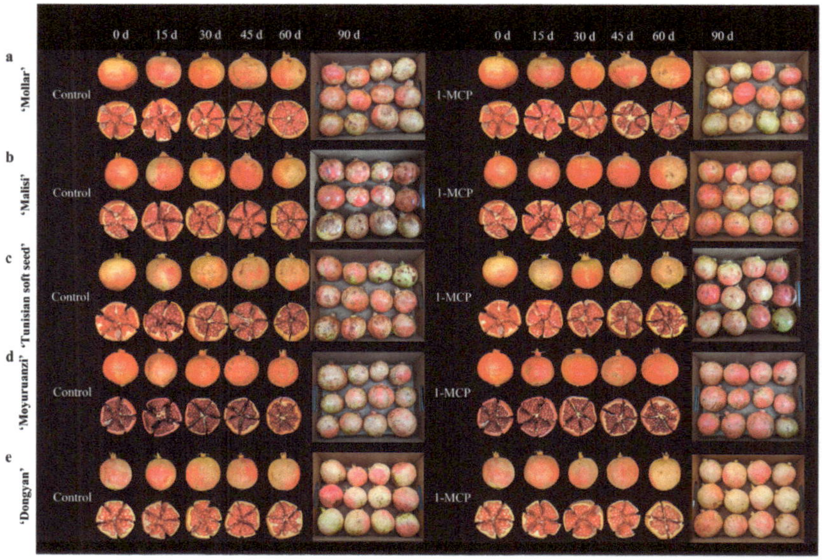

Figure 1. 1–MCP treatment affected the macro performance of the five pomegranate varieties during the low-temperature storage. (**a**) 'Mollar'; (**b**) 'Malisi'; (**c**) 'Tunisian soft seed'; (**d**) 'Moyuruanzi'; (**e**) 'Dongyan'. Control: untreated control; 1–MCP: samples that were treated with 1–MCP stickers. All the fruit were stored at the low temperature of 4 °C and with the relative humidity of 85–90%.

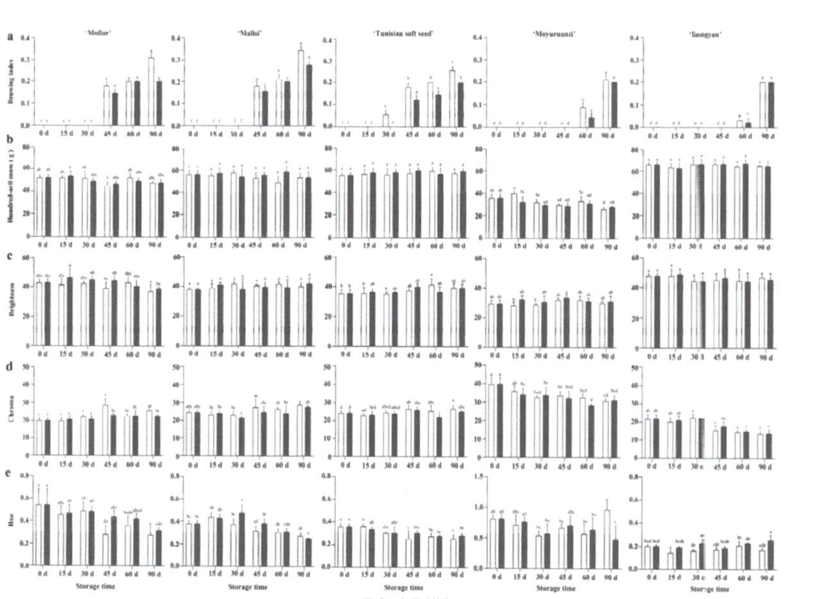

Figure 2. 1–MCP treatment affected the exterior quality index of the five pomegranate varieties during low-temperature storage, which included browning index (**a**), hundred-aril mass (**b**), brightness values (**c**), hue values (**d**), and chroma values (**e**). Control: untreated control; 1–MCP: samples that were treated with 1–MCP stickers. All the fruit were stored at the low temperature of 4 °C with the relative humidity of 85–90%. Data represent mean values of the replications, and the error bars represent standard deviations of the means. Different letters indicate significance differences at a significant level of $p = 0.05$ using Duncan's test.

3.1.2. Color

As shown in Figure 2b, the 1–MCP treatment did not affect the hundred-aril mass of these five varieties during low-temperature storage, with 'Moyuruanzi' being the lowest. The 1–MCP treatment also caused no significant change in aril brightness of these five varieties compared to the untreated control, which maintained a stable level during storage (Figure 2c). However, their chromatic values showed an increase in control arils of three soft-seed varieties but a decrease in two semi-soft-seed varieties during storage. Nevertheless, these values showed no variation after the 1–MCP treatment (Figure 2d). Hue values decreased significantly from harvest with a slight increase at 60–90 d after storage in control soft-seed arils. With the treatment of 1–MCP, hue values were obviously higher at 30–45 d compared to the control. These results indicated that 1–MCP could contribute to maintaining the red color of these pomegranate arils. Additionally, the red color of 'Moyuruanzi' arils was stronger and more saturated for their higher chromatic and hue values than other varieties (Figure 2e).

3.2. Effects of 1–MCP on Interior Quality

3.2.1. Total Soluble Solids and Titratable Acids

Figure 3a,b depicts the variations in total soluble solids and titratable acids of the five pomegranate varieties. Total soluble solids in control arils of all five varieties essentially did not change much during storage; on the contrary, titratable acids showed a substantial downward trend, particularly after 30 d. Total soluble solids and titratable acids of 'Moyuruanzi' retained 16.1–17.7% and 2.41–3.28%, respectively, which were both higher than those of the other varieties. The titratable acids of 'Malisi' were the lowest, ranging between 0.13% and 0.21%. Ratios of total soluble solids and titratable acids in these arils, except 'Moyuruanzi', dramatically rose throughout storage, rising from around 50 at 0 days to about 100 at 90 days. In 'Moyuruanzi' arils, the ratios exhibited a significant increase from 5.2 to 6.1, which might be the result from its attribute of extreme high titratable acids (Figure 3c). With 1–MCP treatment, total soluble solids were unaffected at almost all storage stages; however, the decrease in titratable acids were apparently restrained (Figure 3a,b). At 90 d after storage, the titratable acids were 27.5%, 12.5%, 25.6%, 18.0%, and 18.4% greater in arils of the five varieties with the treatment of 1–MCP than those in controls in the order depicted in Figure 3b. Associated with the slow decrease in titratable acids in 1–MCP-treated arils of the five varieties with storage duration, their ratios of total soluble solid and titratable acid were lower than those in control arils (Figure 3c).

3.2.2. Vc and Anthocyanins

Figure 3d also presented a serious and significant loss of Vc content in arils of the five pomegranate varieties with storage duration. At 0 d, their Vc in control arils ranged from 87 to 116 mg kg^{-1}, but they had decreased to 36–41 mg kg^{-1} at 90 d. Although a slight increase in Vc in arils with 1–MCP treatment were observed at 30 d except for 'Tunisian soft seed' at 60 d, 1–MCP was unable to reverse these decreases finally. Anthocyanins are important phenolic compounds presented in pomegranates that also affect the color of fruits [34]. To verify the effect of 1–MCP on postharvest pomegranate fruit during low-temperature storage, anthocyanin contents in arils were investigated in the current study (Figure 3e). In contrast with declining Vc content, the anthocyanin content in arils displayed a significant increase with the storage period. Higher anthocyanins ranging from 148.2 to 284.8 mg L^{-1} were found in 'Moyuruanzi' arils in comparison with those in other arils. Three soft-seed arils had 17.6–43.7 mg L^{-1} anthocyanins, and 'Dongyan' had the lowest 11.4–15.0 mg L^{-1}. Pomegranate arils of 'Mollar', 'Moyuruanzi', and 'Dongyan' treated with 1–MCP showed a slight increasing anthocyanins after 30 d during storage.

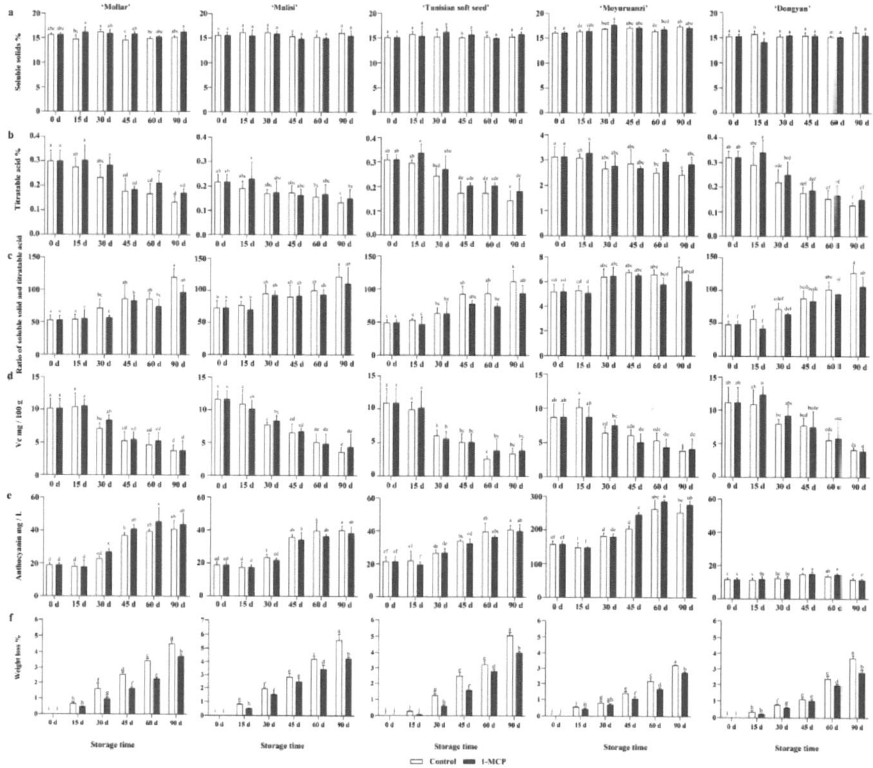

Figure 3. 1–MCP treatment affected the interior quality index of the five pomegranate varieties during low-temperature storage, which included total soluble solid contents (**a**), titratable acid contents (**b**), ratios of soluble solid and titratable acid (**c**), Vc contents (**d**), anthocyanins (**e**), and weight loss (**f**). Control: untreated control; 1–MCP: samples that were treated with 1–MCP stickers. All the fruit were stored at the low temperature of 4 °C with the relative humidity of 85–90%. Data represent mean values of the replications, and the error bars represent standard deviations of the means. Different letters indicate significant differences at a significance level of p = 0.05 using Duncan's test.

3.2.3. Weight Loss

On average, increasing storage time from 0 d to 90 d increased weight loss across all varieties, both of the 1–MCP and control treatments. After the 90 d storage period, the highest (5.33%) and lowest (2.75%) weight losses were recorded in the fruit of 'Malisi' with the control and 'Moyuruanzi' with the 1–MCP treatment, respectively (Figure 3f). However, weight loss was significantly lesser for all fruit treated with 1–MCP, compared to the untreated control. Among the five varieties, the greatest difference of increasing weight loss between the 1–MCP treatment and the control was recorded in 'Malisi' fruits, followed by 'Mollar', and the smallest was in 'Dongyan' (Figure 3f). These analyses demonstrated that 1–MCP was an effective treatment for weight loss.

3.3. *Effects of 1–MCP on Respiration Rate and Ethylene Production*

3.3.1. Respiration Rates of Whole Fruit and Arils

In most cases, the respiration of whole fruit and arils of the five varieties both with the 1–MCP and control treatments displayed increases with fluctuations during storage at the low temperature (Figure 4a,b). Among whole fruit varieties, the lowest respiration rates were recorded in 'Dongyan' with no more than 1.6 mg kg^{-1} h^{-1} (Figure 4a). Compared with the whole fruit of the control, respiration rates were significantly increased by the

1–MCP treatment in soft-seed 'Mollar', 'Malisi', and 'Tunisian soft seed' at 30–45 d after storage, but they were mainly declined in semi-soft-seed 'Moyuruanzi' and 'Dongyan' (Figure 4a). Nevertheless, at 90 d, the increment percentages of respiration rates from 0 d in whole fruit of 'Mollar', 'Malisi', 'Tunisian soft seed', and 'Moyuruanzi' (52.3%, 52.6%, 43.3%, and 25.8%, respectively) were recorded as 42.5%, 56.5%, 25.9%, and 26.2% lower, respectively, in fruits with the 1–MCP treatment than those with the control (94.9%, 109.1%, 69.1%, and 96.3%, respectively). For 'Dongyan', it (14.0%) was 2% higher (16.0%) (Figure 4a). These results indicated that 1–MCP increased whole fruit respiration in the soft-seed varieties during the middle storage. Still, it prevented a subsequent increase in respiration, including the semi-soft-seed 'Moyuruanzi'.

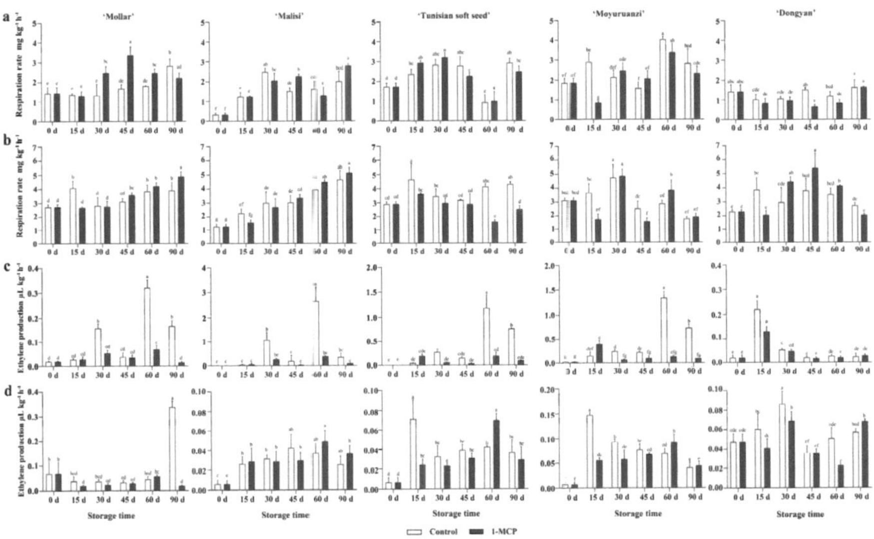

Figure 4. 1–MCP treatment affected respiration rate and ethylene production of the five pomegranate varieties during low-temperature storage. (**a**) Respiration rates of whole fruit; (**b**) respiration rates of arils; (**c**) ethylene production of whole fruit; (**d**) ethylene production of arils. Control: untreated control; 1–MCP: samples that were treated with 1–MCP stickers. All the fruit were stored at the low temperature of 4 °C with a relative humidity of 85–90%. Data represent mean values of the replications, and the error bars represent standard deviations of the means. Different letters indicate significant differences at a significance level of $p = 0.05$ using Duncan's test.

Among arils of varieties, the respiration rates of 'Mollar' and 'Malisi' both with the 1–MCP and control treatments increased progressively from around 2.0 mg kg^{-1} h^{-1} to approximately 4.0 mg kg^{-1} h^{-1}. However, a significant decrease of 34.6% was observed in 'Mollar' arils caused by the 1–MCP treatment at 15 d after storage. Simultaneously, similar decreases of 10%, 24.1%, 52.9%, and 47.2% were observed in 'Malisi', 'Tunisian soft seed', 'Moyuruanzi', and 'Dongyan', respectively (Figure 4b). After that, the respiration rate of 'Tunisian soft seed' arils with 1–MCP treatment was invariably lower than that with control. Although respiration rates of 'Moyuruanzi' and 'Dongyan' exhibited different increases after 15 d, the increment percentages of their respiration rates from 0 d to 90 d after the 1–MCP treatment were 4.5% and 29.5% lower than those after the control treatment, respectively (Figure 4b). Through the analysis above, it was suggested that 1–MCP effectively declined aril respiration rates in all five varieties at the beginning of storage and hindered the rates of aril respiration in 'Tunisian soft seed', 'Moyuruanzi', and 'Dongyan' at the end.

3.3.2. Ethylene Production of Whole Fruit and Arils

The analysis of variance among the experimental effects showed that 1–MCP significantly affected the ethylene production in the treated whole fruit and arils of the five pomegranates, which are referred to as a kind of non-climacteric fruit (Figure 4c). The changed pattern was quite different from that of the respiration rates, especially in whole fruit. The 1–MCP treatment significantly suppressed ethylene production peaks of whole fruit of different varieties, mainly at 30 d, 60 d, and 90 d, except for 'Dongyan', which peaked only at 15 d. At another period, the ethylene production of whole fruit of the control remained at 0.04 μL kg^{-1} h^{-1} and 0.76 μL kg^{-1} h^{-1} during storage; by contrast, it was 0.083 μL kg^{-1} h^{-1} at most after the 1–MCP treatment. Among the whole fruit of five cultivars, the highest ethylene production of 2.67 μL kg^{-1} h^{-1} at 60 d was recorded in 'Malisi' of the control that declined into 0.38 μL kg^{-1} h^{-1} by 1–MCP at 90 d (Figure 4c).

Among arils of varieties both of the 1–MCP and control, ethylene production was always under 0.094 μL kg^{-1} h^{-1} during storage, except with a quite high value of 0.34 μL kg^{-1} h^{-1} in 'Mollar' at 90 d (Figure 4d). Despite the fact that ethylene production displayed quite lower levels in arils than those in whole fruit, some effects were also found in arils by 1–MCP. Compared to the control, the ethylene productions were slowed down in arils of all five varieties of the 1–MCP treatment at 15–45 d after storage. At 45 d, arils in the varieties of the treatment with 1–MCP showed 10.8%, 30.1%, 21.1%, 13.5%, and 5.3% lower ethylene production than those of the control in sequence shown in Figure 4d. On the contrary, except in 'Mollar' and 'Tunisian soft seed', ethylene production was mostly displayed slightly higher in arils with the 1–MCP treatment than those with the control from then. At 90 d, arils of 'Malisi', 'Moyuruanzi', and 'Dongyan' with the 1–MCP treatment presented 44.5%, 11.6%, and 16.2% higher ethylene production than those of the control, respectively (Figure 4d). Overall, the treatment with 1–MCP promoted a reduction of ethylene production and, remarkably declined the peak in whole fruit of all the varieties.

3.4. *Effects of 1–MCP Treatment on Sensory Quality*

The descriptive sensory panelists evaluated overall acceptance, odor, flavor, and color of pericarp and aril at 90 d after the low-temperature storage with the 1–MCP and control treatments (Figure 5). 'Moyuruanzi' outperformed the other investigated varieties in terms of sensory properties and preferences at the end of the low-temperature storage. Regarding overall acceptance, odor, flavor, and color, 'Moyuruanzi' scored the highest, whereas 'Malisi' scored the lowest. 'Mollar' and 'Tunisan soft seed' showed clearly higher scores of overall acceptance with the 1–MCP treatment (6.10 and 5.90) than the control (4.90 and 5.10). Few changes of odor and flavor were perceived in all five varieties by 1–MCP. A better acceptance of fruit color, especially of peel color, were rather recorded in fruit with the 1–MCP treatment compared to the control. The 1–MCP treatment resulted in a pericarp color of 'Mollar', 'Tunisan soft seed', and 'Moyuruanzi' being higher by 30.6%, 23.5%, and 14.8% than those of the control, respectively. Additionally, their aril color scores differed from the control by 9.8%, 5.4%, and 12.6%, respectively. These results showed that there was no negative effect on the sensory quality of 'Malisi' and 'Dongyan', but overall acceptance and color were particularly appreciated in 'Mollar', 'Tunisan soft seed', and 'Moyuruanzi'.

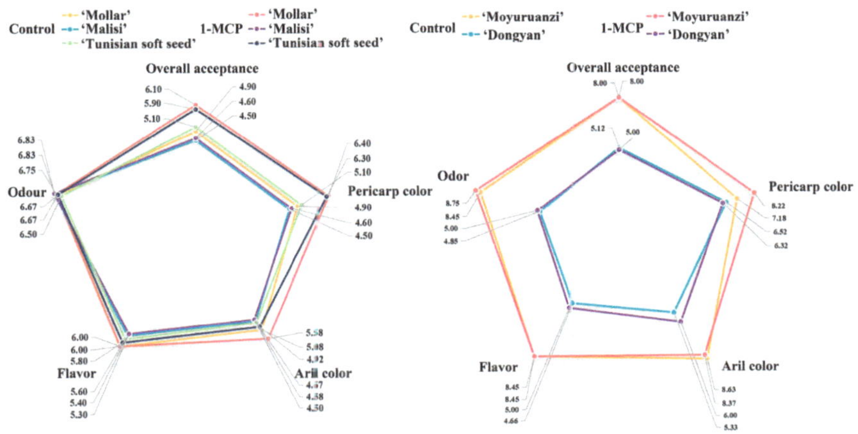

Figure 5. 1–MCP treatment affected sensory properties of the five pomegranate varieties during low-temperature storage. Control: untreated control; 1–MCP: samples that were treated with 1–MCP stickers. All the fruit were stored at the low temperature of 4 °C with the relative humidity of 85–90%.

4. Discussion

In recent years, the soft-seed pomegranate industry has developed rapidly and has become an essential pillar for poverty alleviation and rural revitalization in Yunnan, Sichuan, and Henan in China, especially with the introduction of several new varieties in recent years [14]. Nonetheless, they are all facing a limited postharvest storage life. Pomegranate fruit is prone to several physiological and chemical disorders, such as the major storage problems of water loss and browning symptoms [5–7,33]. As is common knowledge, low-temperature storage is widely used to maintain the nutritional value of fruits and vegetables. Nevertheless, studies have shown that some non-climacteric fruit can benefit somewhat from 1–MCP [23,34]. This study showed that the storage life of five pomegranate varieties treated with 1.0 μL L^{-1} 1–MCP was prolonged compared to that of the control during the low-temperature storage at 4 °C, and the fruit maintained a comparatively lower peel browning and weight loss (especially for three soft-seed varieties) as well as better fruit quality.

4.1. Peel Browning

The browning of pomegranate peels represents a common problem after harvest. Although browning increases under 5 °C, storage at low temperatures is necessary [6]. Symptoms of peel browning include pitting, husk scald, some softening, a higher sensitivity to decay, internal seed discoloration, and browning of chilling injury in postharvest pomegranate fruit during low-temperature storage [29,33,35]. We observed peel browning in the three soft-seed pomegranate fruit ('Mollar', 'Malisi', and 'Tunisian soft seed') from 30–45 d and in the two semi-soft-seed fruit ('Moyuruanzi' and 'Dongyan') from 60 d after the low-temperature storage. The two types of semi-soft- and soft-seed displayed obvious differences with the storage duration (Figures 1 and 2). A similar difference among varieties was even reported on soft-seed 'Tunisia' compared with hard-seed 'Yudazi' [10], suggesting that the different responses to 1–MCP were due to different cultivar traits. Additionally, it can be connected to their various ancestries. From Tunisia and Israel, these three soft-seed types had less endurance in cold temperatures than semi-soft-seed kinds originating in China.

It is reported that 1–MCP can significantly reduce the browning of grape stalks and lychees during postharvest storage [19,21]. 'Wonderful', 'Dahongpao', and 'Tunisian' all had a lower browning index in pomegranate peels with 1–MCP treatment [5,6,27,28]. Similarly, 1–MCP did not delay the occurrence of browning, except in 'Tunisian soft seed'; however, a significant decrease in browning index was recorded, especially in the three soft-seed varieties after the 1–MCP treatment during the low-temperature storage (Figures 1 and 2a) Malonaldehyde (MDA) levels are always elevated, and enzymatic components are always triggered when pomegranates brown [10,32,34,35]. The 1–MCP treatment had been proven to reduce the browning by inhibiting the activities of PPO (polyphenol oxidase) and POD (peroxidase) and lowering MDA levels in 'Tunisian soft seed' fruit (unpublished data). A previous report showed that decreased peel browning in 'Dahongpao' and 'Wonderful by 1–MCP is linked to a decrease in MDA level and PPO activity and an increase in total antioxidant capacity [23,28]. Instead, higher MDA level and PPO activity but lower antioxidant capacity (ascorbate peroxidase and catalase, etc.) have been implicated in aril browning of 'Tunisia' during cold storage, peel browning of 'Baiyushizi', and superficial browning of 'Wonderful' caused by chilling injury [32,36,37].

4.2. Weight Loss

Weight loss indicated by water and carbon losses and caused by transpiration, respiration processes, and a vapor pressure deficit in pomegranate is important in spite of its thick rind and tough leathery outer skin, leading to loss of ethylene gas, as well as aromatic and volatile organics [6,7,34]. The application of modified atmosphere packaging, film wrapping, and coating, such as perforated polypropylene film and acacia gum, can successfully reduce weight loss [5,34,38]. In this present study, compared to the control, retarded weight loss was recorded in pomegranate fruit of the five varieties with the 1–MCP treatment during the low-temperature storage (Figure 3f). Moreover, when weight loss is excessive, it will result in browning of the peel and arils and hardening of the rind [7,34,39].

4.3. Fruit Quality

Regarding quality criteria, the color of pomegranate is an essential attribute affecting marketability, purchasability, and consumer preference [40]. Through the calculation of hue values that measure the degree of saturation of color and the evaluation by sensory panelists, a marginally improved color was shown in the studied pomegranates following 1–MCP treatment, especially in 'Moyuruanzi', in comparison to the control (Figures 2d and 5). Acidity affects the taste and color of pomegranate arils [27,41], and previous studies have reported that organic acids of fruit are substrates consumed during storage in respiratory processes [39]. Here, a preventation of the decrease in titratable acids was found in all the investigated arils with 1–MCP treatment during low-temperature storage (Figure 3b). Furthermore, 1–MCP had no effect on total soluble solids. Obviously, ratios of total soluble solid and titratable acid were undoubtedly reduced by the 1–MCP treatment (Figure 3c). Valdenegro et al. [5] indicated that no significant differences were observed in these parameters for arils of 'Wonderful' with 1–MCP or ethylene treatment during the whole period of cold storage at 2 °C, including 3 days at 20 °C. However, the application of coatings reduces the loss of titratable acids and decreases ratios of total soluble solids and titratable acids during low-temperature storage [39].

Vc is a typical non-enzymatic antioxidant substance with antioxidant and anti-aging effects [10]. As the main component of phenolics, anthocyanin is also an essential anti-aging and antioxidant substance [42]. They also affect the color of fruits [40]. An increasing trend of the anthocyanins but a decrease trend of Vc was observed in the arils of these five pomegranates during low-temperature storage. However, the 1–MCP treatment had a favorable impact on halting the decline of Vc, while it had no discernible impact on the growth of their anthocyanins (Figure 3d,e). Similar results had also been reported in 'Wonderful' treated with 1–MCP during the low-temperature storage at 2 °C and 'Rabbab-e-Neyriz' treated with coating during cold storage (2 ± 0.5 °C) for 45 d [38,39]. Furthermore, the fruit of three soft-seed varieties with the treatment of 1–MCP rather than the two semi-soft-seed varieties effectively maintained the color and total acceptance (Figure 5), further indicating that the various responses to 1–MCP were brought on by various cultivar features.

4.4. Respiration and Ethylene Production

Although pomegranate is a non-climacteric fruit, it was affected by 1–MCP and analogous exogenous ethylene, indicating that ethylene might be involved in its senescence [5]. 1–MCP promotes the postharvest preservation of climacteric or non-climacteric fruits, usually by lowering or delaying the peak of respiration intensity and ethylene production [15,16]. Valdenegro et al. [5] found that exogenous ethylene treatment had only a temporary or no effect on the ethylene production of postharvest pomegranate fruits stored at 2 °C. However, Zhang et al. and Wan et al. [26,28] also demonstrated that 1–MCP may considerably reduce the rate of the pomegranate respiration or ethylene production during cold storage. Therefore, in this study, the effects of 1–MCP treatment on respiration and ethylene production of whole fruit and arils of these five pomegranates during the low-temperature storage were thoroughly examined. The results showed that 1–MCP boosted whole-fruit respiration of the soft-seed varieties at the middle storage (30–45 d), but it blocked their subsequent increase in respiration, including that of 'Moyuruanzi'. Meanwhile, the treatment with 1–MCP triggered a reduction of ethylene production, and it remarkably declined the peak in whole fruit of all the cultivars at 30–60 d after storage, for example of 'Malisi', decreasing the peak of 2.67 μL kg^{-1} h^{-1} at 60 d down to 0.38 μL kg^{-1} h^{-1} at 90 d (Figure 4a,c).

Notably, peel browning began coinciding with increased respiration rates and reduced ethylene generation in the soft-seed whole fruit treated with the 1–MCP compared to control fruits (Figures 1, 2 and 4). According to earlier research, the enhanced respiration and increased energy charge in longan when exposed to pure oxygen are related to their reduced browning level [43]. After short-term anaerobic treatment, a sufficient energy supply reduces the peel browning of post-harvest litchi. In contrast, a shortage of energy is one of the main reasons for longan browning [44,45]. Furthermore, Valdenegro et al. [5] also observed that the pomegranate browning during cold storage is preceded by a spike in ethylene production and an increase in respiration rates and consumption of oxygen by 1–MCP. Therefore, we hypothesized that 1–MCP treatment enhanced the respiration intensity of the three soft-seed pomegranates, allowing for the redistribution of additional materials and energy charges to the peels. This could ensure requirement for resistance to halt browning processes during low-temperature storage, which might be implicated in the ethylene associated pathway induced by the 1−MPC application.

5. Conclusions

Short storage life is the main problem of pomegranate fruit, indicated by peel browning, weight loss, flavor and color loss, and other symptoms. Despite its universal presence in storing and preserving fruits and vegetables, it must be enhanced when applied to various soft-seed and semi-soft-seed pomegranates. This study determined the effects of low-temperature storage and 1–MCP on peel browning and some physiological and biochemical indices of three soft-seed and two semi-soft-seed pomegranate varieties. All fruit studied with the 1–MCP treatment showed less quality degradation than those in the control, as seen by lighter peel browning, less weight loss, fewer decrease in titratable acids and Vc, and better color and acceptability. Although pomegranate is a non-climacteric fruit that always has low levels of ethylene and respiration during ripening, remarkable suppression of ethylene production peaks in all varieties and periodical increase in respiration rates were observed at 30–60 d after the 1–MCP treatment compared to the control. This was especially true in the whole fruit of three soft-seed varieties. Noticeably, this behavior began when 1–MCP started to reduce peel browning, if not earlier. Collectively, our research showed that 1–MCP positively impacted pomegranate fruit quality during the low-temperature storage, which would serve as an important theoretical foundation for postharvest practicality on pomegranate fruit in China.

Author Contributions: Conceptualization, R.W., J.S. (Jinhui Song), Q.H. and Y.C.; formal analysis, R.W., J.S. (Jinhui Song), Z.Y., X.C. and Z.J.; visualization, J.S. (Jinhui Song), Z.L., X.Q., Z.F. and R.T.; investigation, J.S. (Jinhui Song), Z.L., X.Q., Z.F. and J.S. (Jiangli Shi); resources, R.W., Q.H. and Y.C.; writing—original draft, R.W. and J.S. (Jinhui Song); writing—review and editing, R.W., J.S. (Jinhui Song), Q.H. and Y.C. All authors have read and agreed to the published version of the manuscript.

Funding: We appreciate for the support from the Key Science and Technology Project of the Xinjiang Production and Construction Corps (no. 2021AB015).

Data Availability Statement: Not applicable.

Conflicts of Interest: The authors declare no conflict of interest.

References

1. Yuan, Z.H.; Fang, Y.M.; Zhang, T.K.; Fei, Z.; Han, F.; Liu, C.; Liu, M.; Xiao, W.; Zhang, W.; Wu, S.; et al. The pomegranate (*Punica granatum L.*) genome provides insights into fruit quality and ovule developmental biology. *Plant Biotechnol. J* **2018**, *16*, 1363–1374. [CrossRef] [PubMed]
2. Gbinigie, O.A.; Onakpoya, I.J.; Spencer, E.A. Evidence for the effectiveness of pomegranate supplementation for blood pressure management is weak: A systematic review of randomized clinical trials. *Nutr. Res.* **2017**, *46*, 38–48. [CrossRef] [PubMed]
3. Melgarejo-Sánchez, P.; Núñez-Gómez, D.; Martínez-Nicolás, J.J.; Hernández, F.; Legua, P.; Melgarejo, P. Pomegranate variety and pomegranate plant part, relevance from bioactive point of view: A review. *Bioresour. Bioprocess.* **2021**, *8*, 2. [CrossRef]
4. Caleb, O.J.; Mahajan, P.V.; Opara, U.L.; Witthuhn, C.R. Modeling the effect of time and temperature on respiration rate of pomegranate arils (cv. 'Acco' and 'Herskawitz'). *J. Food Sci.* **2012**, *77*, E80–E87. [CrossRef] [PubMed]
5. Valdenegro, M.; Huidobro, C.; Monsalve, L.; Bernales, M.; Fuentes, L.; Simpson, R. Effects of ethrel, 1–MCP and modified atmosphere packaging on the quality of 'Wonderful' pomegranates during cold storage. *J. Sci. Food Agric.* **2018**, *98*, 4854–4865. [CrossRef]
6. Pareek, S.; Valero, D.; Serrano, M. Postharvest biology and technology of pomegranate. *J. Sci. Food Agric.* **2015**, *95*, 2360–2379. [CrossRef]
7. Lufu, R.; Ambaw, A.; Opara, U.L. The contribution of transpiration and respiration processes in the mass loss of pomegranate fruit (cv. Wonderful). *Postharvest Biol. Technol.* **2019**, *157*, 110982. [CrossRef]
8. Liu, Q.; Guo, X.; Du, J.; Guo, Y.; Guo, X.; Kou, L. Comparative analysis of husk microstructure, fruit quality and concentrations of bioactive compounds of different pomegranate cultivars during low temperature storage. *Food Biosci.* **2023**, *52*, 102400. [CrossRef]
9. Belay, Z.A.; Caleb, O.J.; Mahajan, P.V.; Opara, U.L. Response of pomegranate arils (cv. 'Wonderful') to low oxygen stress under active modified atmosphere condition. *J. Sci. Food Agric.* **2019**, *99*, 1088–1097. [CrossRef]
10. Shi, J.; Gao, H.; Wang, S.; Wu, W.; Tong, R.; Wang, S.; Li, M.; Jian, Z.; Wan, R.; Hu, Q.; et al. Exogenous arginine treatment maintains the appearance and nutraceutical properties of hard- and soft-seed pomegranates in cold storage. *Front. Nutr.* **2022**, *9*, 828946. [CrossRef]
11. Chen, Y.; Shi, J.; Wan, R.; Jian, Z.; Hu, Q. Development status and suggestions of soft seed pomegranate industry in China. *Deciduous Fruits* **2020**, *52*, 1–4+79–80. [CrossRef]
12. Guo, X.; Wu, X. Selection and breeding of 'Shaanxi Dazi', a new variety of pomegranate with large fruit and large grain. *China Fruits* **2010**, *04*, 14–15+22+79. [CrossRef]
13. Xu, Z. Early high-yield cultivation technique of 'Yudazi' pomegranate. *Mod. Agric. Res.* **2011**, *06*, 25. [CrossRef]
14. Yao, F.; Cao, S.; Wu, G.; Li, H.; Ma, G.; Zhao, Z. Breeding of a new soft-seeded pomegranate variety 'Malisi'. *North. Hortic.* **2019**, *13*, 172+178–180.
15. Dias, C.; Ribeiro, T.; Rodrigues, A.C.; Ferrante, A.; Vasconcelos, M.W.; Pintado, M. Improving the ripening process after 1–MCP application: Implications and strategies. *Trends Food Sci. Technol.* **2021**, *113*, 382–396. [CrossRef]
16. Hu, B.; Sun, D.-W.; Pu, H.; Wei, Q. Recent advances in detecting and regulating ethylene concentrations for shelf-life extension and maturity control of fruit: A review. *Trends Food Sci. Technol.* **2019**, *91*, 66–82. [CrossRef]
17. Botondi, R.; De Sanctis, F.; Bartoloni, S.; Mencarelli, F. Simultaneous application of ethylene and 1–MCP affects banana ripening features during storage. *J. Sci. Food Agric.* **2014**, *94*, 2170–2178. [CrossRef]
18. Win, N.M.; Yoo, J.; Naing, A.H.; Kwon, J.-G.; Kang, I.-K. 1–Methylcyclopropene (1–MCP) treatment delays modification of cell wall pectin and fruit softening in 'Hwangok' and 'Picnic' apples during cold storage. *Postharvest Biol. Technol.* **2021**, *180*, 111599. [CrossRef]
19. Li, L.; Kaplunov, T.; Zutahy, Y.; Daus, A.; Porat, R.; Lichter, A. The effects of 1-methylcyclopropane and ethylene on postharvest rachis browning in table grapes. *Postharvest Biol. Technol.* **2015**, *107*, 16–22. [CrossRef]
20. Si, W.; Li, K.; Jiang, Y.; Zi, W.; Kai, M.; Sha, H.; Yi, G.; Xin, L. Effect of 1–MCP treatment on quality of winter jujube during low-temperature storage. *Food Res. Dev.* **2021**, *42*, 64–70. [CrossRef]
21. De Reuck, K.; Sivakumar, D.; Korsten, L. Integrated application of 1–methylcyclopropene and modified atmosphere packaging to improve quality retention of litchi cultivars during storage. *Postharvest Biol. Technol.* **2009**, *52*, 71–77. [CrossRef]

22. Aguayo, E.; Jansasithorn, R.; Kader, A.A. Combined effects of 1–methylcyclopropene, calcium chloride dip, and/or atmospheric modification on quality changes in fresh-cut strawberries. *Postharvest Biol. Technol.* **2006**, *40*, 269–278. [CrossRef]
23. Li, L.; Lichter, A.; Chalupowicz, D.; Gamrasni, D.; Goldberg, T.; Nerya, O.; Ben-Arie, R.; Porat, R. Effects of the ethylene-action inhibitor 1–methylcyclopropene on postharvest quality of non-climacteric fruit crops. *Postharvest Biol. Technol.* **2016**, *111*, 322–329. [CrossRef]
24. Gamrasni, D.; Gadban, H.; Tsvilling, A.; Goldberg, T.; Neria, O.; Ben-Arie, R.; Wolff, T.; Stern, Y.J. 1–MCP improves the quality of stored 'Wonderful' pomegranates. *Acta Hortic.* **2015**, *1079*, 229–234. [CrossRef]
25. Yan, X.; Zhang, R.; Liang, Q.; Guo, X.; Yao, G.; Li, Y.; Zhang, Y. Effects of low temperature combined with 1–MCP on postharvest quality of Tunisia soft-seed pomegranate. *Food Ferment. Ind.* **2021**, *47*, 147–155. [CrossRef]
26. Wan, R.; Song, J.; Lv, Z.; Qi, X.; Han, X.; Guo, Q.; Wang, S.; Shi, J.; Jian, Z.; Hu, Q.; et al. Genome-wide identification and comprehensive analysis of the *AP2/ERF* gene family in pomegranate fruit development and postharvest preservation. *Genes* **2022**, *13*, 895. [CrossRef]
27. Guo, Q. Effect of 1–MCP and Potassium Permanganate on Storage and Aging Mechanism of 'Tunisian Soft-Seeded' Pomegranate. Master's Thesis, Henan Agricultural University, Zhengzhou, China, 2019. [CrossRef]
28. Zhang, L.; Zhang, Y.; Li, L.; Li, Y. Effects of 1–MCP on peel browning of pomegranates. *Acta Hortic.* **2008**, *774*, 275–282. [CrossRef]
29. Adams-Phillips, L.; Barry, C.; Giovannoni, J. Signal transduction systems regulating fruit ripening. *Trends Plant Sci.* **2004**, *9*, 331–338. [CrossRef]
30. Matityahu, I.; Marciano, P.; Holland, D.; Ben-Arie, R.; Amir, R. Differential effects of regular and controlled atmosphere storage on the quality of three cultivars of pomegranate (*Punica granatum* L.). *Postharvest Biol. Technol.* **2016**, *115*, 132–141. [CrossRef]
31. Chen, Y.; Gao, H.; Wang, S.; Liu, X.; Hu, Q.; Jian, Z.; Wan, R.; Song, J.; Shi, J. Comprehensive evaluation of 20 pomegranate (*Punica granatum* L.) cultivars in China. *J. Integr. Agric.* **2022**, *21*, 434–445. [CrossRef]
32. Shi, J.; Wang, S.; Tong, R.; Wang, S.; Chen, Y.; Wu, W.; He, F.; Wan, R.; Jian, Z.; Hu, Q.; et al. Widely targeted secondary metabolomics explored pomegranate aril browning during cold storage. *Postharvest Biol. Technol.* **2022**, *186*, 111839. [CrossRef]
33. Osondu, H.A.A.; Akinola, S.A.; Shoko, T.; Pillai, S.K.; Sivakumar, D. Coating properties, resistance response, molecular mechanisms and anthracnose decay reduction in green skin avocado fruit ('Fuerte') coated with chitosan hydrochloride loaded with functional compounds. *Postharvest Biol. Technol.* **2022**, *186*, 111812. [CrossRef]
34. Zhang, R.; Guo, X. Pomegranate storage preservation technology. *Northwest Hortic.* **2017**, *03*, 14–15.
35. Maghoumi, M.; Amodio, M.L.; Fatchurrahman, D.; Cisneros-Zevallos, L.; Colelli, G. Pomegranate husk scald browning during Storage: A review on factors involved, their modes of action, and its association to postharvest treatments. *Foods* **2022**, *11*, 3365. [CrossRef] [PubMed]
36. Mishra, V.; Kaplan, Y.; Ginzberg, I. Mitigating chilling injury of pomegranate fruit skin. *Sci. Hortic.* **2022**, *304*, 111329. [CrossRef]
37. Qi, X.; Zhao, J.; Jia, Z.; Cao, Z.; Liu, C.; Li, J.; Su, Y.; Pan, Y.; He, C.; Xu, Y.; et al. Potential metabolic pathways and related processes involved in pericarp browning for postharvest pomegranate fruits. *Horticulturae* **2022**, *8*, 924. [CrossRef]
38. Valdenegro, M.; Fuentes, L.; Bernales, M.; Huidobro, C.; Monsalve, L.; Hernández, I.; Schelle, M.; Simpson, R. Antioxidant and fatty acid changes in pomegranate peel with induced chilling injury and browning by ethylene during long storage times. *Front. Plant Sci.* **2022**, *13*, 771094. [CrossRef]
39. Kawhena, T.G.; Opara, U.L.; Fawole, O.A. Effects of gum arabic coatings enriched with lemongrass essential oil and pomegranate peel extract on quality maintenance of pomegranate whole fruit and arils. *Foods* **2022**, *11*, 593. [CrossRef]
40. Lei, Y.; Niu, H.; Yun, Y.; Tian, J.; Lao, F.; Liao, X.; Gao, Z.; Ren, D.; Zhou, L. Analysis of coloration characteristics of Tunisian soft-seed pomegranate arils based on transcriptome and metabolome. *Food Chem.* **2022**, *370*, 131270. [CrossRef]
41. Maghoumi, M.; Fatchurrahman, D.; Amodio, M.L.; Quinto, M.; Cisneros-Zevallos, L.; Colelli, G. Is pomegranate husk scald during storage induced by water loss and mediated by ABA signaling? *J. Sci. Food Agric.* **2023**, *103*, 2914–2925. [CrossRef]
42. Ali, H.M.; Almagribi, W.; Al-Rashidi, M.N. Antiradical and reductant activities of anthocyanidins and anthocyanins, structure-activity relationship and synthesis. *Food Chem.* **2016**, *194*, 1275–1282. [CrossRef] [PubMed]
43. Su, X.; Jiang, Y.; Duan, X.; Liu, H.; Li, Y.; Lin, W.; Zheng, Y. Effects of pure oxygen on the rate of skin browning and energy status in longan fruit. *J. Food Technol. Biotechnol.* **2005**, *43*, 359–365.
44. Lin, Y.; Lin, Y.; Chen, Y.; Wang, H.; Shi, J.; Lin, H. Hydrogen peroxide induced changes in energy status and respiration metabolism of harvested longan fruit in relation to rericarp browning. *J. Agric. Food Chem.* **2016**, *64*, 4627–4632. [CrossRef] [PubMed]
45. Ting, L.; Zheng, Q.; En, Y.; Fu, W.; Hong, Q.; Yue, J. Respiratory activity and energy metabolism of harvested litchi fruit and their relationship to quality deterioration. *J. Fruit Sci.* **2010**, *27*, 946–951. [CrossRef]

horticulturae

Article

Pomegranate Cultivars with Diverse Origins Exhibit Strong Resistance to Anthracnose Fruit Rot Caused by *Colletotrichum gloeosporioides*, a Major Disease in Southeast United States

Alexander Schaller [1], John M. Chater [2], Gary E. Vallad [3], Jeff Moersfelder [4], Claire Heinitz [4] and Zhanao Deng [1,*]

1 Department of Environmental Horticulture, Gulf Coast Research and Education Center, University of Florida, IFAS, 14625 County Road 672, Wimauma, FL 33598, USA; aschaller@ufl.edu
2 Department of Horticultural Sciences, Citrus Research and Education Center, University of Florida, IFAS, 700 Experiment Station Rd., Lake Alfred, FL 33850, USA; jchater@ufl.edu
3 Department of Plant Pathology, Gulf Coast Research and Education Center, University of Florida, IFAS, 14625 County Road 672, Wimauma, FL 33598, USA; gvallad@ufl.edu
4 National Clonal Germplasm Repository, USDA-ARS, University of California, Davis, CA 95616, USA; jeff.moersfelder@usda.gov (J.M.); claire.heinitz@usda.gov (C.H.)
* Correspondence: zdeng@ufl.edu

Abstract: Pomegranate, a pivotal fruit that is well recognized globally and a rapidly emerging crop in the southeastern United States and other subtropical regions, faces a formidable challenge from *Colletotrichum* spp., a fungal pathogen causing anthracnose fruit rot, which leads to severe to complete premature fruit drop. The development and use of disease-resistant cultivars are considered the most cost-effective and sustainable approach to managing this disease. Identifying sources of resistance is essential for developing new cultivars with improved resistance to this disease This project aimed to expand the scope of evaluation through a 2-year field study in central Florida, examining fruit from 35 cultivars from diverse origins using both artificial inoculation at the petal dehiscent stage and natural infection. Lesion size on the fruit was measured during the growing season in a field setting. Subsequently, seven cultivars were selected for further testing by inoculating detached mature fruit and measuring lesion size to confirm observed resistance and determine the correlation between resistance observed in planta in the field and on detached fruit in the laboratory. The field study revealed significant genetic differences among pomegranate cultivars in susceptibility to naturally occurring and induced anthracnose fruit rot and classified cultivars into five resistance or susceptibility classes. Five cultivars that originated from different regions of the world, including 'Azadi', showed consistent resistance to anthracnose fruit rot in the field. Resistance remained strong on detached mature fruit. A strong positive correlation existed between resistance levels on in-planta fruit and on detached mature fruit, suggesting a possible simple, efficient approach to screening breeding populations for anthracnose fruit rot resistance in pomegranate. These findings represent an important step toward developing new anthracnose-resistant cultivars and understanding and improving disease resistance in this increasingly important fruit crop in the world.

Keywords: *Punica granatum*; disease resistance; fungal pathogen; *Colletotrichum*; breeding; germplasm; genetic diversity

Citation: Schaller, A.; Chater, J.M.; Vallad, G.E.; Moersfelder, J.; Heinitz, C.; Deng, Z. Pomegranate Cultivars with Diverse Origins Exhibit Strong Resistance to Anthracnose Fruit Rot Caused by *Colletotrichum gloeosporioides*, a Major Disease in Southeast United States. *Horticulturae* **2023**, *9*, 1097. https://doi.org/10.3390/horticulturae9101097

Academic Editors: Yuepeng Han, Zhaohe Yuan, Gaihua Qin and Julián Bartual

Received: 7 May 2023
Revised: 21 September 2023
Accepted: 27 September 2023
Published: 3 October 2023

1. Introduction

The pomegranate (*Punica granatum* L.) is a subtropical fruit tree with a long history of cultivation across the world. Commercial pomegranate orchards can be found in the Middle East and Caucasus region, North and tropical Africa, the Indian subcontinent, Central Asia, Southeast Asia, the Mediterranean Basin, North and South America, and Australia [1]. World pomegranate production is estimated to be well above 300,000 ha [2] and an estimated 3 million metric tons annually as of 2017 [3]. In recent years, consumer demand for pomegranate has been increasing worldwide [4–6] due to its multiple health benefits [7,8], including strong antimicrobial and antiviral activities [9]. Research on increasing the production and supply of quality pomegranate fruit is much needed to meet consumer demand for this superfood. Destructive diseases have been the most important constraint for successful commercial production of pomegranate in many countries.

Spanish settlers introduced pomegranates to the southern United States and Mexico [10,11] when they first colonized North America in the 1700s. Current pomegranate production in the United States remains relatively small compared to other tree fruit crops, with 12,736 ha of production, 98% of which is in California as of 2017 [12]. The crop has garnered increasing interest in the United States [13,14], including in Florida [15,16]. Pomegranates have been cultivated in Florida since their arrival with Spanish settlers, but commercial production remains very limited. Early research in Florida revealed that pomegranates can grow well in the state, but fruit production is threatened by the high incidence of fungal diseases, particularly anthracnose fruit rot caused by *Colletotrichum* species [17–19]. Anthracnose in pomegranates appears on leaves as small circular leaf spots with yellow halos and on fruit as brown lesions that progress through the fruit causing external and internal rot. In many cases, disease pressure is so intense that the fruit succumbs to rot well before maturity, resulting in up to 100% fruit loss (Figure 1).

Figure 1. Anthracnose symptoms on pomegranate fruit. The symptoms typically begin at the calyx and work their way up until the fruit is completely rotted.

Colletotrichum has been reported worldwide in pomegranates [20–25] and thrives in the higher temperature and humidity in subtropical environments such as Florida, Southeast Asia, and India. While some fungicides have been researched and approved for fungal control in pomegranates in multiple countries [26–29], disease-resistant cultivars are considered essential for commercial production in subtropical regions including Florida. Use of disease-resistant cultivars represents a much more sustainable and cost-effective option for managing pomegranate diseases not only in Florida but also globally.

Identifying sources of disease resistance is the first and most critical step for developing new cultivars with greater disease resistance. With hundreds of cultivars worldwide, there is a vast range of potential diversity for screening for disease resistance [30,31]. However, very limited research has been published on pomegranate cultivars that are resistant to

Colletotrichum spp., and of the cultivars that have been screened, very few show strong resistance. Joshi et al. (2014) [32] screened cultivars for *Colletotrichum gloeosporioides* fruit rot resistance and found that two popular Indian cultivars, 'Arakta' and 'Bhagwa', were susceptible. However, two local cultivars, 'Yarcud Local' and 'Bedana', showed resistance to *C. gloeosporioides* isolates. Jayalakshmi et al. (2015) [33] investigated nineteen pomegranate cultivars for resistance to *C. gloeosporioides* using detached leaves but found no resistance, with 'Arakta', 'Ganesh', and 'Kesar' showing a higher level of susceptibility. Yu et al. 2018 [34] also reported 'Arakta' and 'Bhagwa' as being highly susceptible to *C. gloeosporioides* in a detached leaf assay, while a local Florida cultivar 'Cedar Key Sunset' exhibited moderate resistance. These previous results indicated that resistance to *Colletotrichum* might exist within the pomegranate germplasm. Our hypothesis for this study was that by expanding screening efforts to include a larger number of cultivars from a diverse background, more useful sources of resistance to *Colletotrichum* could be identified for pomegranate breeding.

The objectives of this study were to (1) evaluate a subset of the USDA pomegranate germplasm collection and a group of heritage Florida cultivars for resistance to fruit rot caused by *C. gloeosporioides* in the 'real world' under natural disease pressure, (2) confirm identified resistance through artificial inoculation of the pathogen, and (3) determine the correlation between the resistance levels observed on in-planta fruit in the field and on detached mature fruit in the laboratory. The evaluation of these cultivars for resistance to this highly destructive fungal disease is important for developing new cultivars with greater disease resistance and expanding our understanding of disease resistance in pomegranate. The results revealed remarkable genetic diversity among the cultivars in terms of susceptibility to the disease and identified five cultivars with strong resistance. Interestingly and unexpectedly, these sources of resistance have a diverse origin in the world. Strong fruit rot resistance was observed to be expressed on detached mature fruit. These new findings can play an important role not only for breeding new pomegranate cultivars with resistance to anthracnose fruit rot but also for managing other important diseases in pomegranate production in the world.

2. Materials and Methods

2.1. Experimental Site and Pomegranate Cultivars

The experiments were conducted over two years, from 2021 and 2022, at the University of Florida's Gulf Coast Research and Education Center in central Florida. The region's climate is characterized by hot, humid summers with frequent rains from May through September and warm, dry winters (Figure 2). This weather pattern creates a conducive environment for fungal growth during late spring into summer when pomegranate trees are producing flowers and young fruit.

The experimental pomegranate orchard was established in 2015 and was grouped by cultivar with two or three plants per cultivar. Before each evaluation season, the plants were subjected to defoliation utilizing ethephon to encourage earlier blooming. No fungicides were applied during the two growing seasons to ensure high pressure of natural fruit rot disease. In the 2022 season, a series of freezes in January and February caused production issues (few flowers) in some cultivars that limited the number of fruit available for evaluation. Thirty-five pomegranate cultivars were included in 2021, and 27 cultivars in 2022, due to some trees not producing enough fruit for evaluation during that growing season (Table 1). These cultivars originated from five different regions of the world, including the southeastern United States (Florida and Georgia), western United States (California), Turkmenistan and adjacent region, the former Soviet Union, and India [35–37]. The United States Department of Agriculture (USDA) germplasm accession numbers (DPUN) for all cultivars are included in Supplementary Table S1.

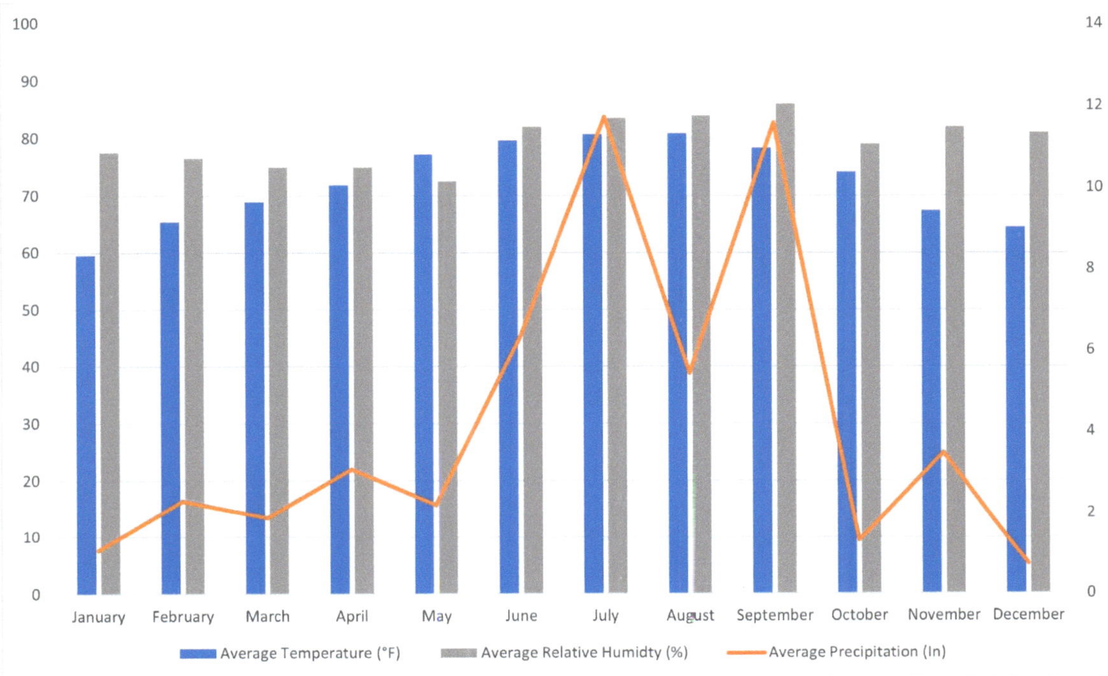

Figure 2. The average temperature, relative humidity, and precipitation in central Florida where the experiments were conducted. Data were collected from Florida Automated Weather Network (FAWN).

Table 1. Average fruit rot severity rating for fruit that were naturally infected or artificially inoculated for 35 pomegranate cultivars. The disease severity rating is on a scale of 0–6, with 6 being the most susceptible fruit. The top three most susceptible and most resistant cultivars for each treatment are bolded. The standard error is included for each cultivar.

Cultivar	Artificial Inoculation 2021		Natural Infection 2021		Artificial Inoculation 2022		Natural Infection 2022	
Afganski	2.1 ± 0.71	bc	2.3 ± 0.37	a–i	4.0 ± 0.38	abcd	3.7 ± 0.37	a–f
Al-Sirin-Nar	3.0 ± 0.52	abc	4.0 ± 0.29	abcd	**5.3 ± 0.24**	**ab**	**5.0 ± 0.20**	**ab**
Ambrosia	**4.0 ± 0.47**	**ab**	3.7 ± 0.28	a–e	4.5 ± 0.27	abc	4.9 ± 0.18	abc
Angel Red	1.3 ± 0.56	bc	1.8 ± 0.48	c–i			3.3 ± 0.52	a–f
Apseronski Krasnyj	3.6 ± 0.18	ab	3.9 ± 0.33	abcd				
Arakta	1.4 ± 0.98	bc	**0.4 ± 0.56**	**hi**	0.8 ± 0.68	gh	**0.8 ± 0.14**	**gh**
Azadi	**0.1 ± 0.12**	**c**	**0.1 ± 0.08**	**i**	**0.1 ± 0.07**	**h**	**0.3 ± 0.29**	**h**
Bala Miursal	2.1 ± 0.63	bc	2.8 ± 0.46	a–g	1.4 ± 0.41	efgh	2.3 ± 0.29	defg
Christina	1.6 ± 0.94	bc	1.4 ± 0.45	d–i	**0.2 ± 0.20**	**h**	2.8 ± 0.95	c–g
Cranberry	1.2 ± 0.68	bc	3.8 ± 0.30	abcd	4.3 ± 1.09	abcd	1.6 ± 0.46	fgh
Desertnyi	3.3 ± 0.29	ab	**4.4 ± 0.17**	**ab**	4.9 ± 0.18	ab	4.3 ± 0.26	abcd
Don Somner South	1.1 ± 0.71	bc	1.9 ± 0.62	c–i	3.1 ± 0.33	cde	3.9 ± 0.43	a–e
Eve	2.9 ± 0.64	abc	**4.3 ± 0.38**	**abc**	**5.5 ± 0.11**	**a**	**5.2 ± 0.24**	**ab**

Table 1. *Cont.*

Cultivar	Artificial Inoculation 2021		Natural Infection 2021		Artificial Inoculation 2022		Natural Infection 2022	
Eversweet	2.0 ± 1.39	bc	1.2 ± 0.54	e–i	**0.7 ± 0.45**	**h**	1.6 ± 0.51	fgh
Fleischman	**1.0 ± 0.52**	**bc**	**0.4 ± 0.20**	**ghi**	1.1 ± 0.60	fgh	**0.5 ± 0.18**	**gh**
Gainey Sweet	1.2 ± 0.64	bc	0.7 ± 0.38	ghi				
Girkanets	1.7 ± 0.65	bc	2.2 ± 0.49	b–i	2.3 ± 0.41	defg	2.6 ± 0.35	defg
Gissarskii Rozovyi	1.3 ± 0.58	bc	1.3 ± 0.39	d–i				
Grenada	1.6 ± 0.68	bc	2.8 ± 0.45	a–g	3.1 ± 0.58	cde	3.9 ± 0.27	a–e
Jimmy Roppe	1.4 ± 0.59	bc	0.6 ± 0.21	ghi	2.0 ± 0.60	d–h	2.0 ± 0.66	d–h
Kaim-anor	3.2 ± 0.64	ab	3.3 ± 0.51	a–f	4.5 ± 0.32	abc	4.8 ± 0.26	abc
Kazake	**5.2 ± 0.38**	**a**	**4.6 ± 0.00**	**a**	**5.3 ± 0.16**	**ab**	**5.3 ± 0.12**	**a**
Larkin	3.8 ± 0.88	ab	2.5 ± 0.15	a–h	2.7 ± 0.45	def	2.6 ± 0.45	defg
Medovyi Vahsha	3.1 ± 0.47	abc	1.8 ± 0.59	c–i				
Molla Nepes	**4.4 ± 0.00**	**ab**	1.2 ± 0.39	e–i				
Nikitski Ranni	3.4 ± 0.52	ab	3.3 ± 0.44	a–e	5.2 ± 0.17	ab	4.7 ± 0.15	abc
Parfianka	3.2 ± 0.64	ab	1.9 ± 0.40	c–i	3.7 ± 0.24	bcd	3.4 ± 0.28	a–f
Rose	3.2 ± 1.02	abc	1.9 ± 0.76	c–i				
Sakerdze	1.5 ± 0.75	bc	2.0 ± 0.41	c–i	2.5 ± 0.39	def	2.0 ± 0.68	d–h
Salavatski	2.6 ± 0.87	bc	2.3 ± 0.38	a–i	1.6 ± 0.41	efgh	1.9 ± 0.27	efgh
Sin Pepe	1.9 ± 0.66	bc	1.5 ± 0.45	d–i				
Sirenevyi	2.2 ± 1.27	bc	0.6 ± 0.47	ghi				
Surh-Anor	**1.0 ± 0.54**	**bc**	0.9 ± 0.48	fghi	3.0 ± 0.81	cde	3.2 ± 0.30	b–f
Sweet	1.6 ± 1.02	bc	2.3 ± 0.59	a–i	4.1 ± 0.42	abcd	4.0 ± 0.31	abcd
Vkusnyi	1.6 ± 0.60	bc	1.7 ± 0.34	d–i	1.8 ± 0.37	efgh	2.3 ± 0.44	d–h

Letters represent the differences among cultivars. In cases where more than 4 characters are present, a dash is used as a shorthand, e.g., a–i is abcdefghi.

2.2. Evaluation of Fruit Rot Severity under Natural Disease Pressure

Twelve or eighteen young fruit per cultivar were randomly selected and tagged in May 2021 and May 2022. Fruit was examined weekly for a period of eight weeks from May to July. When fruit rot appeared, the size of the rotted area (lesion size) on each fruit was manually measured and rounded to the nearest whole numbers in cm. Lesion size measurements were then converted to a 0 to 6 disease severity scale: 0 = no disease; 1 = lesions only occurring on the calyx; 2 = lesion of 1 or 2 cm; 3 = lesion of 3 or 4 cm; 4 = lesion of 5 or 6 cm; 5 = lesion 7 cm or greater; and 6 = the fruit had dropped from the tree due to fruit rot disease.

2.3. Preparation of Fungal Spore Suspension for Artificial Inoculation

Fungal isolate C30 for *C. gloeosporioides* was cultured on a potato dextrose agar (PDA) and incubated at (30 °C) for 12–15 days. A spore suspension was prepared by flooding the plates with 6 mL of autoclaved distilled water and scraping the agar surface. The spore suspension was then filtered through cheesecloth into a 50 mL Falcon tube. Spore density was determined using a hemacytometer and adjusted with autoclaved distilled water to a final suspension of 1×10^5 conidia/mL.

2.4. Inoculation of Fungal Spores to Open Flowers and Young Fruit on Plants in the Field

Artificial inoculation was achieved by applying 500 μL of 1×10^5 conidia/mL *C. gloeosporioides* inoculum into open hermaphroditic flowers that were at the anther dehiscence stage or the young fruitlet stage. Inoculated flowers and fruit were each enclosed in a mesh bag and a brown paper bag for 24 h to provide a high humidity environment for promoting fungal infection. After 24 h, the paper bag was removed, but the mesh bag was left in place to protect the inoculated fruit. In each growing season, 14 to 36 young fruit per cultivar were inoculated. For each cultivar, mock inoculations were performed on three to five fruit using sterile deionized water (SDW).

2.5. Measuring Lesion Size and Disease (Fruit Rot) Development after Artificial Inoculation of In-Planta Fruit the Field

All inoculated fruit were tagged and examined for fruit rot lesion size every week. The examinations continued for 8 weeks, from May into early July 2021 or mid-July 2022. Lesion measurements were then converted to a 0 to 6 ranking scale as described above.

2.6. Ranking of Fruit Rot Resistance Levels

Pomegranate cultivars were ranked for fruit rot resistance based on an empirical method that considered the average disease severity rating from both the natural infection and artificial inoculation over two years and the percent of fruit that had symptoms that were at level 5 or 6 on the above ranking scale. Only cultivars with complete data for both artificial inoculation and natural infection over the two years were ranked.

2.7. Re-Isolation of Fungal Pathogen from Inoculated Fruit

During the final three weeks of the evaluation, ten fruit showing symptoms of *Colletotrichum* fruit rot were collected at random from the field for a total of 30 fruit each year. Isolates were recovered from the fruit following the protocol of Xavier et al. (2019) [17]. After 10 days, the fungal cultures were visually examined to count the number of isolates that were *Colletotrichum*.

2.8. Evaluation of Detached Mature Fruit for Resistance to Fruit Rot

Mature fruit for many pomegranate cultivars were not available in Florida, so mature fruit were harvested from an experimental orchard at the USDA National Clonal Germplasm Repository (NCGR) in Davis, California, in the fall of 2022 and shipped to Florida. Seven available pomegranate cultivars with varying levels of fruit rot resistance were selected, including 'Afganski' (susceptible), 'Al-Sirin-Nar' (highly susceptible), 'Azadi' (highly resistant), 'Eversweet' (resistant), 'Fleishman' (resistant), 'Kazake' (highly susceptible), and 'Nikitski Ranni' (highly susceptible).

Fruit received from NCGR were first washed to remove any soil or debris, surface-sterilized by soaking in 0.0025% sodium hypochlorite for 30 min, and then air dried in the laboratory under ambient conditions. Individual surface-sterilized fruit were wounded to a 5 mm depth with a 3 mm diameter nail and sterilized with 70% ethanol after each use. Four wounds were made at equal distance apart around the center of each fruit. To each wound, 30 μL of 1×10^5 *C. gloeosporioides* spore inoculum was applied, allowing five minutes for the inoculum to absorb into the wound before sealing with petroleum jelly. Mock inoculations were made using sterile deionized water (SDW). Inoculated fruit were placed into a clear plastic container lined with damp paper towels and incubated for 14 days in a 24 °C growth chamber with a 12 h photoperiod. Starting six days after inoculation, fruit lesions were measured every two days. The detached fruit inoculation experiment was repeated four times, each time with four fruit per cultivar and four inoculated sites per fruit.

2.9. Statistical Analysis

Analysis of variance was conducted to test the disease responses of different pomegranate cultivars. Mean separation procedures among cultivars were conducted using a Tukey

HSD test. Pearson's correlation between infection methods and between years was tested using the cor.test function. All statistical analysis was performed in R (version 4.2.3) and the 'agricolae' [38] package was used for the HSD comparison.

3. Results

3.1. Fruit Rot under Natural Disease Pressure

Fruit rot began to appear on young fruit in May in both years and continued to enlarge as the fruit developed. Under natural infection, different patterns of disease development were observed over the 8 weeks in 2021 (Supplementary Figure S1). For the cultivars that were highly susceptible, disease progression happened even at the early fruit development stage and almost linearly, and rapidly increased over the season, especially as conditions for fungal growth improved with the beginning of the rainy season (Figure 3). For some of the more resistant cultivars such as 'Azadi' and 'Fleishman', disease progression happened at a much slower pace over the observation period and disease symptoms did not tend to expand on the fruit. A few cultivars showed little disease at the early phase but consistently experienced a sharp increase in disease symptoms later in the season when the conditions for fungal growth were more optimal. These cultivars included 'Eversweet', 'Girkanets', and 'Parfianka'.

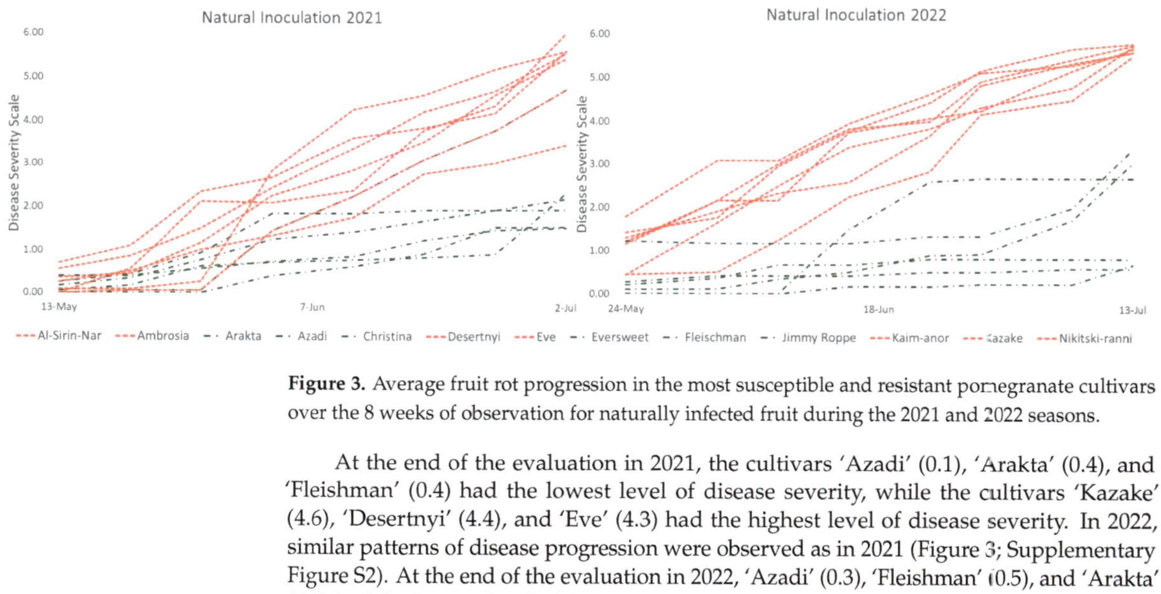

Figure 3. Average fruit rot progression in the most susceptible and resistant pomegranate cultivars over the 8 weeks of observation for naturally infected fruit during the 2021 and 2022 seasons.

At the end of the evaluation in 2021, the cultivars 'Azadi' (0.1), 'Arakta' (0.4), and 'Fleishman' (0.4) had the lowest level of disease severity, while the cultivars 'Kazake' (4.6), 'Desertnyi' (4.4), and 'Eve' (4.3) had the highest level of disease severity. In 2022, similar patterns of disease progression were observed as in 2021 (Figure 3; Supplementary Figure S2). At the end of the evaluation in 2022, 'Azadi' (0.3), 'Fleishman' (0.5), and 'Arakta' (0.8) had the lowest level of disease severity at the end of the evaluation, while 'Kazake' (5.3), 'Eve' (5.2), and 'Al-Sirin-Nar' (5.0) had the highest level of disease severity.

3.2. Fruit Rot after Artificial Inoculation

Under artificial inoculation during the 2021 season, fruit rot progressed linearly across the 8 weeks for all cultivars (Figure 4; Supplementary Figure S3). During the 2022 season, a similar trend was observed for all but five cultivars, including 'Eversweet' and 'Girkanets', which experienced a sharp increase in disease symptoms towards the end of the 8 weeks (Figure 4; Supplementary Figure S4).

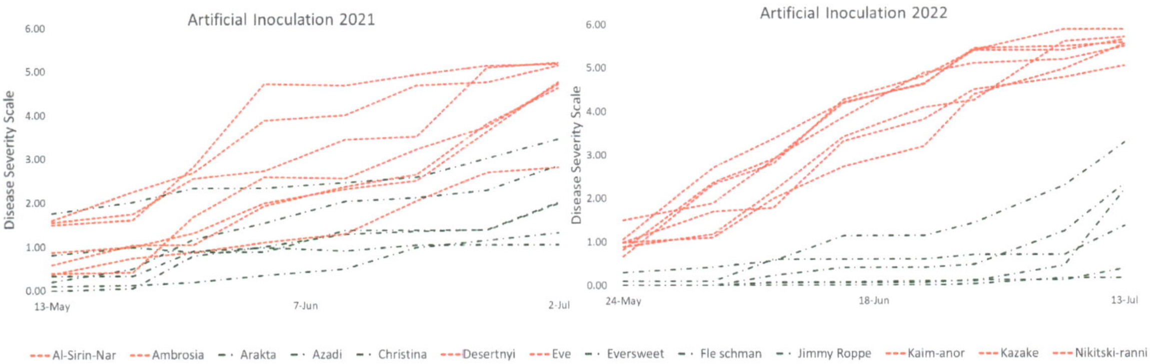

Figure 4. Average fruit rot progression of the most susceptible and resistant individuals over the 8 weeks of observation of artificially inoculated fruit for the 2021 and 2022 seasons.

There were statistically significant differences among cultivars in disease severity in both 2021 and 2022 (Table 1) for the artificially inoculated fruit. The cultivars 'Azadi' (0.1), 'Fleishman' (1.0), and 'Surh-Anor' (1.0) had the lowest level of disease severity in 2021, while the cultivars 'Kazake' (5.2), 'Molla Nepes' (4.4), and 'Ambrosia' (4.0) had the highest severity ratings. In 2022, the cultivars 'Azadi' (0.1), 'Christina' (0.2), and 'Eversweet' (0.7) had the lowest disease severity ratings in 2022, while 'Eve' (5.5), 'Al-Sirin-Nar' (5.3), and 'Kazake' (5.4) had the highest severity ratings.

At the end of the growing season, inoculated fruit were collected to re-isolate the pathogen. During the 2021 evaluation, out of the 30 fruits selected from the field, *Colletotrichum* was re-isolated from infected fruit tissue 16/30 times. During the 2022 evaluation, *Colletotrichum* was re-isolated from infected fruit tissue 18/30 times.

3.3. Ranking of Resistance Level and Correlation of Inoculation Methods and Years

Out of the 35 cultivars evaluated, 27 had data for all four categories over the two years and were included in the ranking of resistance. Of those 27 cultivars, one was highly resistant ('Azadi'), five were resistant, eight were moderately resistant, six were susceptible, and seven were highly susceptible (Table 2). The average percent of fruit that had a ranking of 5 or 6 at the end of the year was included for each class of phenotype, along with the average disease rating over the two years between the two infection methods.

Table 2. The 27 pomegranate cultivars that were included in ranking of resistance.

Categories of Fruit Rot Resistance Level	Cultivars	% Fruit Rot Mean	Disease Rating Mean (0–6 Scale)
Highly Resistant	Azadi	2	0.2
Resistant	Arakta, Christina, Eversweet, Fleishman, Jimmy Roppe,	25 ± 1	1.2 ± 0.15
Moderately Resistant	Angel Red, Bala Miursal, Cranberry, Don Somner South, Sakerdze, Salavatski, Surh-Anor, Vkusnyi	39 ± 2	2.2 ± 0.10
Susceptible	Afganski, Girkanets, Grenada, Larkin, Parfianka, Sweet	61 ± 2	2.8 ± 0.12
Highly Susceptible	Al-Sirin-Nar, Ambrosia, Desertnyi, Eve, Kaim-anor, Kazake, Nikitski Ranni	87 ± 2	4.4 ± 0.13

There was a significant positive correlation between the inoculation methods in both 2021 (r = 0.653, $p < 0.0001$) (Figure 5A) and 2022 (r = 0.876, $p < 0.0001$) (Figure 5B) and both years (Figure 5C) (r = 0.794, $p < 0.0001$). There was also significant correlation between the two years (r = 0.694, $p < 0.0001$) (Figure 5D). These significant correlations suggest that the rankings of fruit rot resistance level among pomegranate cultivars in different years were consistent and under genetic control.

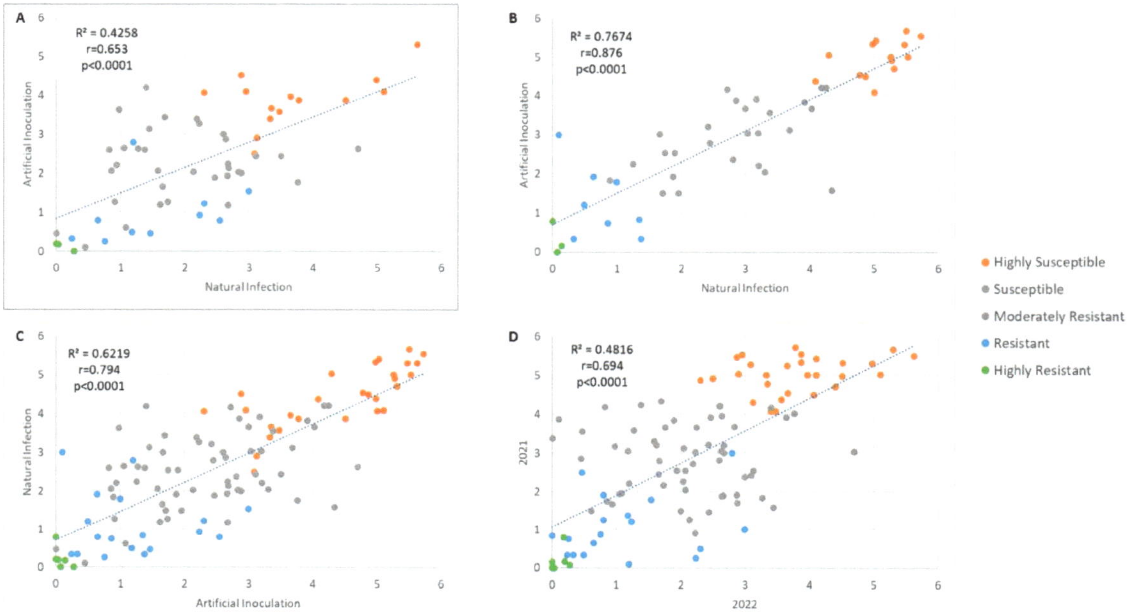

Figure 5. Correlation of disease severity rating between artificial inoculation and natural infection in 2021 (**A**), 2022 (**B**), and both years (**C**), and disease severity rating for both artificial inoculation and natural infection between years (**D**). Data points are color coded by ranking assigned in Table 2.

3.4. Detached Fruit Assay

There were significant cultivar differences in all four experiments, and overall, with 'Azadi' consistently having a smaller lesion diameter in comparison to all other cultivars (Figure 6 and Table 3). 'Afganski' and 'Kazake' were the most susceptible cultivars and had the largest lesions in all experiments except Experiment 2 (Table 3).

Table 3. Anthracnose fruit rot severity (lesion size) (diameter, in mm) for seven pomegranate cultivars after artificial inoculation of detached mature fruit in four replicated experiments. 'Fleishman' was not included in the fourth experiment due to a shortage of fruit available for inoculation. Letters represent the differences among cultivars. The standard error is included for each cultivar.

	Experiment								
Cultivar	**1**		**2**		**3**		**4**		**Average**
Afganski	22.1 ± 1.9	a	20.6 ± 3.8	abc	20.3 ± 1.6	a	12.8 ± 2.8	abc	19.6
Al-Sirin-Nar	17.8 ± 1.1	a	21.2 ± 2.8	ab	9.4 ± 2.0	bc	8.3 ± 1.9	bcd	14.9
Azadi	7.6 ± 1.7	b	8.9 ± 1.9	c	4.7 ± 0.9	c	3.0 ± 0.0	d	6.0
Eversweet	21.7 ± 4.5	a	27.5 ± 4.0	a	8.3 ± 1.5	bc	4.0 ± 0.5	cd	15.5
Fleishman	13.8 ± 2.2	ab	18.8 ± 3.7	abc	5.1 ± 1.2	bc			12.7
Kazake	22.0 ± 1.9	a	12.0 ± 1.7	bc	20.8 ± 1.4	a	18.5 ± 1.5	a	18.4
Nikitski Ranni	15.3 ± 2.1	ab	23.8 ± 3.0	a	11.7 ± 2.1	b	12.9 ± 1.9	ab	15.4

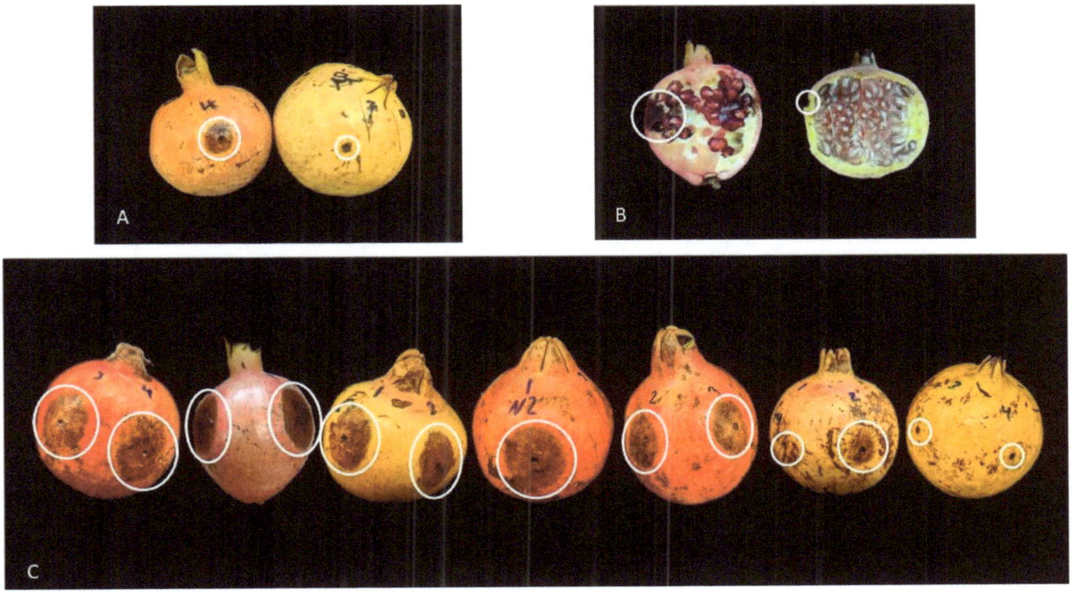

Figure 6. Lesion size comparison among detached mature fruit of pomegranate cultivars. (**A**,**B**) compare two representative cultivars, 'Kazake' on the left and 'Azadi' on the right, 14 days post-inoculation, externally and internally. (**C**) shows lesion sizes of seven cultivars, from the largest average lesion size to the smallest lesion size ('Afganski', 'Kazake', 'Eversweet', 'Nikiski Ranni', 'Al-Sirin-Nar', 'Fleishman', and 'Azadi'). Lesions are indicated with white circles.

4. Discussion

Resistance to anthracnose fruit rot is a much-needed trait for pomegranates, especially in the southeastern United States and other subtropical regions in the world where the environmental conditions are ideal for the growth and spread of the causal fungal pathogen *C. gloeosporioides*. Our results indicate that there are very valuable sources of genetic resistance to this pathogen in the current USDA pomegranate germplasm collection as well as in local heirloom varieties. 'Azadi' seems to be the most promising cultivar and has strong resistance to *C. gloeosporioides*, and a handful of other cultivars also have useful resistance.

In many studies investigating disease resistance in different cultivars, repeatability and differences between years, artificial inoculation, and natural infection, as well as inoculation methods, are often problematic [39–41]. In-field evaluations were further complicated in this experiment by the conducive environment for pathogen spread and the high number of pathogens that infect pomegranates present in Florida [19]. While re-isolation of *Colletotrichum* from infected fruit in the field was possible, multiple other fungal pathogens were also isolated from infected tissues collected from the field. This pathogen pressure meant that many of the mock inoculated fruit showed symptoms of fungal infection due to natural infection occurring, particularly in cultivars with higher susceptibility levels. Despite the many factors that could affect infection in the field study portion, there was statistical significance as well as a strong magnitude of correlation when looking at the correlation of different inoculation methods within years and between years. This would suggest that the rankings of resistance for cultivars were consistent, regardless of the inoculation method (artificial or natural infection) or year, despite the varying environmental effects.

Evaluation of the fruit over the season seems to suggest that in a few cultivars, there was an increase in susceptibility as the fruit matured. For example, 'Eversweet' and 'Girkanets' showed an increase in lesion size late in the season as the fruit approached maturity. These results suggest that a fruit's age may play a role in its resistance to fruit rot. Among the other six cultivars tested in the detached fruit study, the results were consistent with field data. Such consistency may suggest a possibility of screening pomegranate cultivars for fruit rot resistance by inoculating detached fruit in the laboratory rather than inoculating open flowers or young fruit in the field, which is much more challenging, time-consuming, and complicated. Disease resistance screening based on detached fruit would be much easier to implement and better to control environmental variables such as temperature and relative humidity, resulting in an enhanced selection efficiency.

'Arakta' has been evaluated for disease resistance in other studies. Joshi et al. 2014 [32] assessed it for fruit rot resistance, and Jayalakshmi et al. (2015) [33] for leaf spot resistance. Both studies rated it as highly susceptible to *Colletotrichum*. However, in our study, we found 'Arakta' to be resistant with low disease incidence in both artificial inoculations as well as natural infections. These differences may be due to factors such as different pathogen populations, the environment, or interactions among these factors.

Interestingly, of the cultivars that showed resistance or strong resistance to anthracnose fruit rot in this study, all but 'Arakta' have a yellow to light pink peel and light pink arils. In other crops, disease resistance has been found to increase as fruit gain more red color. These pomegranate cultivars do not seem to follow that trend [42,43]. It will be interesting to find out whether fruit peel or aril color is associated with fruit rot resistance in pomegranate. If such an apparent association does exist in pomegranate, it may provide an easy-to-use, visual marker for screening pomegranate breeding populations for anthracnose fruit rot resistance. On the other hand, such an association may slow down the progress for developing new anthracnose-resistant cultivars with deep red fruit peel and arils, which are more popular than yellow to light pink peel or light pink arils among consumers.

Four of the resistant cultivars ('Eversweet', 'Fleishman', 'Christina', and 'Jimmy Roppe') originated within the United States, with the first two from California and the last two from northern Florida/Southern Georgia. 'Azadi' and 'Arakta', on the other hand, originated in Turkmenistan and India, respectively. It is extremely interesting that 'Azadi', originating from a region where anthracnose is not a common problem, has evolved strong resistance to anthracnose. These cultivars may be worth being evaluated for resistance to local *C. gloeosporioides* isolates in other tropical and subtropical regions where anthracnose fruit rot is a major disease.

These anthracnose-resistant cultivars have been crossed with other cultivars with other desirable traits including high yield and appealing external and internal color to develop new cultivars with anthracnose resistance, high yield, and superior quality. Resistant cultivars with different origins have also been intercrossed with the hope that their progeny may have stronger or broader-spectrum resistance than their parents. Genome and transcriptome sequencing is underway to identify candidate genes and develop molecular markers for this disease resistance trait so that large breeding populations can be screened more efficiently using molecular markers.

In recent years, a number of highly destructive diseases have been reported in pomegranate [20,22,24,44–48]. Resistance to diseases has become a much-needed trait for new pomegranate cultivars, and more research has been devoted to find sources of disease resistance, understand their genetic and molecular mechanisms, and develop new tools to incorporate them into new cultivars. For example, Kumari and Ram (2015) [49] evaluated 63 cultivars for resistance to *Coniella granati* and found six cultivars with moderate resistance to this pathogen. Jabnoun-Khiareddine et al. (2018) [50] evaluated nine cultivars and their response to *Coniella granati*; however, they found that all cultivars showed some level of susceptibility. Mincuzzi et al. (2020) [51] revealed that the cultivar 'Wonderful' had a higher resistance to *Coniella granati* than 'Mollar de Elche' and an up-regulation of genes associated with chitinase, phenylalanine ammonia-lyase (PAL), and peroxidase genes as well as phenolic compounds. Priya et al. (2016) [52] discovered five pomegranate genotypes with resistance to the highly destructive disease bacterial blight (BB) caused by *Xanthomonas axonopodis* pv. *punicae*. Kumar et al. (2021) [53] identified three pomegranate accessions with strong tolerance to BB and found that resistant accessions had an up-regulation of phenylalanine ammonia-lyase, callose synthase-3 (CS3), chitinase, pathogenesis-related protein-1 (PR1), and pathogenesis-related protein-10 (PR10) genes. A number of simple sequence repeat (SSR) markers have been associated with BB resistance or tolerance in pomegranate [54]. Overall, our knowledge of disease resistance traits and the availability of genetic, genomic, and molecular tools for these traits in pomegranate are very limited compared to what have been developed in other fruit crops. It is expected that as more genomic resources become available [55–58], they will accelerate the genetic improvement of disease resistance in this important crop.

5. Conclusions

Anthracnose is extremely destructive to pomegranate fruit, particularly in subtropical regions where it is warm, humid, and rainy, and the environmental conditions are ideal for the disease. This study represents a crucial breakthrough toward the development of new *Colletotrichum*-resistant pomegranate cultivars. 'Azadi' and five other cultivars that have demonstrated resistance or high resistance to *Colletotrichum* fruit rot merit further horticultural tests for potential commercial production in these regions. These sources of resistance are playing a pivotal role in developing new pomegranate cultivars for Florida and elsewhere where *Colletotrichum* anthracnose fruit rot is prevalent.

Disease resistance has become a very important objective in pomegranate breeding programs. Our studies indicate that pomegranate germplasm from different countries may harbor highly valuable sources of resistance for major diseases. Preserving such germplasm, including local or heirloom varieties, deserves more attention as climate change and highly destructive diseases become more prevalent in the pomegranate-producing areas in the world.

To enhance our understanding of this vital resistance trait and efficiently utilize it in breeding efforts, future studies need to investigate the inheritance of fruit rot resistance and its genetic relationship with other traits including fruit skin colors and phenolic compounds, identify and locate the gene loci for the observed resistance in 'Azadi' and other cultivars, and develop molecular and genomic selection tools.

Our study reveals the existence of resistance, even strong resistance, to destructive diseases in the pomegranate germplasm that originated from different parts of the world. This exciting finding suggests tremendous potential to improve pomegranate disease resistance and protect this vital crop against the devastating impacts of anthracnose.

Supplementary Materials: The following supporting information can be downloaded at: https://www.mdpi.com/article/10.3390/horticulturae9101097/s1, Table S1: Cultivars and their Origin, Figure S1: Disease progression of natural infection 2021. Figure S2: Disease progression of natural infection 2022. Figure S3: Disease progression of artificial inoculation 2021. Figure S4: Disease progression of artificial inoculation 2022.

Author Contributions: A.S.: formal analysis, investigation, methodology, project administration, writing—original draft, and writing—review and editing. J.M.C.: conceptualization, funding acquisition, and writing—reviewing and editing. G.E.V.: methodology, resources, and writing—reviewing and editing. J.M.: resources. C.H.: resources. Z.D.: conceptualization, funding acquisition, project administration, resources, supervision, and writing—reviewing and editing. All authors have read and agreed to the published version of the manuscript.

Funding: This project was supported by the USDA Agricultural Marketing Service (AMS) MultiState Specialty Crop Block Grant through the California Department of Food and Agriculture project number 19-1043-002-SF, the AMS Specialty Crop Block Grant through the Florida Department of Agriculture and Consumer Services (FDACS) Specialty Crop Block Grant Program (Contract #00099169), and the former Florida Pomegranate Association.

Data Availability Statement: Not applicable.

Acknowledgments: The authors are grateful to J. Velte, C. Willborn, and S. Ledon for their assistance with field data collection; J. Jones, G. Bowman, K. Druffel, and G. Garcia Garcia for maintaining the pomegranate orchard; and C. Weistein, R. Bonsteel, J. Alexander, and D. Bice for providing pomegranate plants and their support.

Conflicts of Interest: The authors declare no conflict of interest.

References

1. Spragnoli, F.; Shirazi, R.; Shirazi, Z.; Saeidi Ghavi Andam, S.; Djamali, M. Archaeology, history and symbolism. In *The Pomegranate. Botany, Production and Uses*; Sarkhosh, A., Alimohammad, M.Y., Zamani, Z., Eds.; CABI: Oxfordshire, UK, 2021; pp. 1–14.
2. Melgarejo-Sánchez, P.; Núñez-Gómez, D.; Martínez-Nicolás, J.J.; Hernández, F.; Legua, P.; Melgarejo, P. Pomegranate variety and pomegranate plant part, relevance from bioactive point of view: A review. *Bioresour. Bioprocess* **2021**, *8*, 2. [CrossRef]
3. Kahramanoglu, I. Trends in pomegranate sector: Production, postharvest handling and marketing. *Int. J. Agric. For. Life Sci.* **2019**, *3*, 239–246.
4. Bartual, J.; Zamudio, M.; Gómez, M. Situation of the production, research and economics of the pomegranate industry in Spain. *Acta Hortic.* **2015**, *1089*, 345–349. [CrossRef]
5. Marty, A.; Castillo, A.; Zoppolo, R. Pomegranate: A growing alternative for fruit production in Uruguay. *Acta Hortic.* **2015**, *1089*, 351–355. [CrossRef]
6. Saroj, P.L.; Kumar, R. Recent advances in pomegranate production in India—A review. *Ann. Hortic.* **2019**, *12*, 1–10. [CrossRef]
7. Giménez-Bastida, J.A.; Ávila-Gálvez, M.Á.; Espín, J.C.; González-Sarrías, A. Evidence for health properties of pomegranate juices and extracts beyond nutrition: A critical systematic review of human studies. *Trends Food Sci. Technol.* **2021**, *114*, 410–423. [CrossRef]
8. Asgary, S.; Azadeh, Z.; Dodman, S.; Soleymani, A.; Mirjalili, M.H.; Farzaei, M.H.; Hoseinzadeh-chahkandak, F.; Hozeifi, S.; Jasemi, E.; Keshvari, M.; et al. Pomegranate bioactive compounds and health. In *The Pomegranate: Botany, Production and Uses*; Sarkhosh, A., Alimohammad, M.Y., Zamani, Z., Eds.; CABI: Oxfordshire, UK, 2021; pp. 504–547.
9. Howell, A.; D'Souza, D. The pomegranate: Effects on bacteria and viruses that influence human health. *Evid. Based Complement Alternat. Med.* **2013**, *2013*, 606212. [CrossRef]
10. LaRue, J.H. *Growing Pomegranates in California*; Leaflet 2459. 1980. Available online: http://ucanr.edu/sites/Pomegranates/files/122804.pdf (accessed on 20 March 2023).
11. Stover, E.; Mercure, E.W. The pomegranate: A new look at the fruit of paradise. *HortScience* **2007**, *42*, 1088–1092. [CrossRef]
12. USDA National Agricultural Statistics Service, 2017 Census of Agriculture. Available online: www.nass.usda.gov/AgCensus (accessed on 14 March 2023).
13. Chater, J.M.; Merhaut, D.J.; Jia, Z.; Mauk, P.A.; Preece, J.E. Fruit quality traits of ten California-grown pomegranate cultivars harvested over three months. *Sci. Hortic.* **2018**, *237*, 11–19. [CrossRef]
14. Hooks, T.; Niu, G.; Masabni, J.; Sun, Y.; Ganjegunte, G. Performance and phytochemical content of 22 pomegranate (*Punica granatum*) varieties. *HortScience* **2021**, *56*, 217–225. [CrossRef]
15. Castle, W.S.; Baldwin, J.C.; Singh, M. Pomegranate in Florida for commercial enterprises and homeowners. *Proc. Fla. State Hortic. Soc.* **2011**, *124*, 33–40.
16. Deng, Z.; Castle, W.; Vallad, G.E.; Agehara, S.; Thetford, M.; Díaz-Pérez, J.C. Pomegranate: An emerging fruit crop in southeast United States? *Acta Hortic.* **2019**, *1254*, 149–156. [CrossRef]
17. Xavier, K.; Kc, A.N.; Peres, N.A.; Deng, Z.; Castle, W.; Lovett, W.; Vallad, G.E. Characterization of *Colletotrichum* species causing anthracnose of pomegranate in the southeastern United States. *Plant Dis.* **2019**, *103*, 2771–2780. [CrossRef]
18. Kc, A.N.; Vallad, G. Monitoring pomegranate pathogens towards developing effective disease management program. *Phytopathology* **2016**, *106*, 94.
19. Xavier, K.; Kc, A.N.; Vallad, G. Diseases of pomegranate (*Punica granatum*) in Florida: PP349, 10/2019. *EDIS* **2019**, *5*. [CrossRef]
20. Rahimlou, S.; Babaeizad, V.; Sayari, M. First report of fruit spot of pomegranate caused by *Colletotrichum gloeosporioides* in Iran. *J. Plant Pathol.* **2014**, *96*, 605.
21. Thomidis, T. Fruit rots of pomegranate (cv. Wonderful) in Greece. *Australas. Plant Pathol.* **2014**, *43*, 583–588. [CrossRef]
22. Munhuweyi, K.; Lennox, C.L.; Meitz-Hopkins, J.; Caleb, O.J.; Opara, U.L. Major diseases of pomegranate (*Punica granatum* L.), their causes and management—A review. *Sci. Hortic.* **2016**, *211*, 126–139. [CrossRef]
23. Uysal, A.; Kurt, Ş. *Colletotrichum gloeosporiodes* causing anthracnose on pomegranate in Turkey. *Australas. Plant Dis. Notes* **2018**, *13*, 19. [CrossRef]
24. Silva-Cabral, J.R.A.; Batista, L.R.L.; Costa, J.F.d.O.; Ferro, M.M.d.M.; Silva, S.J.C.; Lima, G.S.d.A.; Assunção, I.P. First report of *Colletotrichum tropicale* causing anthracnose on pomegranate in Brazil. *Plant Dis.* **2018**, *103*, 583. [CrossRef]
25. Patel, D.S. Chemical management of fruit spot of pomegranate caused by *Colletotrichum gloeosporioides* Penz. and Sacc. *Indian Phytopathol.* **2009**, *62*, 252–253.
26. Dev, D.; Narendrappa, T. In vitro evaluation of fungicides against *Colletotrichum gloeosporioides* (Penz.) Penz and Sacc. causing anthracnose of pomegranate (*Punica granatum* L.). *J. Appl. Nat. Sci.* **2016**, *8*, 2268–2272. [CrossRef]

27. Golakiya, B.; Akbari, L.F.; Marakna, N.M. In vitro evaluation of different fungicides against pomegranate anthracnose caused by *Colletotrichum gloeosporioides*. *Int. J. Chem. Stud.* **2020**, *8*, 3669–3674. [CrossRef]
28. Xavier, K.; Vallad, G.E. Efficacy of biological and conventional fungicide programs for foliar disease management on pomegranate (*Punica granatum*) in Florida. *Plant Health Prog.* **2020**, *21*, 199–204. [CrossRef]
29. Xavier, K.; Kc, A.N.; Vallad, G.E. Fungicide application timing essential for the management of leaf spot and fruit rot on pomegranate (*Punica granatum* L.) in Florida. *Plant Dis.* **2020**, *104*, 1629–1637. [CrossRef]
30. Sarkhosh, A.; Zamani, Z.; Fatahi, R.; Ranjbar, H. Evaluation of genetic diversity among Iranian soft-seed pomegranate accessions by fruit characteristics and RAPD markers. *Sci. Hortic.* **2009**, *121*, 313–319. [CrossRef]
31. Chandra, R.; Vilas Jadhav, T.; Sharma, J. Global scenario of pomegranate (*Punica granatum* L.) culture with special reference to India. *Fruit Veg. Cereal Sci. Biotechnol.* **2010**, *4*, 7–18.
32. Joshi, M.S.; Sawant, D.M.; Gaikwad, A.P. Isolate variations in *Colletotrichum gloeosporioides* infecting pomegranate. *J. Plant Prot. Sci.* **2014**, *6*, 21–26.
33. Jayalakshmi, K.; Nargund, V.B.; Raju, J.; Benagi, V.I.; Raghu, S.; Giri, M.S.; Basamma, R.B.; Priti, S.; Rajput, R.B. Pomegranate anthracnose caused by *Colletotrichum gloeosporioides*: A menace in quality fruit production. *J. Pure Appl. Microbiol.* **2015**, *9*, 3093–3097.
34. Yu, X.; Xavier, K.; Vallad, G.E.; Deng, Z. Diseases resistance in pomegranates: Importance, sources, breeding approaches, and progress. *Proc. Fla. State Hortic. Soc.* **2018**, *131*, 1–5. Available online: https://journals.flvc.org/fshs/article/view/114705/110032 (accessed on 20 May 2023).
35. Castle, W. Pomegranates for Now-Accessions. 2022. Available online: https://crec.ifas.ufl.edu/extension/pomegranates/accessions/ (accessed on 20 May 2023).
36. Chater, J.M.; Yavari, A.; Sarkhos, A.; Jia, Z.; Merhaut, D.J.; Preece, J.E.; Cossio, F.; Qin, G.; Liu, C.; Li, J.; et al. World pomegranate cultivars. In *The Pomegranate: Botany, Production and Uses*; Sarkhosh, A., Alimohammad, M.Y., Zamani, Z., Eds.; CABI: Oxfordshire, UK, 2021; pp. 157–199.
37. GRIN-Global. U.S. National Plant Germplasm System. Available online: https://npgsweb.ars-grin.gov/gringlobal/search.aspx (accessed on 20 May 2023).
38. de Mendiburu, F.; de Mendiburu, M.F. Package'Agricolae'; R Package Version; 2021; pp. 3–5. Available online: https://CRAN.R-project.org/package=agricolae (accessed on 14 April 2023).
39. Phillips, D.A.; Harmon, P.F.; Olmstead, J.W.; Peres, N.A.; Munoz, P.R. Screening for susceptibility to anthracnose stem lesions in southern highbush blueberry. *HortScience* **2018**, *53*, 920–924. [CrossRef]
40. Mangandi, J.; Peres, N.A.; Whitaker, V.M. Identifying resistance to crown rot caused by *Colletotrichum gloeosporioides* in strawberry. *Plant Dis.* **2015**, *99*, 954–961. [CrossRef] [PubMed]
41. Grice, K.R.E.; Bally, I.S.E.; Wright, C.L.; Maddox, C.; Ali, A.; Dillon, N.L. Mango germplasm screening for the identification of sources of tolerance to anthracnose. *Australas. Plant Pathol.* **2023**, *52*, 27–41. [CrossRef]
42. Jia, H.; Zhang, C.; Pervaiz, T.; Zhao, P.; Liu, Z.; Wang, B.; Wang, C.; Zhang, L.; Fang, J.; Qian, J. Jasmonic acid involves in grapefruit ripening and resistant against *Botrytis cinerea*. *Funct. Integr. Genom.* **2016**, *16*, 79–94. [CrossRef] [PubMed]
43. Sivankalyani, V.; Feygenberg, O.; Diskin, S.; Wright, B.; Alkan, N. Increased anthocyanin and flavonoids in mango fruit peel are associated with cold and pathogen resistance. *Postharvest Biol. Technol.* **2016**, *111*, 132–139. [CrossRef]
44. Gat, T.; Liarzi, O.; Skovorodnikova, Y.; Ezra, D. Characterization of *Alternaria alternata* causing black spot disease of pomegranate in Israel using a molecular marker. *Plant Dis.* **2012**, *96*, 1513–1518. [CrossRef]
45. Sharma, K.; Sharma, J.; Jadhav, V. Recent developments in bacterial blight of pomegranate and its management. In *Recent Advances in the Diagnosis and Management of Plant Diseases*; Awasthi, L.P., Ed.; Springer: New Delhi, India, 2015; pp. 119–126, ISBN 978-81-322-2571-3.
46. Mirabolfathy, M.; Groenewald, J.Z.; Crous, P.W. First report of *Pilidiella granati* causing dieback and fruit rot of pomegranate (*Punica granatum*) in Iran. *Plant Dis.* **2011**, *96*, 461. [CrossRef]
47. Peduto Hand, F.; Choudhury, R.A.; Gubler, W.D. First report of *Cytospora punicae* causing wood canker and branch dieback of pomegranate (*Punica granatum*) in the United States. *Plant Dis.* **2014**, *98*, 853. [CrossRef]
48. Palavouzis, S.C.; Tzamos, S.; Paplomatas, E.; Thomidis, T. First report of *Neofusicoccum parvum* causing shoot blight of pomegranate in northern Greece. *New Dis. Rep.* **2015**, *32*, 10. [CrossRef]
49. Kumari, N.; Ram, V. Evaluation of pomegranate germplasm for resistance against leaf spot and dry fruit rot (*Coniella granati*). *Int. J. Farm Sci.* **2015**, *5*, 97–104.
50. Jabnoun-Khiareddine, H.; Aydi Ben Abdallah, R.; Daami-Remadi, M.; Mars, M. Response of Tunisian pomegranate (*Punica granatum* L.) cultivars and several plant hosts to *Coniella granati* (Saccardo). *J. Hortic.* **2018**, *5*, 1000245. [CrossRef]
51. Mincuzzi, A.; Ippolito, A.; Brighenti, V.; Marchetti, L.; Benvenuti, S.; Ligorio, A.; Pellati, F.; Sanzani, S.M. The effect of polyphenols on pomegranate fruit susceptibility to *Pilidiella granati* provides insights into disease tolerance mechanisms. *Molecules* **2020**, *25*, 515. [CrossRef] [PubMed]
52. Priya, B.T.; Murthy, B.N.S.; Gopalakrishnan, C.; Artal, R.B.; Jagannath, S. Identification of new resistant sources for bacterial blight in pomegranate. *Eur. J. Plant Pathol.* **2016**, *146*, 609–624. [CrossRef]
53. Kumar, P.; Dashyal, M.S.; Doddaraju, P.; Meti, B.S.; Girigowda, M. Differential gene responses in different varieties of pomegranate during the pathogenesis of *Xanthomonas axonopodis* pv. *punicae*. *3 Biotech.* **2021**, *11*, 180. [CrossRef]

54. Singh, N.V.; Abburi, V.L.; Ramajayam, D.; Kumar, R.; Chandra, R.; Sharma, K.K.; Sharma, J.; Babu, K.D.; Pal, R.K.; Mundewadikar, D.M.; et al. Genetic diversity and association mapping of bacterial blight and other horticulturally important traits with microsatellite markers in pomegranate from India. *Mol. Genet. Genom.* **2015**, *290*, 1393–1402. [CrossRef]
55. Usha, T.; Middha, S.K.; Babu, D.; Goyal, A.K.; Das, A.J.; Saini, D.; Sarangi, A.; Krishnamurthy, V.; Prasannakumar, M.K.; Saini, D.K.; et al. Hybrid assembly and annotation of the genome of the Indian *Punica granatum*, a superfood. *Front. Genet.* **2022**, *13*, 786825. [CrossRef]
56. Roopa Sowjanya, P.; Shilpa, P.; Patil, G.P.; Babu, D.K.; Sharma, J.; Sangnure, V.R.; Mundewadikar, D.M.; Natarajan, P.; Marathe, A.R.; Reddy, U.K.; et al. Reference quality genome sequence of Indian pomegranate cv. 'Bhagawa' (*Punica granatum* L.). *Front. Plant Sci.* **2022**, *13*, 947164. [CrossRef]
57. Luo, X.; Li, H.; Wu, Z.; Yao, W.; Zhao, P.; Cao, D.; Yu, H.; Li, K.; Poudel, K.; Zhao, D.; et al. The pomegranate (*Punica granatum* L.) draft genome dissects genetic divergence between soft- and hard-seeded cultivars. *Plant Biotechnol. J.* **2020**, *18*, 955–968. [CrossRef]
58. Yuan, Z.; Fang, Y.; Zhang, T.; Fei, Z.; Han, F.; Liu, C.; Liu, M.; Xiao, W.; Zhang, W.; Wu, S.; et al. The pomegranate (*Punica granatum* L.) genome provides insights into fruit quality and ovule developmental biology. *Plant Biotechnol. J.* **2018**, *16*, 1363–1374. [CrossRef]

horticulturae

Article

Exploring MicroRNAs Associated with Pomegranate Pistil Development: An Identification and Analysis Study

Yujie Zhao [1,2], Jingyi Huang [3], Ming Li [1,2], Hongfang Ren [3], Jian Jiao [1,2], Ran Wan [1,2], Yu Liu [1,2], Miaomiao Wang [1,2], Jiangli Shi [1,2] Kunxi Zhang [1,2], Pengbo Hao [1,2], Shangwei Song [1,2], Tuanhui Bai [1,2] and Xianbo Zheng [1,2,*]

1 College of Horticulture, Henan Agricultural University, Zhengzhou 450046, China; z1184985369@163.com (Y.Z.); liya1702@163.com (M.L.); jiaojian@henau.edu.cn (J.J.); wanxayl@henau.edu.cn (R.W.); yuliu@henau.edu.cn (Y.L.); wmm2018@henau.edu.cn (M.W.); sjli30@henau.edu.cn (J.S.); kunxi66@163.com (K.Z.); hao_pb@henau.edu.cn (P.H.); songshw959@163.com (S.S.); tuanhuibai88@163.com (T.B.)

2 Henan Province International Joint Laboratory of Horticultural Plant Biology, Henan Agricultural University, Zhengzhou 450046, China

3 College of Forestry, Nanjing Forestry University, Nanjing 210037, China; 19959932089@163.com (J.H.); 17503482685@163.com (H.R.)

* Correspondence: xbzheng@henau.edu.cn

Abstract: The interaction between miRNAs (microRNAs) and target genes plays an important role in plant pistil development. MiRNAs related to pistils were explored in pomegranate. The differentially expressed miRNAs were screened at different developmental stages of pomegranate pistils, and their target differentially expressed mRNAs were further identified to clarify the regulatory effect of miRNAs on pistil development. In our study, 61 conserved miRNAs were identified in 30 families, including miR395, miR394, miR393, miR161, miR162, and miR168. Among them, miR156, miR157, miR159, miR160, miR164, miR165, miR166, miR167, miR169, and miR172 were involved in the development of flower organs. Eight miRNAs were randomly selected and verified for qRT-PCR analysis. The result analysis indicated that miR160, miR164, and miR172 might be positive factors in the regulation of pomegranate pistil development. MiR156 and miR166 might be involved in regulation of pomegranate pistil development as negative factors.

Keywords: pomegranate; pistil; miRNA; correlation analysis

Citation: Zhao, Y.; Huang, J.; Li, M.; Ren, H.; Jiao, J.; Wan, R.; Liu, Y.; Wang, M.; Shi, J.; Zhang, K.; et al. Exploring MicroRNAs Associated with Pomegranate Pistil Development: An Identification and Analysis Study. *Horticulturae* **2024**, *10*, 85. https://doi.org/10.3390/horticulturae10010085

Academic Editor: Xuewu Duan

Received: 30 November 2023
Revised: 10 January 2024
Accepted: 12 January 2024
Published: 16 January 2024

1. Introduction

miRNAs are the class of non-coding small RNAs of eukaryotes genes, most of which are 21–24 nt in length [1,2]. In the nucleus, RNA polymerase II transcribes miRNA genes to generate pri-miRNAs with a stem loop structure and then generates miRNA/miRNA double strands under the action of Dicer enzyme cleavage. Finally, the miRNA strands combine with proteins such as AGO in the cytoplasm to form RNA-induced silencing complexes, which in turn regulate target genes. miRNAs regulate target genes at the post-transcriptional level in two main ways: degrading mRNA or inhibiting protein translation. If the miRNA is fully complementary and paired with its target gene mRNA, the AGO protein bound to miRNA cleaves and degrades the mRNA, resulting in mRNA that cannot be translated [3]. If miRNAs are not highly complementary to the target mRNA, miRNAs bind incompletely to mRNA and inhibit mRNA translation [4]. Plant miRNAs were first reported in *Arabidopsis thaliana* in 2002 [5]. In plants, most miRNAs are exactly matched to target genes, so degradation of mRNAs is the primary way in which miRNAs regulate target genes. miRNAs play important regulatory roles at the post-transcriptional level and participate in the regulation of plant growth and development, including flowering, megasporogenesis, inflorescence, and ovule development [2,6,7]. The *miR2118* mutant leads to complete male and female sterility [8]. MiR167 regulates its target gene ARFs (auxin response factors), which plays an important role in the development of pistil and stamen groups [9], and miR167 is also involved in regulating the fertility of male and female flowers in *Arabidopsis thaliana* [10]. MiR156 is involved in regulating plant growth cycle transitions [11] and directly inhibits the expression of members of SQUAMOSA promoter binding protein-like (SPL) family, thereby inhibiting the transition from vegetative growth to reproductive growth [12,13]. MiR164 regulates the formation of flower organ boundaries and the boundary formation of lateral organs [14]. The sequences of *miR159* and *miR319* are similar, and the target genes are MYB and TCP transcription factor families, respectively MiR159-MYB33-ABI5 synergistically regulates the transition of the plant vegetative growth stage [15].

Pomegranate trees produce large numbers of both bisexual flowers that produce fruit and functional male flowers that typically drop and fail to set fruit. Bisexual flowers have a discoid stigma covered with copious exudate, elongated stigmatic papillae, a single elongate style, and numerous and anatropous ovules. In contrast, functional male flowers have reduced female parts and exhibit shortened pistils of variable height. The outer and inner integument primordia form in bisexual flowers with a vertical diameter of 8.1–10.0 mm, and the ovule grows parallel to the nucellus through anticlinal cell division and elongation. However, the integument primordia are not observed in functional male flowers. When the vertical diameter is 10.1–13.0 mm, the outer integument grows rapidly and completely encloses the inner integument in bisexual flowers. Functional male flowers have sterile pistils that show abnormal ovule development. This result indicates that the vertical diameter of 8.1–13.0 mm is a critical stage for pomegranate ovule development [16]. In our study, pomegranate miRNAs of 'Taishanhong' bisexual flowers and functional male flowers were sequenced at the critical stages of pomegranate ovule development. miRNAs related to pistil development were mined at the post-transcriptional regulatory level, which laid the foundation for exploring the development mechanism of pomegranate ovules.

2. Materials and Methods

2.1. Plant Materials

According to Zhao's study [16], pomegranate ovule development was divided into three stages (initial ovule development stage: 5.0–10.0 mm; critical ovule abortion stage: 10.1–13.0 mm; ovule maturity stage: 13.1–18.0 mm). The pistils of bisexual and functional male flowers with vertical bud diameters of 5.0–10.0 mm (I), 10.1–13.0 mm (II), and 13.1–18.0 mm (III) were used as the test materials for miRNAs sequencing. The calyx, petal, and stamen tissues were removed, and only the female organs (ovary, style, and stigma) were retained for mixed-pool transcriptome and miRNA sequencing. The transcriptome data (PRJNA754480) were reported in our previous study [16]. Three biological replicates (18 samples in total) were obtained for each test sample. BF1, BF2, and BF3 represented bisexual flowers' pistils when their vertical diameters were 5.0–10.0 mm, 10.1–13.0 mm, and 13.1–18.0 mm, respectively. Similarly, MF1, MF2, and MF3 were used to represent functional male flowers' pistils when their vertical diameters were 5.0–10.0 mm, 10.1–13.0 mm, and 13.1–18.0 mm, respectively.

2.2. Sequencing and Data Analysis

2.2.1. Library Preparation and Sequencing

Total RNA was extracted from pomegranate samples. A total amount of 3 µg of RNA per sample was used as input material for the small RNA library. Sequencing libraries were generated using the NEBNext® Multiplex Small RNA Library Prep Set for Illumina® (NEB, Ipswich, MA, USA) following the manufacturer's recommendations and index codes were added to attribute sequences of each sample. First strand cDNA was synthesized using M-MuLV Reverse Transcriptase (RNase H^-). PCR amplification was performed using LongAmp Taq 2X Master Mix, SR Primer for illumina, and index (X) primer. PCR products were purified on 8% polyacrylamide gel (100 V, 80 min). DNA fragments corresponding to 140–160 bp (the length of small non-coding RNA plus the 3′ and 5′ adaptors) were recovered and dissolved in 8 μL of elution buffer. Lastly, library quality was assessed on the Agilent Bioanalyzer 2100 system using DNA High-Sensitivity Chips.

After the library was constructed, Qubit 2.0 was used for preliminary quantification, and the insert size of the library was then detected with Agilent 2100 before the effective concentration of the library was further accurately quantified (>2 nM). The library preparations were sequenced on an Illumina Hiseq 2000 platform and 50 bp single-end reads were generated.

2.2.2. Comparison and Analysis of Raw Data

Clean reads were obtained by deleting the raw reads with splices and low quality. Then, sRNAs in the 18–30 base range were screened from clean reads for subsequent analysis. The small RNA tags were mapped to the pomegranate genome (ASM286412v1) using Bowtie without mismatches to analyze their expression and distribution on the pomegranate genome.

2.2.3. Identification of Conservative miRNA and Novel miRNA

The reads on the pomegranate reference genome were mapped and compared in the miRBase database to obtain the known miRNA secondary structure, sequence, and bases number of sRNA matched on each sample.

miREvo (linux version) [17] and mirdeep2 [18] software were integrated to perform predictive analysis of novel miRNAs in pomegranate.

2.2.4. miRNA Expression and Differential Analysis

The expression levels of known and novel miRNAs in each sample were counted and normalized with TPM (transcripts per million reads) [19]. Normalization formula: normalized expression = mapped readcount/total reads × 1,000,000.

The sample data analysis was firstly based on the negative binomial distribution of DESeq2 [20], and the difference expression analysis was then performed using the DEGseq R package (1.8.3) [21]. The *p*-values were adjusted using the Benjamini and Hochberg method. A corrected *p*-value of 0.05 was set as the threshold for significantly different expression by default.

The heatmap of differential expression of miRNAs was constructed with $\log_2$(TPM) values using online software (http://www.heatmapper.ca/expression/, accessed on 30 November 2023).

2.2.5. Prediction and Enrichment Analysis of miRNA Target Genes

psRobot_tar in psRobot [22] and targetFinder were used to predict the target genes, and the correspondence between conserved and novel miRNAs and the target genes was analyzed. Gene ontology (GO) and KEGG enrichment analysis were further performed on the target genes.

2.2.6. Correlation Analysis of Sequencing Results

Firstly, the differentially expressed miRNAs were identified, and information regarding the relationship between miRNAs and target genes was further obtained. The differential expression of miRNA and mRNA was analyzed to identify key miRNAs and genes, and the regulatory relationship between miRNAs and target genes was directly displayed through the miRNA–target genes network regulation map. Through the integrated analysis of transcriptome and miRNA sequencing data, the miRNAs involved in regulating the development of pomegranate ovules and their target genes were mined.

2.2.7. qRT-PCR Verification of Sequencing Results

Mature miRNAs were used as the template, *PgActin* was used as a normalizer gene, and the specific primers are shown in Supplemental Table S1 for fluorescence quantitative verification of sequencing results. The remaining RNA from miRNA sequencing was used for qRT-PCR. Reverse transcription was performed using the PrimeScriptTM RT Reagent Kit with gDNA Eraser (Perfect Real Time) (TaKaRa, Osaka, Japan). The primer was designed and synthesized according to the report of Chen et al. [23]. After the primers were mixed, the temperature was set according to Tang et al. [24]. qRT-PCR was performed using SYBR® Premix Ex TAQTii (Tli RNaseH Plus) (TaKaRa, Osaka, Japan). Finally, the PCR analyses were performed on an Applied Biosystems 7500 and the thermal cycler was set as follows: pre-denaturation at 95 °C for 30 s, denaturation at 95 °C for 5 s, and denaturation at 60 °C for 34 s for 40 cycles, with fluorescence then acquired at the second step of each cycle. Dissolution curves were gained as follows: 95 °C for 15 s, 60 °C for 60 s, and 95 °C for 15 s. Three biological and technical replicates were designed for each miRNA. The data were quantitatively analyzed using the $2^{-\Delta\Delta CT}$ method [25]. Data were analyzed using SPSS software 22.0 (IBM, Armonk, NY, USA).

3. Results and Analysis

3.1. Sequencing Results

In this study, the pistils of bisexual flowers and functional male flowers were used as samples to extract RNA. After quality detection, a small RNA library was constructed for sequencing. The raw data obtained by Illumina HiSeqTM2500/MiSeq sequencing were uploaded to the NCBI database (PRJNA793612). After filtering, a total of 18 libraries were obtained with the lowest number of clean reads (12,326,514) and the highest number (22,788,789). The GC content ranged from 48 to 54%. In total, 93–97% of the filtered data fragments could be compared to the reference genome (Table 1), indicating that the sequencing results met the requirements for subsequent analysis.

Table 1. Small RNA sequencing data quality and comparison rate statistics.

Sample	Raw Reads	Clean Reads	GC Content (%)	Q20/Q30 (%)	Total sRNA	Mapped sRNA
BF1_1	17,367,419	17,051,495 (98.18%)	49.96	99.04/96.85	14,117,927	94.72%
BF1_2	13,704,093	13,340,571 (97.35%)	50.11	99.39/97.63	8,884,127	94.02%
BF1_3	13,401,810	12,885,356 (96.15%)	50.21	99.29/97.67	7,375,939	95.61%
BF2_1	21,426,062	20,858,435 (97.35%)	49.25	99.31/97.59	17,745,249	94.55%
BF2_2	13,406,968	12,787,925 (95.38%)	49.50	98.48/95.09	10,792,373	96.84%
BF2_3	14,948,932	14,681,792 (98.21%)	49.16	99.05/96.87	11,500,997	94.93%
BF3_1	21,966,339	19,930,652 (90.73%)	50.56	99.15/97.22	14,925,094	96.23%
BF3_2	18,298,780	15,436,529 (84.36%)	50.91	99.24/97.59	5,175,907	97.24%
BF3_3	21,665,923	20,317,773 (93.78%)	50.01	98.07/94.05	18,571,396	95.28%
MF1_1	21,655,951	21,171,252 (97.76%)	48.37	99.41/97.73	16,395,117	92.75%
MF1_2	23,257,849	22,788,789 (97.98%)	49.29	99.41/97.76	19,202,540	93.99%
MF1_3	13,748,317	13,525,096 (98.38%)	49.36	99.33/97.61	10,193,022	94.36%
MF2_1	14,418,240	13,551,334 (93.99%)	51.05	99.43/97.80	6,959,440	95.06%
MF2_2	12,606,612	12,326,514 (97.78%)	50.73	98.21/94.51	6,948,964	93.62%
MF2_3	18,135,526	17,633,277 (97.23%)	49.33	99.11/97.08	14,981,807	96.68%
MF3_1	21,572,493	18,175,437 (84.25%)	49.52	99.13/97.20	15,960,976	94.39%
MF3_2	19,010,757	17,214,821 (90.55%)	51.60	98.91/96.99	10,989,846	97.12%
MF3_3	17,985,939	16,567,891 (92.12%)	53.63	99.31/97.74	4,197,113	96.47%

3.2. Identification and Analysis of Conserved miRNAs and Novel miRNAs

Clean reads were screened, and small RNAs with lengths of 18–30 nt were analyzed, with miRNAs then concentrated at 21–22 nt. A total of 61 conserved miRNAs and 348 novel miRNAs were identified. The 61 conserved miRNAs were identified in 30 families, including *miR395*, *miR394*, *miR393*, *miR161*, *miR162*, and *miR168*. *miR156*, *miR157*, *miR159*, *miR160*, *miR164*, *miR165*, *miR166*, *miR167*, *miR169*, and *miR172* were involved in the development process of plant flower organs (Supplemental Table S2).

Pomegranate miRNA precursors could form the typical stem ring secondary structure, but the number of stem rings formed varies. The first base of most mature sequences was U, the length of which was concentrated at 20–21 nt, and a few newly identified miRNAs were 24 nt. There were differences in the location of the precursor sequence, such as *miR156*, *miR157*, *novel 104*, and *novel 105* at the 3′ end arm and *miR159*, *miR160*, *novel 10*, *novel 100*, *novel 101*, *novel 102*, *novel 107*, and *novel 108* at the 5′ end arm (Figure 1). The results showed that the mature bodies of the same family members of pomegranates had the same conserved sequence, and different positions of mature bodies might determine the performance of different functions [26], indicating that different miRNAs members of the same family of pomegranates were functionally conserved and diverse.

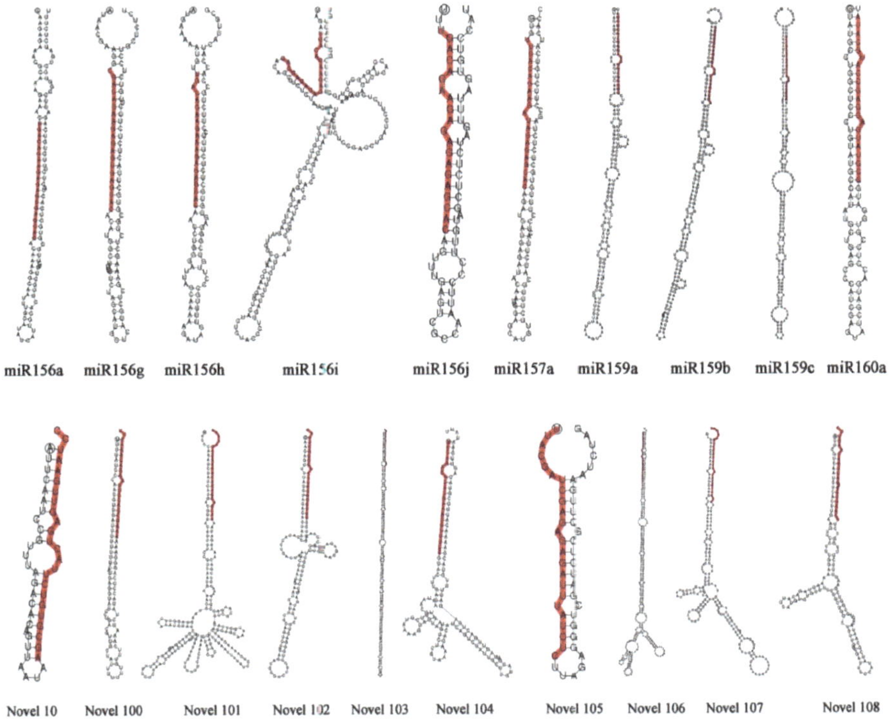

Figure 1. Secondary structure diagram of partial conserved miRNA and novel miRNA. Note: miR represents known microRNA, Novel represents new microRNA. The sequence is the precursor of miRNA, and the red part is the mature sequence.

3.3. *miRNA Differential Expression Analysis*

Among all pomegranate miRNAs, 76 miRNAs were differentially expressed in the pistil development of bisexual and functional male flowers, including 22 conserved miRNAs and 54 novel miRNAs. A total of 22 miRNAs were differentially expressed at the 5.0–10.0 mm (BF1 vs. MF1) stage, and 51 miRNAs were differentially expressed at the 10.1–13.0 mm (BF2 vs. MF2) stage of pomegranate flowers. A total of 98 miRNAs were expressed differently at different developmental stages of the pistils of bisexual flowers, among which 49 miRNAs were expressed differently between BF2 and BF3. The expression of 26 miRNAs showed significant differences during functional male flower pistil development (Figure 2).

At different pomegranate pistil developmental stages, differentially expressed miRNAs had different expression patterns (Figure 3). Conserved *miR156*, *miR157*, *miR159*, *miR160*, *miR164*, *miR165*, *miR166*, *miR167*, *miR169*, and *miR172* and *novel 41*, *novel 95*, *novel 111*, *novel 178*, *novel 312*, *novel 391*, *novel 437*, and *novel 472* were significantly differentially expressed in pomegranate pistil development (Figure 4). The expression levels of *miR159*, *miR160*, *miR164*, *miR167a*, *miR167d*, and *miR172* were significantly higher in bisexual flowers than in functional male flowers. These results indicated that these miRNAs were involved in pomegranate pistil development. However, the expression levels of *miR156* *miR166a-5p*, *novel 312*, and *novel 437* were significantly higher in functional male flowers than in bisexual flowers, suggesting that these miRNAs might play an important role in pomegranate pistil abortion.

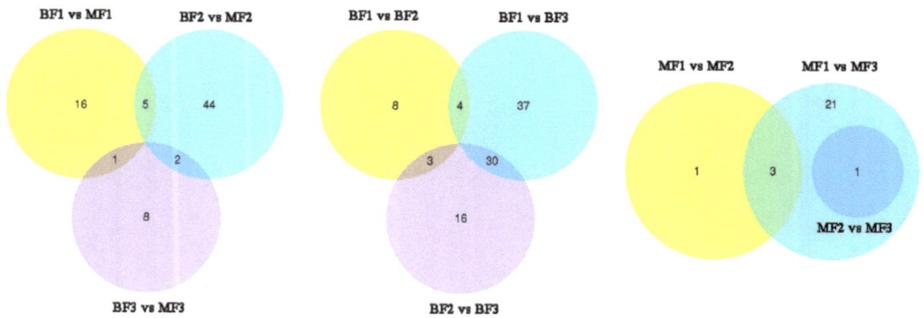

Figure 2. Venn diagram of the differential expression of miRNAs.

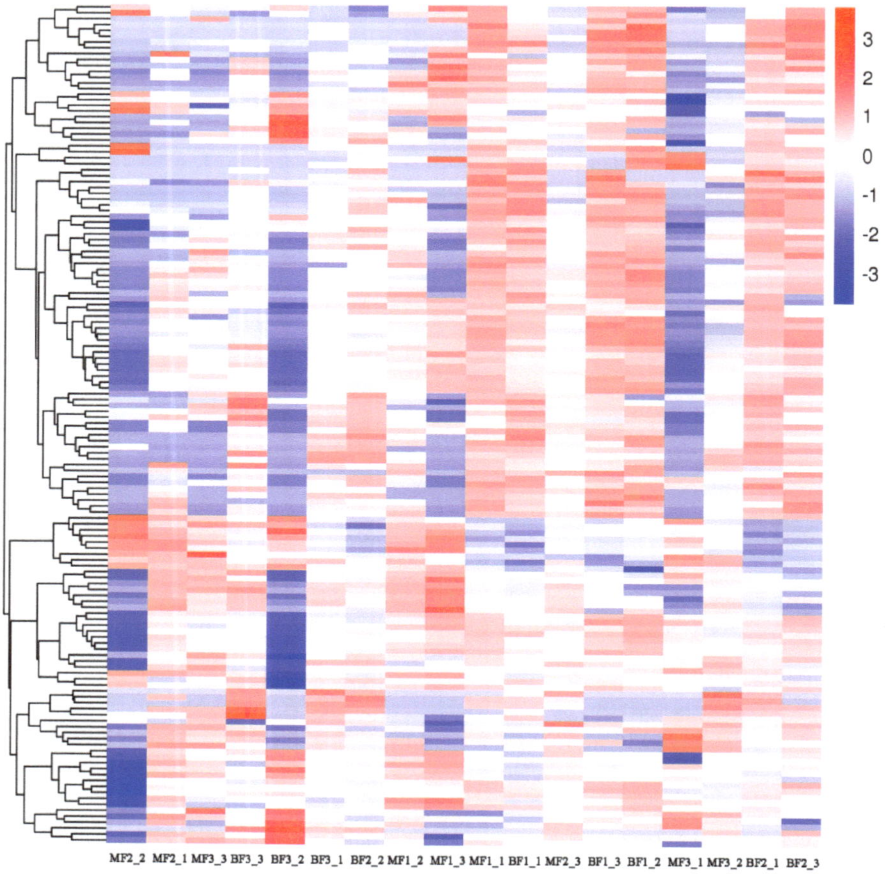

Figure 3. Heat map of differential expression of miRNAs in pomegranate pistils. Note: _1, _2, and _3 (such as BF1_1, BF1_2, and BF1_3) represent the three replicates of the sample.

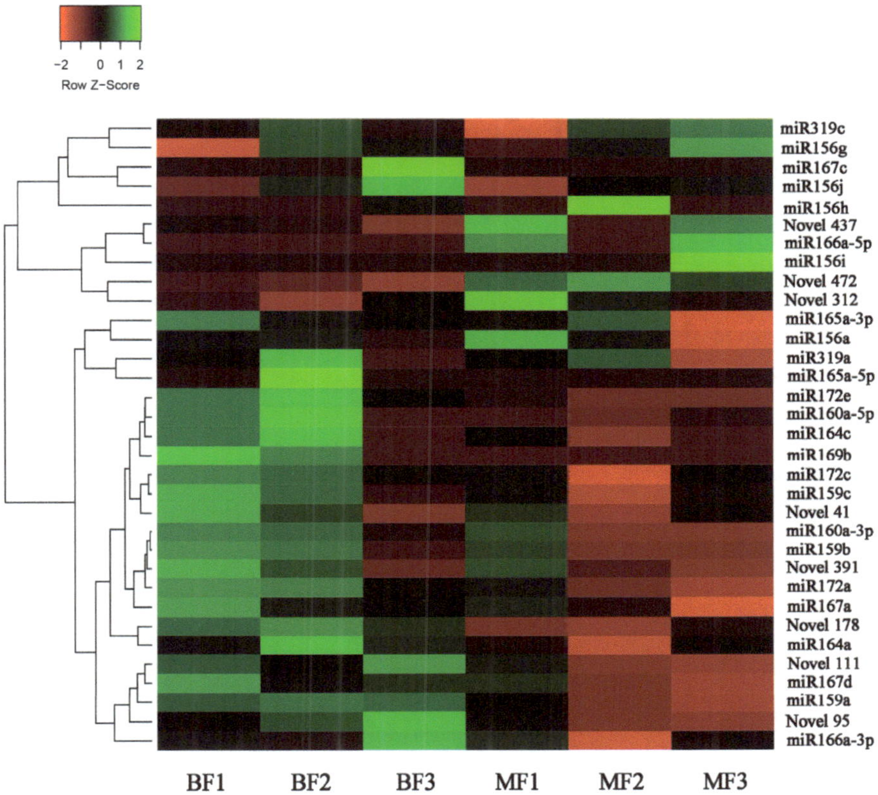

Figure 4. Heat map of conserved miRNAs and novel miRNAs expression patterns in pomegranate pistils.

3.4. miRNA Target Gene Prediction

Pomegranate's 61 known miRNAs and 348 novel miRNAs predicted 4952 and 6932 target genes, respectively. The results of differential expression analysis of miRNAs and target genes indicated that 76 differentially expressed miRNAs predicted 3539 target genes, and some conserved miRNAs and novel miRNAs targeted the same gene (Supplemental Table S3). One miRNA can target multiple target genes, ranging from a few to dozens [27,28]. Our study found that *miR156a-5p* and *miR157a-5p* could target *gene1341*, *gene9689*, *gene2311*, *gene26316*, *gene1095*, and *gene2300*, while *novel 356* and *novel 326* could target *gene26639*. Both *novel 251* and *novel 171* targeted *gene26063*, while *miR172* and *novel 77* were found to target *gene24967*. *PgmiRNA167* identified three target genes (*PgARF6a*, *PgARF6b*, and *PgARF6c*), and *PgARF6a* had a directly targeted regulatory relationship with *PgmiR167a* in pomegranate [29]. According to previous research results, the relationship between pomegranate miRNAs and the target genes will be confirmed in further research.

GO and KEGG function enrichment were performed on the target genes of 76 differentially expressed miRNAs to obtain annotation information for target genes (Supplemental Figure S1). The target genes of differentially expressed miRNAs were mainly annotated to biological processes and molecular functions, including biological regulatory processes (GO:0065007), metabolic processes (GO:0019222), gene expression regulatory biological processes (GO:0010468), protein-binding molecular functions (GO:0005515), and anion binding (GO:0043168). The KEGG function was used to enrich plant hormone signaling, auxin biosynthesis, and BR biosynthesis (Supplemental Figure S2).

3.5. Correlation Analysis of miRNAs and mRNAs

miRNAs regulate target gene expression by binding to complementary sites of target genes to degrade target mRNAs or inhibit their translation [30,31], indicating that miRNAs have a negative correlation with target genes. In our study, correlation analysis of miRNA sequencing and RNA-seq sequencing data was conducted to analyze the expression trend and targeted regulatory relationship between differentially expressed miRNAs and target genes. The statistics of differentially expressed genes as acquired by transcriptome sequencing are shown in Figure 5. The total number of differentially expressed genes was 1722, with 424 genes upregulating expression and 1298 genes downregulating expression. In stage I, 118 genes upregulated and 613 genes downregulated expression. In stage II, 661 genes upregulated and 916 genes downregulated expression. In stage III, 3721 genes upregulated and 3065 genes downregulated expression. As shown in Figure 5, the total number of differentially expressed miRNAs was 76, with 53 miRNAs upregulating expressions and 23 miRNAs downregulating expression. In stage I of pomegranate flower development, 9 miRNAs were upregulated and 22 miRNAs were downregulated. In stage II, 53 miRNAs were upregulated and 5 miRNAs were downregulated. In stage III, 9 miRNAs were upregulated and 9 miRNAs were downregulated.

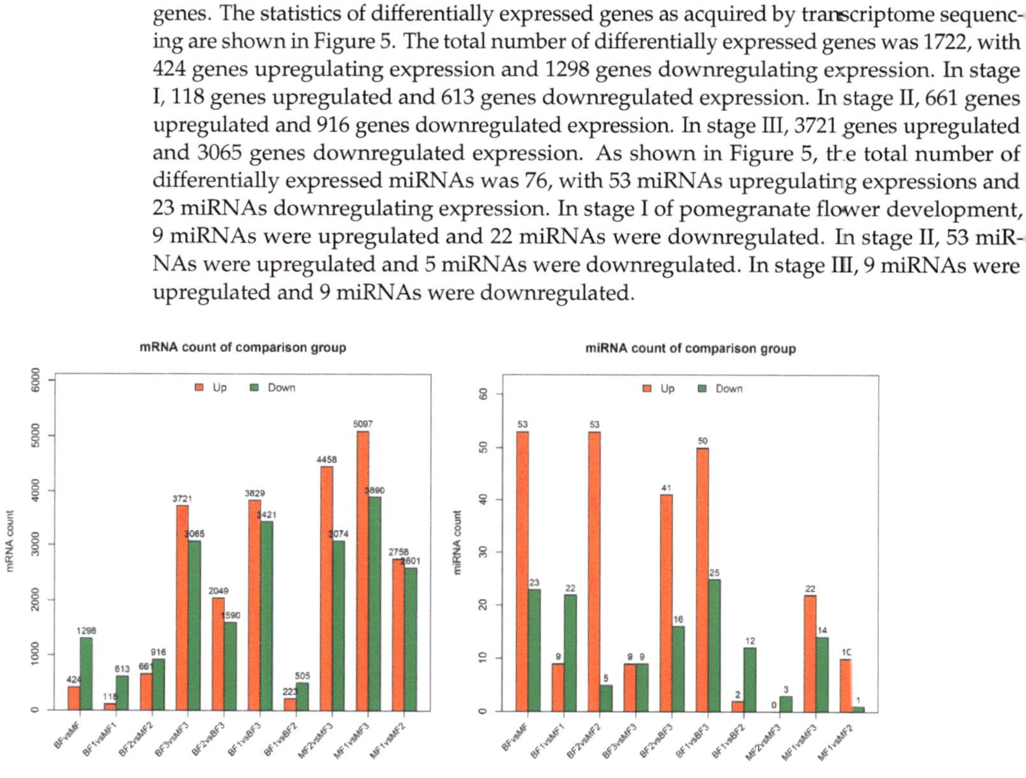

Figure 5. The statistical results of differential expression of mRNA–miRNA in comparison groups. Note: The x-coordinate represents the comparison combination of samples, and the y-coordinate represents the number of differentially expressed mRNA and miRNA in different comparison groups.

Between BF1 and MF1, *miR167a-5p* inhibited the expression of *gene6546*, *gene508*, *gene20506*, and *gene797*, while *miR165a-3p* inhibited *gene8460*. *miR172e-3p* inhibited the expression of *gene9394*, *gene7494*, and *gene22758*. *miR8175* inhibits *gene20953* and *gene20954* expression while promoting *gene1116* expression (Supplemental Figure S3). Between BF2 and MF2, *miR159a*, *miR159b-3p*, and *miR159c* co-inhibited the expression of *gene2996* and *gene10808*. *miR172a*, *miR172c*, and *miR172e-3p* inhibited the expression of *gene10725*, *gene15501*, *gene14955*, and *gene24967*, whereas *miR172a* and *miR172c* promoted *gene8013* expression (Figure 6). *miR164a* and *miR164c-5p* inhibited *gene1312*, *gene18425*, *gene24433*, and *gene25847* expression while promoting the expression of *gene4878*, *gene23729*, *gene21039*, and *gene12163*. During the maturation and development of pomegranate flowers' pistils (13.1–18.0 mm), *novel 111* inhibited the expression of *gene12900*, *gene10203*, *gene3421*, *gene20233*, *gene5397*, *gene10767*, and *gene6293* while promoting *gene6798* expression. Association analysis indicated that *miR164a*, *miR164c-5p*, *miR167a-5p*, *miR172a*, *miR172c*, *miR172e-3p*, and their target genes might be involved in regulating pomegranate pistil development.

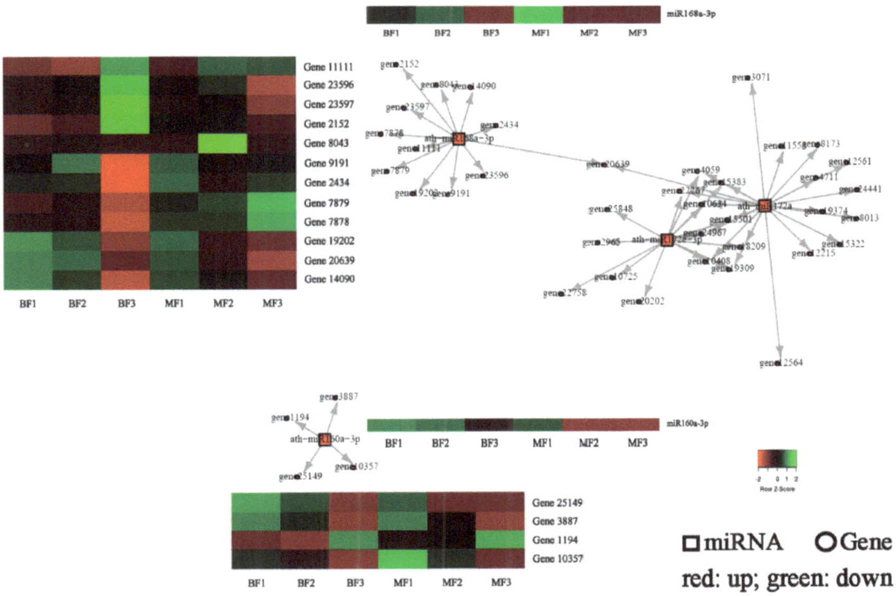

Figure 6. Analysis of the relationship between differentially expressed miR158/miR160/miR172 and target genes. Note: The square represents miRNAs, and the circles represent target genes. Red indicates up-regulated expression, while green represents down-regulated expression.

3.6. qRT-PCR Validation of Differential Transcripts

The qRT-PCR results of miRNA sequencing are shown in Figure 7. In functional male flowers, the expression level of *Pgnovel472* in stage III was higher than that in stages I and II, and the expression level in stage I was the lowest. *Pgnovel472* expression levels in stage III of bisexual flowers were higher than in stages I and II. In functional male flowers, the expression level of *Pgnovel437* in stage I was higher than that in stage II, and the expression of stage III was the lowest. The expression of *Pgnovel437* in stage III of bisexual flowers was five times higher than that in stage I. *Pgnovel178* had a lower expression level in stage II of bisexual flowers. In functional male flowers, the expression of *Pgnovel178* in stage III was higher than that in stages I and II, and the expression in stage I was the lowest.

In bisexual flowers, the expression levels of *PgmiR159a* in stages I and II were lower than in stage III, while the expression in stage II was the lowest. In functional male flowers, *PgmiR159a* expression levels in stages II and III were higher than that in stage I, with the expression in stage II being the highest. The expression level of *PgmiR160a* in stage III of bisexual flowers was higher than that in stage I, and the expression in stage II was the lowest. In functional male flowers, the expression level of *PgmiR160a* at stage II was higher than that at stage I. *PgmiR164c* in bisexual flowers had the highest expression at stage III. The expression level of *PgmiR164c* was the highest at stage II of functional male flower development. The expression of *PgmiR167d* gradually increased in the development of bisexual flowers, with the highest expression level found at stage III. The expression level of *PgmiR167d* was the highest at stage II of functional male flowers. In functional male flowers, the expression level of *PgmiR172e* in stage II was higher than that of stages I and III, with expression at stage I being the lowest. The expression level of *PgmiR172e* was the highest in stage III of bisexual flowers.

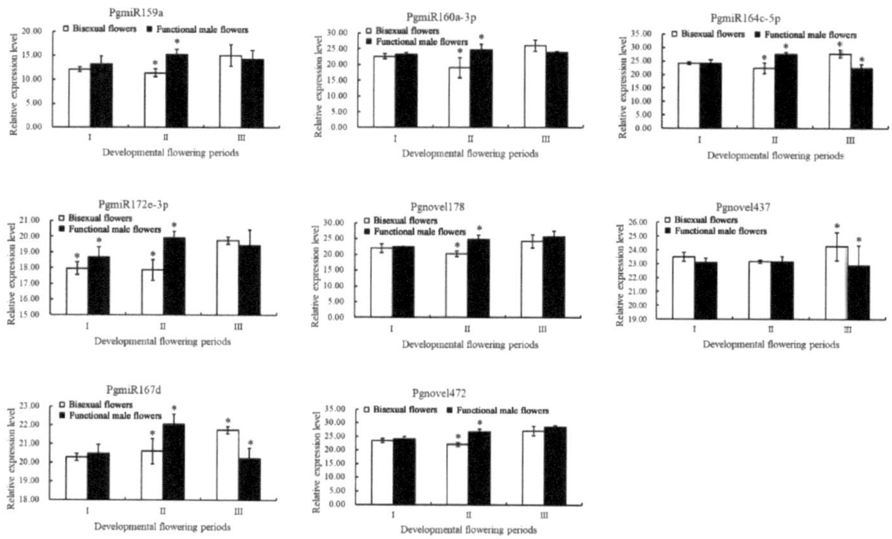

Figure 7. qRT-PCR analysis verified the result of miRNA sequencing. Note: The white bars represent bisexual flowers, and the black bars represent functional male flowers. Data were means ± SD of three technical replicates. * represents a significance level of $p < 0.05$ in independent sample *t*-tests.

4. Discussion

miRNAs are involved in flower development processes such as flowering regulation, flower organ morphogenesis, flower organ size and shape, ovule development, and flower organ polarity [7,32–34]. *miR156* can directly inhibit the expression of SPL family members, which control the transition from vegetative stage to reproductive stage. Overexpression of *miR159* and *miR319* causes flower development disorders such as delayed flowering [35]. *miR172* targets *AP2* to control flower organ development [36,37]. Cotton flower organs with *miRNA157* overexpression become smaller with a decreased number of ovules [38]. *miR167* regulates pistil and stamen development in Arabidopsis by targeting *ARF6* and *ARF8* [9]. In our study, a total of 61 conserved miRNAs and 348 novel miRNAs were discovered in the pistils of bisexual flowers and functional male flowers, among which 22 conserved miRNAs and 54 novel miRNAs were differentially expressed. The results of differential expression analysis showed that *miR156, miR157, miR159, miR160, miR164, miR165, miR166, miR167, miR169, miR172, novel 41, novel 95, novel 111, novel 178, novel 312, novel 391, novel 437*, and *novel 472* were significantly differentially expressed in the pistil development of bisexual pomegranate flowers and functional male flowers. *Novel 312, novel 437*, and *novel 472* were highly expressed in functional male flowers. *Novel 41, novel 95, novel 111*, and *novel 178* exhibited higher expression in bisexual flowers than in functional male flowers. These results suggest that these differentially expressed novel miRNAs may be involved in regulating pomegranate pistil development.

Apple *mdm-miR156h* was overexpressed in Arabidopsis, resulting in a prolonged juvenile period, increased leaf number, abnormal flower organ development, short horn fruit, and partial seed abortion [13]. Overexpression of populus *miR156j* promoted the development of Arabidopsis rosette leaves, resulting in delayed flowering and negatively regulated target genes *SPL6, SPL9*, and *SPL11* [39]. After overexpression of *miRl56*, plants showed delayed flowering and decreased fertility under short-day conditions [11]. Our study found that the expression levels of pomegranate *miR156a, miR156h*, and *miR156i* in the pistils of functional male flowers were higher than in bisexual flowers, which was consistent with the expression of chestnut *cmo-miR156* in male flower clusters and stamens [40]. However, the expression level of *miR156j* in the pistils of bisexual flowers was higher than that in functional male flowers. These results indicated that *miR156* is involved in regulating the development of pomegranate pistils.

miR160 targets *ARFs* in ovule development and pollen wall formation [41,42]. Grape *vvi-miR160c/d/e* target *VvARF18* to participate in regulating seed development [43]. Transgenic plants with overexpression of *sly-miR160a* produced tomato fruits with abnormal shape, demonstrating *sly-miR160a* affects early fruit development in tomato by regulating *SlARF10a/10b/17* expression [44,45]. It has been reported that *sly-miR160* regulates the expression of *ARFs* to affect ovary development by regulating auxin polar transport [45]. In our study, it was found that the expression levels of *miR160a-3p* and *miR160a-5p* in the pistils of bisexual flowers were higher than that in functional male flowers, and they were not expressed in the range of 13.1 mm to 18.0 mm. Bisexual flower ovules showed normal development, and functional male flower ovules showed abortion, indicating that *miR160* might be involved in regulating pomegranate ovule development.

During post-harvest storage of strawberry fruits, the expression levels of *fan-miR164d* and *fan-miR164e* were significantly increased, while the expression of *NAC*, their target gene, was downregulated [46,47]. The petals of tomato plants with overexpression of *sly-miR164* did not fall off normally and fruit was seedless [14]. The expression levels of *miR164a* and *miR164c* in the pistils of bisexual flowers were significantly higher than those in functional male flowers, while ovule development was normal in bisexual flowers. These data showed that *miR164* was expressed in pomegranate ovules to maintain normal ovule development.

miR167 targets *ARF6* and *ARF8*, which play an important role in the regulation of the development and maturity of pistil and stamen groups [9]. Overexpression of *miR167* reduces ovule maturity [48]. Our study found that *miR167a* and *miR167d* were expressed in the initial stage of ovule development in bisexual flowers and functional male flowers, while the expression of *miR167c* in the mature stage of the pistils of bisexual flowers was significantly higher than in other developmental stages and in functional male flowers. miRNA target gene prediction showed that *PgmiR167a* and *PgmiR167d* have binding sites on the *PgARF6s* gene sequence [29]. The results showed that *PgmiR167s* regulated *PgARF6s* expression to participate in ovule abortion in pomegranate.

In Arabidopsis, *miR172* regulates plant flowering time by regulating the expression of *AP2* (APETALA2), which in turn affects flower organ determination and flower morphology [32,49]. *AP2* sequence mutation, which occurs at the *miR172* binding site, results in severe defects in the development of *Arabidopsis* organs [50]. Overexpression of *miR172* in rice causes spikelet loss, flower organ developmental malformation, and decreased fertility [51]. Apple fruit size in transgenic overexpressed *mdm-miR172* plants was significantly reduced [52]. The deletion of *ppe-miR172* binding sites on the peach *AP2* sequence increases the number of peach petals and stamens [53]. The expression of *rch-miR172* was significantly downregulated in the petals, pistils, and stamens of Chinese rose, suggesting that *rch-miR172* may negatively regulate the expression of target gene *AP2* during the development of Chinese rose [54]. *Ach-miR172* targets and regulates *AP2* expression, and the function loss of *ach-miR172* leads to abnormal flower organ development in kiwifruit [55]. The above results show that *miR172* targets *AP2* in the regulation of flower development. Our study found that the expression of *miR172a*, *miR172c*, and *miR172e* in bisexual flowers was higher than in functional male flowers, and they were not expressed at the critical stage of ovule abortion in functional male flowers (10.1 mm–13.0 mm) nor in the mature development of functional male flowers (13.1 mm–18.0 mm). Correlation analysis showed that *miR172* was associated with target genes, which were differentially expressed in pomegranate flowers. These results suggest that *miR172* was involved in regulating the normal development of ovules in pomegranate pistils.

miR167 and *miR165/166* have been shown to be required for integument growth. *miR165/166* is closely related to the formation of meristems in flower organs and in the regulation of meristem activity [56]. Overexpression of *miR165/166* affects flower organ development, such as overexpression of *miR166* in *men1* and *jba-1D* mutants where the pistils are small and the number of carpels is reduced. *PHB* is involved in ovule primordium morphology and capsular development, and *miR166/165* regulates ovule development by regulating *PHB* expression in the inner ovule primordium [57,58]. *Pg-miR166a-3p* showed significantly higher expression in the pistils of functional male flowers of 'Tunisian soft seed' pomegranate than in bisexual flowers. The seed pods of 35S::*Pg-miR166a-3p* transgenic *Arabidopsis thaliana* became smaller, the number of seeds decreased, and the number of flower primordium and plant branches increased [59]. The results showed that the expression of *miR166a-5p* in functional male flowers was higher than that in bisexual flowers, and it was expressed at the critical stage of ovule development. These results suggested that *miR166a-5p* might be involved in regulating pomegranate ovule abortion.

5. Conclusions

After miRNA sequencing and analysis of the three developmental stages of bisexual and functional male pomegranate flowers, it was found that *miR156*, *miR157*, *miR159*, *miR160*, *miR164*, *miR165*, *miR166*, *miR167*, *miR169*, and *miR172* and *novel 41*, *novel 95*, *novel 111*, *novel 178*, *novel 312*, *novel 391*, *novel 437*, and *novel 472* were expressed differently during the pistil development of pomegranate. Target gene prediction, functional enrichment analysis, expression trends, and association analysis of differentially expressed miRNAs showed that *novel 41*, *novel 95*, *novel 111*, *novel 178*, *miR160*, *miR164*, and *miR172* were important regulators involved in the pistil development of pomegranate. *miR160*, *miR164*, and *miR172* might be positive factors in regulation of the pistil development of pomegranate. *miR156* and *miR166* might be involved in regulation of pistil development in pomegranate as negative factors.

Supplementary Materials: The following supporting information can be downloaded at: https://www.mdpi.com/article/10.3390/horticulturae10010085/s1, Supplemental Figure S1: Gene ontology-based term classification of different expressed miRNA targets (BF1 vs. MF1); Supplemental Figure S2: Gene ontology-based term classification of different expressed miRNA targets (BF2 vs. MF2 and BF3 vs. MF3); Supplemental Figure S3: Analysis of the relationship between differentially expressed miRNAs and target genes; Supplemental Table S1: The qRT-PCR primer of miRNAs; Supplemental Table S2: Molecular characteristics of partial known and novel miRNAs; Supplemental Table S3: Information for the target genes of partial miRNAs; Supplemental Table S4: Annotation information for differential expression genes of pomegranate.

Author Contributions: Writing—original draft preparation, Y.Z.; writing—review and editing, Y.Z. and J.H.; methodology, M.L.; software, Y.L. and M.W.; validation, J.J. and R.W.; formal analysis, S.S. and T.B.; resources, J.S.; data curation, P.H. and K.Z.; visualization, H.R.; funding acquisition, X.Z. All authors have read and agreed to the published version of the manuscript.

Funding: This work was funded by the fund for modern agricultural industrial technology systems of Henan province (HARS-22-09-Z2).

Data Availability Statement: The transcriptome data (PRJNA754480) and microRNA sequencing data (PRJNA793612) can be downloaded from the NCBI database.

Conflicts of Interest: The authors declare no conflicts of interest.

References

1. Bartel, D.P. MicroRNAs: Genomics, biogenesis, mechanism, and function. *Cell* **2004**, *116*, 281–297. [CrossRef] [PubMed]
2. Voinnet, O. Origin, Biogenesis, and Activity of Plant MicroRNAs. *Cell* **2009**, *136*, 669–687. [CrossRef] [PubMed]
3. Arribas-Hernández, L.; Kielpinski, L.J.; Brodersen, P. mRNA decay of most *Arabidopsis* miRNA targets requires slicer activity of AGO1. *Plant Physiol.* **2016**, *171*, 2620–2632. [CrossRef] [PubMed]
4. Singh, N.K. microRNAs databases: Developmental methodologies, structural and functional annotations. *Interdiscip. Sci.* **2017**, *9*, 357–377. [CrossRef] [PubMed]
5. Reinhart, B.J.; Weinstein, E.G.; Rhoades, M.W.; Bartel, B.; Bartel, D.P. MicroRNAs in plants. *Genes Dev.* **2002**, *16*, 1616–1626. [CrossRef]
6. Khraiwesh, B.; Zhu, J.K.; Zhu, J.H. Role of miRNAs and siRNAs in biotic and abiotic stress responses of plants. *Biochim. Biophys. Acta(BBA)-ene Regul. Mech.* **2012**, *1819*, 137–148. [CrossRef]

7. Petrella, R.; Cucinotta, M.; Mendes, M.A.; Underwood, C.J.; Colombo, L. The emerging role of small RNAs in ovule development, a kind of magic. *Plant Reprod.* **2021**, *34*, 335–351. [CrossRef]
8. Araki, S.; Le, N.T.; Koizumi, K.; Villar-Briones, A.; Nonomura, K.-I.; Endo, M.; Inoue, H.; Saze, H.; Komiya, R. miR2118-dependent U-rich phasiRNA production in rice anther wall development. *Nat. Commun.* **2020**, *11*, 3115. [CrossRef]
9. Wu, M.F.; Tian, Q.; Reed, J.W. Arabidopsis microRNA167 controls patterns of ARF6 and ARF8 expression, and regulates both female and male reproduction. *Development* **2006**, *133*, 4211–4218. [CrossRef]
10. Tan, C.; Zhang, Z.M.; Liu, H.J.; Gao, J.; Rong, Y.Z.; Pan, G.T. Advances on microRNAs regulated flower development of higher plant. *J. Agric. Biotechnol.* **2011**, *19*, 938–952.
11. Schwab, R.; Palatnik, J.F.; Riester, M.; Schommer, C.; Schmid, M.; Weigel, D. Specific effects of microRNAs on the plant transcriptome. *Dev. Cell* **2005**, *8*, 517–527. [CrossRef] [PubMed]
12. Klein, J.; Saedler, H.; Huijser, P. A new family of DNA binding proteins includes putative transcriptional regulators of theAntirrhinum majus floral meristem identity geneSQUAMOSA. *Mol. Genet. Genom.* **1996**, *250*, 7–16. [CrossRef]
13. Sun, C.; Zhao, Q.; Liu, D.D.; You, C.X.; Hao, Y.J. Ectopic expression of the apple Md-miRNA156h gene regulates flower and fruit development in Arabidopsis. *Plant Cell* **2013**, *112*, 343–351. [CrossRef]
14. Yang, C.W. MiR164 is Required for Tomato Flower Initiation and Fruit Development. Master's Thesis, Chongqing University, Chongqing, China, 2012.
15. Guo, C.; Jiang, Y.; Shi, M.; Wu, X.; Wu, G. *ABI5* acts downstream of miR159 to delay vegetative phase change in Arabidopsis. *New Phytol.* **2021**, *231*, 339–350. [CrossRef] [PubMed]
16. Zhao, Y.; Wang, Y.; Yan, M.; Liu, C.; Yuan, Z. *BELL1* interacts with *CRABS CLAW* and *INNER NO OUTER* to regulate ovule and seed development in pomegranate. *Plant Physiol.* **2023**, *191*, 1066–1083. [CrossRef] [PubMed]
17. Wen, M.; Shen, Y.; Shi, S.; Tang, T. miREvo: An integrative microRNA evolutionary analysis platform for next-generation sequencing experiments. *BMC Bioinform.* **2012**, *13*, 140. [CrossRef]
18. Friedländer, M.R.; Mackowiak, S.D.; Li, N.; Chen, W.; Rajewsky, N. miRDeep2 accurately identifies known and hundreds of novel microRNA genes in seven animal clades. *Nucleic Acids Res.* **2011**, *40*, 37–52. [CrossRef]
19. Zhou, L.; Chen, J.; Li, Z.; Li, X.; Hu, X.; Huang, Y.; Zhao, X.; Liang, C.; Wang, Y.; Sun, L.; et al. Integrated Profiling of MicroRNAs and mRNAs: MicroRNAs Located on Xq27.3 Associate with Clear Cell Renal Cell Carcinoma. *PLoS ONE* **2010**, *5*, e15224. [CrossRef]
20. Love, M.I.; Huber, W.; Anders, S. Moderated estimation of fold change and dispersion for RNA-seq data with DESeq2. *Genome Biol.* **2014**, *15*, 550. [CrossRef]
21. Wang, L.; Feng, Z.; Wang, X.; Wang, X.; Zhang, X. DEGseq: An R package for identifying differentially expressed genes from RNA-seq data. *Bioinformatics* **2010**, *26*, 136–138. [CrossRef]
22. Wu, H.J.; Ma, Y.-K.; Chen, T.; Wang, M.; Wang, X.J. PsRobot: A web-based plant small RNA meta-analysis toolbox. *Nucleic Acids Res.* **2012**, *40*, 22–28. [CrossRef]
23. Chen, C.; Ridzon, D.A.; Broomer, A.J.; Zhou, Z.; Lee, D.H.; Nguyen, J.T.; Barbisin, M.; Xu, N.L.; Mahuvakar, V.R.; Andersen, M.R.; et al. Real-time quantification of microRNAs by stem-loop RT-PCR. *Nucleic Acids Res.* **2005**, *33*, e179. [CrossRef] [PubMed]
24. Tang, F.; Hajkova, P.; Barton, S.C.; Lao, K.; Surani, M.A. MicroRNA expression profiling of single whole embryonic stem cells. *Nucleic Acids Res.* **2006**, *34*, e9. [CrossRef] [PubMed]
25. Livak, K.J.; Schmittgen, T.D. Analysis of relative gene expression data using realtime quantitative PCR and the 2-ΔΔCT method. *Methods* **2001**, *25*, 402–408. [CrossRef]
26. Mi, S.; Cai, T.; Hu, Y.; Chen, Y.; Hodges, E.; Ni, F.; Wu, L.; Li, S.; Zhou, H.; Long, C.; et al. Sorting of small RNAs into *Arabidopsis* Argonaute complexes is directed by the 5′ terminal nucleotide. *Cell* **2008**, *133*, 116–127. [CrossRef] [PubMed]
27. Jones-Rhoades, M.W.; Barte, D.P. Computational identification of plant microRNAs and their targets, including a stress-induced miRNA. *Mol. Cell* **2004**, *14*, 787–799. [CrossRef]
28. Jones-Rhoades, M.W.; Bartel, D.P.; Bartel, B. MicroRNAs and their regulatory roles in plants. *Annu. Rev. Plant Biol.* **2006**, *57*, 19–53. [CrossRef]
29. Zhao, Y.; Wang, Y.; Zhao, X.; Yan, M.; Ren, Y.; Yuan, Z. ARF6s identification and function analysis provide insights into flower development of *Punica granatum* L. *Front. Plant Sci.* **2022**, *13*, 833747. [CrossRef]
30. Kurihara, Y.; Watanabe, Y. *Arabidopsis* micro-RNA biogenesis through Dicer-like 1 protein functions. *Proc. Natl. Acad. Sci. USA* **2004**, *101*, 12753–12758. [CrossRef]
31. Filipowicz, W.; Bhattacharyya, S.N.; Sonenberg, N. Mechanisms of post-transcriptional regulation by microRNAs: Are the answers in sight? *Nat. Rev. Genet.* **2008**, *9*, 102–114. [CrossRef]
32. Allen, R.S.; Li, J.; Stahle, M.I.; Dubroué, A.; Gubler, F.; Millar, A.A. Genetic analysis reveals functional redundancy and the major target genes of *Arabidopsis* miR159 family. *Proc. Natl. Acad. Sci. USA* **2007**, *104*, 16371–16376. [CrossRef] [PubMed]
33. Aukerman, M.J.; Sakai, H. Regulation of flowering time and floral organ identity by a microRNA and its APETALA2-like target genes. *Plant Cell* **2003**, *15*, 2730–2741. [CrossRef] [PubMed]
34. Zhang, Q.W.; Yang, X.H.; Li, F.; Deng, Y.T. Advances in miRNA-mediated growth and development regulation in horticultural crops. *Acta Hortic. Sin.* **2022**, *49*, 1145–1161.

35. Palatnik, J.F.; Wollmann, H.; Schommer, C.; Schwab, R.; Boisbouvier, J.; Rodriguez, R.; Warthmann, N.; Allen, E.; Dezulian, T. Huson, D.; et al. Sequence and expression differences underlie functional specialization of *Arabidopsis* MicroRNAs miR159 and miR319. *Dev. Cell* **2007**, *13*, 115–125. [CrossRef] [PubMed]
36. Wu, G.; Poethig, R.S. Temporal regulation of shoot development in Arabidopsis thaliana by Mir156 and its target SPL3. *Development* **2006**, *133*, 3539–3547. [CrossRef] [PubMed]
37. Wu, G.; Park, M.Y.; Conway, S.R.; Wang, J.-W.; Weigel, D.; Poethig, R.S. The sequential action of miR156 and miR172 regulates developmental timing in *Arabidopsis*. *Cell* **2009**, *138*, 750–759. [CrossRef] [PubMed]
38. Liu, N.; Tu, L.; Wang, L.; Hu, H.; Xu, J.; Zhang, X. MicroRNA 157-targeted SPL genes regulate floral organ size and ovule production in cotton. *BMC Plant Biol.* **2017**, *17*, 7. [CrossRef]
39. Duan, Z.X. Expression Pattern and Functional Analysis of microRNA Peu-miR156j and Peu-miR169 from *Populus euphratica* Master's Thesis, Beijing Forestry University, Beijing, China, 2012.
40. Chen, G.; Li, J.; Liu, Y.; Zhang, Q.; Gao, Y.; Fang, K.; Cao, Q.; Qin, L.; Xing, Y. Roles of the GA-mediated *SPL* gene family and miR156 in the floral development of Chinese chestnut (*Castanea mollissima*). *Int. J. Mol. Sci.* **2019**, *20*, 1577. [CrossRef]
41. Liu, X.D.; Zhang, H.; Zhao, Y.; Feng, Z.Y.; Li, Q.; Yang, H.Q.; Luan, S.; Li, J.M.; He, Z.H. Auxin controls seed dormancy through stimulation of abscisic acid signaling by inducing ARF-mediated ABI3 activation in Arabidopsis. *Proc. Natl. Acad. Sci. USA* **2013**, *110*, 15485–15490. [CrossRef]
42. Wu, S.C. Functional Characterization of mir160 in Cotton Ovule Development. Master's Thesis, Huazhong Agricultural University, Wuhan, China, 2016.
43. Bai, Y.H.; Wang, W.R.; Dong, T.Y.; Guan, L.; Su, Z.W.; Jia, H.F.; Fang, J.G.; Wang, C. vvi-miR160s in mediating vvARF18 response to gibberellin regulation of grape seed development. *Sci. Agric. Sin.* **2020**, *53*, 1890–1903.
44. Hendelman, A.; Buxdorf, K.; Stav, R.; Kravchik, M.; Arazi, T. Inhibition of lamina outgrowth following Solanum lycopersicum auxin response factor 10 (SlARF10) derepression. *Plant Mol. Biol.* **2012**, *78*, 561–576. [CrossRef] [PubMed]
45. Damodharan, S.; Zhao, D.Z.; Arazi, T. A common mi RNA160-based mechanism regulates ovary patterning, floral organ abscission and lamina outgrowth in tomato. *Plant J.* **2016**, *86*, 458–471. [CrossRef] [PubMed]
46. Li, J.; Lai, T.; Song, H.; Xu, X. MiR164 is involved in delaying senescence of strawberry (*Fragaria ananassa*) fruit by negatively regulating NAC transcription factor genes under low temperature. *Russ. J. Plant Physiol.* **2017**, *64*, 251–259. [CrossRef]
47. Zhang, X.; Wang, M.; Gan, C.; Ren, Y.; Zhao, X.; Yuan, Z. Riboflavin application delays senescence and relieves decay in harvested strawberries during cold storage by improving antioxidant system. *LWT-Food Sci. Technol.* **2023**, *182*, 114810. [CrossRef]
48. Ru, P.; Xu, L.; Ma, H.; Huang, H. Plant fertility defects induced by the enhanced expression of microRNA167. *Cell Res.* **2006**, *16*, 457–465. [CrossRef]
49. Jung, J.H.; Seo, Y.H.; Seo, P.J.; Reyes, J.L.; Yun, J.; Chua, N.-H.; Park, C.-M. The GIGANTEA-regulated microRNA172 mediates photoperiodic flowering Independent of CONSTANS in Arabidopsis. *Plant Cell* **2007**, *19*, 2736–2748. [CrossRef]
50. Grigorova, B.; Mara, C.; Hollender, C.; Sijacic, P.; Chen, X.; Liu, Z. LEUNIG and SEUSS co-repressors regulate *miR172* expression in *Arabidopsis* flowers. *Development* **2011**, *138*, 2451–2456. [CrossRef]
51. Zhu, Q.-H.; Upadhyaya, N.M.; Gubler, F.; Helliwell, C.A. Over-expression of miR172 causes loss of spikelet determinacy and floral organ abnormalities in rice (*Oryza sativa*). *BMC Plant Biol.* **2009**, *9*, 149. [CrossRef]
52. Yao, J.; Xu, J.; Cornille, A.; Tomes, S.; Karunairetnam, S.; Luo, Z.; Bassett, H.; Whitworth, C.; Rees-George, J.; Ranatunga, C.; et al. A microRNA allele that emerged prior to apple domestication may underlie fruit size evolution. *Plant J.* **2015**, *84*, 417–427. [CrossRef]
53. Gattolin, S.; Cirilli, M.; Pacheco, I.; Ciacciulli, A.; Linge, C.D.S.; Mauroux, J.; Lambert, P.; Cammarata, E.; Bassi, D.; Pascal, T.; et al. Deletion of the miR172 target site in a TOE-type gene is a strong candidate variant for dominant double-flower trait in Rosaceae. *Plant J.* **2018**, *96*, 358–371. [CrossRef]
54. Sui, M.J.; Yan, H.J.; Wang, Z.Z.; Qiu, X.Q.; Jian, H.Y.; Wang, Q.G.; Chen, M.; Zhang, H.; Tang, K.X. Identification of microRNA associated with flower organ development in Rosa chinensis 'Viridiflora'. *Plant Sci. J.* **2019**, *37*, 37–46.
55. Varkonyi-Gasic, E.; Lough, R.H.; Moss, S.M.A.; Wu, R.; Hellens, R.P. Kiwifruit floral gene APETALA2 is alternatively spliced and accumulates in aberrant indeterminate flowers in the absence of miR172. *Plant Mol. Biol.* **2012**, *78*, 417–429. [CrossRef]
56. Zhang, X.; Henderson, I.R.; Lu, C.; Green, P.J.; Jacobsen, S.E. Role of RNA polymerase IV in plant small RNA metabolism. *Proc. Natl. Acad. Sci. USA* **2007**, *104*, 4536–4541. [CrossRef]
57. Sieber, P.; Gheyselinck, J.; Gross-Hardt, R.; Laux, T.; Grossniklaus, U.; Schneitz, K. Pattern formation during early ovule development in Arabidopsis thaliana. *Dev. Biol.* **2004**, *273*, 321–334. [CrossRef]
58. Hashimoto, K.; Miyashima, S.; Sato-Nara, K.; Yamada, T.; Nakajima, K. Functionally diversified members of the miR165/6 gene family regulate ovule morphogenesis in Arabidopsis thaliana. *Plant Cell Physiol.* **2018**, *59*, 1017–1026. [CrossRef]
59. Chen, L.N. Identification of microRNAs Involved in Pomegranate Female Sterility and Functional Analysis on the Regulation of Pg-miR166a-3p on Pistil Development. Ph.D. Thesis, Chinese Academy of Agricultural Sciences, Beijing, China, 2020.

horticulturae

Article

Phenotypic Diversity of Pomegranate Cultivars: Discriminating Power of Some Morphological and Fruit Chemical Characteristics

Mira Radunić [1,2], Maja Jukić Špika [2,3] and Jelena Gadže [4,*]

1 Department of Plant Sciences, Institute for Adriatic Crops and Karst Reclamation Split, Put Duilova 11, 21000 Split, Croatia; mira.radunic@krs.hr
2 Centre of Excellence for Biodiversity and Molecular Plant Breeding, Svetošimunska cesta 25, 10000 Zagreb, Croatia; maja.jukic.spika@krs.hr
3 Department of Applied Science, Institute for Adriatic Crops and Karst Reclamation Split, Put Duilova 11, 21000 Split, Croatia
4 Department for Pomology, Division of Horticulture and Landscape Architecture, Faculty of Agriculture, University of Zagreb, Svetošimunska 25, 10000 Zagreb, Croatia
* Correspondence: jgadze@agr.hr

Abstract: In modern agricultural production, where a small number of commercial cultivars dominate, the collection, evaluation, and preservation of germplasm are important tasks to reduce the erosion of genes and preserve biodiversity. The aim of this study is to characterize the morphological and fruit chemical properties of the pomegranate germplasm grown on the East Adriatic coast, including the commercial cultivars 'Hicaznar', 'Granada', and 'Wonderful', and to highlight the characteristics with the greatest discriminating power. The characterization of the tree, leaf, flower, arils, seed, and juice was carried out using the UPOV descriptor. The colors of the peel, arils, and juice were analyzed according to the CIEL*a*b* method, total soluble solids were measured using refractometers, and total acidity was determined by titration with 0.1 M NaOH. The research results showed significant diversity between the cultivars, which were grouped into several clusters using an unsupervised analysis technique. Factors such as plant vigor, plant growth habit, predominant number of leaves per node on young shoots, crown type, fruit shape, fruit shape in cross-section, peel weight, total aril weight, aril weight, number of arils per fruit, seed length and width, seed yield, total acidity, TSS/TA ratio, and color parameters of the peel, arils, and juice showed high variability, indicating their strong discriminating power in determining the phenotypic diversity of pomegranate.

Keywords: pomegranate; germplasm; ex situ collection; qualitative and quantitative markers; color; acidity; fruit shape; crown type fruit shape in cross-section; juice yield

Citation: Radunić, M.; Jukić Špika, M.; Gadže, J. Phenotypic Diversity of Pomegranate Cultivars: Discriminating Power of Some Morphological and Fruit Chemical Characteristics. *Horticulturae* **2024**, *10*, 563. https://doi.org/10.3390/horticulturae10060563

Academic Editors: Zhaohe Yuan, Gaihua Qin, Julián Bartual and Daniele Bassi

Received: 12 April 2024
Revised: 21 May 2024
Accepted: 24 May 2024
Published: 28 May 2024

1. Introduction

The pomegranate is a flavorful fruit that is native to the Mediterranean. Growing public awareness of the health benefits of pomegranate has led to an increasing demand for its cultivation and commercialization. Numerous cultivars have descended from wild populations, making the pomegranate a prime example of the domestication of a wild fruit tree [1]. Fundamental botanical knowledge, diverse morphological variation within a plant species, and an understanding of growing conditions and physiological needs are essential for germplasm identification and for growers to optimize crop production and management. More than 500 pomegranate cultivars have been identified worldwide, showing their great genetic variability, but only about 50 are cultivated for commercial purposes [2]. This selective cultivation has significantly reduced the genetic richness. Establishing repositories comprising germplasm from wild, semi-wild, and less popular cultivars in different geographical regions is crucial to maintaining the genetic basis for future breeding and improvement of pomegranate cultivars, as well as enhancing crop quality [3].

Mediterranean countries, such as Spain, Morocco, Tunisia, Greece, Turkey, and Egypt, have established local collections of pomegranate germplasm [4], which play a central role in maintaining genetic diversity and provide valuable breeding material. To avoid redundancy and effectively assess the diversity within these collections, it is essential to thoroughly characterize each cultivar, not only by its physical characteristics but also by comprehensive genetic analysis [5]. While geneticists deal with molecular diversity, agronomists focus on observable morphological variation for sustainable breeding methods, as Hawkes [6] discussed in 1991. The use of germplasm collections to study phenotypic variation not only improves our understanding of basic plant biology but also holds promise for studying the adaptive responses of long-lived woody plants to changing environmental conditions [7].

As with many other fruit species, the nomenclature of pomegranate cultivars is confusing, and numerous synonyms and homonyms further complicate their characterization. Fruit quality assessment is a multi-layered approach that takes into account external appearance, morphological characteristics, chemical composition, post-harvest characteristics, and microbiological and chemical safety. The most important parameters include the visual aspects of the fruit, such as size, shape, and color, as well as chemical characteristics such as acidity, sweetness, and antioxidant activity. All these factors together influence the commercial value of pomegranate products, market perception, and consumer preferences [8].

Ex situ germplasm collections of pomegranate cultivars grown on the East Adriatic coast exist in Croatia, but the lack of morphological and molecular studies prevents a comprehensive assessment of the preserved genetic diversity, indicating the need for further investigations aimed at qualitative use in breeding and/or breeding programs.

The main objectives of this research were to evaluate the morphological and some fruit chemical characteristics of native and introduced pomegranate cultivars grown on the eastern coast of the Adriatic Sea. In addition, based on the obtained results, the aim was to determine which of the characteristics assessed have the greatest discriminating power in determining the diversity of pomegranate cultivars. The detailed characterization, evaluation, and documentation of the studied germplasm are invaluable for the further development of pomegranate production, genetic conservation, and for facilitating future breeding programs aimed at sustainable improvement; this underscores the crucial role of living collections for the conservation of genetic diversity.

2. Materials and Methods

2.1. Site Characteristics, Environmental, Conditions and Plant Material

The evaluation was carried out on fruit samples from nineteen cultivars, consisting of sixteen native pomegranate cultivars traditionally grown on the eastern Adriatic coast ('Barski slatki', 'Bokežan', 'Dividiš', 'Domaći kiseli', 'Dubrovački kasni', 'Glavaš', 'Konjski zub', 'Kristal', 'Medunac', 'Mojdiški sitnozrni', 'Pastun', 'Sladun', 'Slatki crveni', 'Slatki tankokorac', 'Šerbetaš', and 'Zamorac') and three introduced cultivars ('Granada', 'Hicaznar', and 'Wonderful'). The native cultivars were collected along the eastern Adriatic coast, propagated by cuttings, and planted in 2011 in the Gene Bank of Mediterranean Fruit Species located at the Institute for Adriatic Crops and Karst Reclamation, Split, Croatia (43°30′22″ N; 16°29′47″ E; 60 m a.s.l.). The trees were healthy and bush-shaped with three to four main trunks, planted 4.5 × 3 m apart. The orchard was drip-irrigated and the usual cultivation practices were applied (pruning, fertilization, and plant protection).

The collection was grown in a Mediterranean climate, defined as a Csa climate type according to the Koppen–Geiger climate classification [9]. The average annual temperature was 17.5 °C, with mild winters (absolute minimum 2 °C in January) and hot summers (absolute maximum 33.2 °C in August). The annual precipitation at this location was 754 mm during vegetation (April to October, 50% of annual precipitation). The average duration of sunshine was 2742 h (Croatian Meteorological and Hydrological Service).

2.2. Sampling

Twenty-three qualitative and fifty-eight quantitative characteristics of the tree, leaf, flower, fruit, aril, juice, and seed were evaluated on three trees per cultivar in four consecutive years (2017–2020). In all years studied, samples of 30 mature leaves and 25 bell-shaped flowers (fertile flowers) per tree were randomly collected in the canopy. The leaves were taken from the outer part of the tree at the end of August. The flowers were collected at the time of full bloom during the second half of May. A total of 36 fruits (3 fruits per tree × 3 trees × 4 years), which had developed from the flowers in the first part of flowering were collected randomly around the canopies for each cultivar. The fruits were harvested when they were ripe according to local practices (fruit size and external color) between 15–30 October (depending on the year and cultivar). The samples were taken to the Pomology Laboratory of the Institute for Adriatic Crops and Karst Reclamation, Split, Croatia, where the morphological characterization and chemical analyses of the juice were performed.

2.3. Morphological Characteristics

The morphological characteristics were assessed according to the UPOV (International Union for the Protection of New Varieties of Plants) descriptors for pomegranates [10].

2.3.1. Qualitative Characteristics

Tree characteristics were described in terms of the vigor of the tree, the growth habit, the intensity of the gray color on the main branches, the number of one-year-old shoots ending in thorns, and the predominant number of leaves per node on young shoots. The shape of the leaf blade (except for the apex), the anthocyanin coloration of the petiole, the intensity of the green coloration of the leaf blade, the calyx color, the color of the corolla, the surface of the petals, the predominant number of flowers on the one-year-old shoot, the predominant type of arrangement of the flowers, the fruit shape, the shape of the fruit base, the shape of the fruit apex, the shape in cross-section, the crown type, the fruit overcolor, the extent of the aril, the aril color, the juice color, and the hardness of the seed were described on a scale from 1 to 9 according to the UPOV descriptor. The evaluation was conducted by a five-member panel with extensive experience in the morphological characterization of fruit species, specializing in pomegranates.

2.3.2. Quantitative Characteristics

Leaves and flowers were scanned using an Epson Perfection V700 photo scanner; the leaf blade length and width, leaf blade length and width ratio, leaf area, the petiole length, the calyx length, the calyx width, the ratio of the calyx length and width, the petal length, the petal width, the petal length and width ratio, petal form coefficient, the petal area, and the petal perimeter were measured using WinFOLIA Pro 2014a software (Regent Instruments Inc., Quebec, QC, Canada). The number of sepals and petals was determined by counting.

The weight of the individual fruits, peel, aril, and seed weight, as well as the total aril weight, were measured using an electronic scale ±0.01 g (Shimadzu, Kyoto, Japan). Fruit length without crown, fruit diameter, crown length, crown diameter, and peel thickness were determined using digital micrometers. For the fruit and crown diameter, two perpendicular measurements were taken around the equatorial plane, while for peel thickness, two measurements were taken from opposite sides of the fruit. The fruit form index (percentage ratio of equatorial fruit diameter/fruit length excluding crown), fruit crown index (percentage ratio of crown length/total fruit length), and fruit peel thickness index (percentage ratio of peel thickness/fruit diameter) were calculated. The aril yield was calculated as the ratio between aril weight and fruit weight multiplied by 100. The number of arils per fruit was estimated by counting the number of arils in a 100 g sample and extrapolating the number of arils based on the total weight of arils per fruit. Aril length, aril width, seed length, and seed width were measured on 30 arils per year for each cultivar using the Epson Perfection V700 photo scanner and WinSEEDLE Pro 2019a analysis software (Regent

Instruments Inc., Quebec, QC, Canada). Seed yield was determined as the ratio of seed weight per 50 g of arils, multiplied by 100.

The color parameters of the fruit peel were measured in the middle part of the fruit twice at different points of each fruit. After measuring the peel color, the fruits were opened and the arils were removed. Thirty arils were randomly taken from each cultivar for the measurements. Moreover, 200 g of the arils from three fruits per tree were pressed through four layers of gauze cloth and used to measure the juice color. The Chroma meter CR-400 (Konica Minolta, Osaka, Japan) was used to measure the color of the peel, arils, and juice (CIE L*a*b* method) of the pomegranate, expressed in the parameters L*, a*, b*, C*, and h°. The color parameter L* indicates the lightness of the peel, arils, or juice, and ranges from 0 (opaque) to 100 (completely transparent). Value a* stands for redness and ranges from negative values for green to positive values for red. Value b* denotes 'yellowing' and ranges from negative values for blue to positive values for yellow [11]. Chroma (C*) and hue angle (h°) denote the visual color appearances. The hue angle represents the visual experience according to which the color is evaluated with the following values: 0–90°: red–violet, 90–180°: yellow, 180–270°: blue–green, and 270–360°: blue. The C* value stands for the color intensity. These parameters are calculated using the following equation:

$$(180 - h^\circ)/(L^* + C^*)$$

to obtain what we call the 'color index' [12].

2.4. *Juice Yield, Total Soluble Solids, and Total Acidity Content*

Fruit juice content was obtained by extracting 50 g of arils per fruit by squeezing through four layers of gauze cloth. The juice yield was determined as the ratio of milliliters of juice per 50 g of arils sampled, multiplied by 100. Total soluble solid (TSS) and total acidity (TA) were analyzed to classify cultivars into sweet, sweet–sour or sour groups. Total soluble solid (TSS) and total acidity (TA) analyses were performed. TSS was determined using a digital refractometer (Mettler Toledo, Greifensee, Switzerland, calibrated with distilled water) at 21 °C and TA was determined by titrating to pH 8.1 with a 0.1 M NaOH solution, expressed as g citric acid per 100 g juice [13].

2.5. *Statistical Analysis*

The approach used in this study (to detect the associated characteristics and distribution of samples in 19 pomegranate cultivar datasets) included univariate and multivariate statistical algorithms. For the generated dataset, descriptive statistics were generated for morphological parameters of trees, leaves, flowers, fruit, and arils, as well as color parameters of fruits, arils, juice, and fruit chemical traits. To determine the differences between cultivars, significant differences at the 5% level between means were determined using a non-parametric test (Kruskal–Wallis) for qualitative morphological traits and a one-way ANOVA analysis test for other traits, followed by Tukey's honest significant difference (HSD) test. The minimum, maximum, mean, standard deviation (SD), and coefficient of variation (CV) were calculated and used as indicators of variability.

An unsupervised analysis technique, utilizing a heatmap created through clustering using the Ward method, was employed to analyze data, facilitating the identification of patterns, and enhancing the understanding of interrelationships or similarities. This analysis focused on studied pomegranate cultivars (19), which were clustered based on 44 traits specifically chosen for their coefficient of variation exceeding 15%. The Pearson correlation coefficient analysis was performed to examine and quantify the relationships between the analyzed morphological and chemical characteristics, thereby elucidating the extent and nature of their interdependencies within the dataset. The correlation was considered significant at a value of $p \leq 0.05$.

The analyses were performed with Statistica 14.0.0.15 (Tibco Software Inc., Palo Alto, CA, USA, 2020), whereas RStudio was applied for Heatmap (RStudio 2023.09.1, Posit Software, PBC 2009–2023).

3. Results

3.1. Morphological Characteristics

Qualitative characteristics of the studied pomegranate cultivars are shown in Table 1 while the distribution frequency for these qualitative characteristics is shown in Table 2. For 10 characteristics, significant differences were found between cultivars (e.g., plant growth habit and fruit shape ($p < 0.001$), as well as the predominant number of leaves per node, shape of the fruit apex, fruit extent of overcolor, fruit shape in cross-section, seed hardness, petal surface, flower arrangement, and petiole anthocyanin coloration, which were predominant ($p \leq 0.05$).

3.1.1. Tree

Seven tree characteristics were part of the first group of characteristics used to evaluate the cultivars (Tables 1 and 2). Plant vigor, plant growth habit, and the predominant number of leaves per node on young shoots showed high variability. The vigor of the cultivars studied was generally medium (53%) and strong (42%), except for 'Konjski zub', which was weak. The growth habit predominantly spread (63%), while 37% of the cultivars exhibited an upright growth habit. In this study, 47% of cultivars had a medium-intensity gray color on the main branches, while all cultivars had some annual shoots ending in thorns. Cultivars were divided into three groups based on the predominant number of leaves per node on young shoots: 52% had more than three leaves per node ('Bokežan', 'Dividiš', 'Domaći kiseli', 'Dubrovački kasni', 'Glavaš', 'Medunac', 'Sladun', 'Slatki crveni', 'Zamorac', and 'Hicaznar'), 32% with two leaves ('Barski slatki', "Konjski zub', 'Kristal', 'Mojdiški sitnozrni', 'Slatki tankokorac', and 'Granada'), and 16% ('Pastun', 'Šerbetaš', and 'Wonderful') with three leaves per node. In 63% of the cultivars, the flowers were arranged in an inflorescence, while the cultivars 'Dividiš' 'Dubrovački kasni', 'Konjski zub', 'Medunac', 'Mojdiški sitnozrni', 'Šerbetaš', 'and 'Zamorac' had a solitary flower arrangement.

3.1.2. Leaf and Flower

Leaf characteristics were part of the second group of characteristics used for the study. The mean values between the studied cultivars for all evaluated characteristics differed significantly ($p \leq 0.05$) and the coefficient of variation was between 6.65% and 10.97% (Table 3).

Leaf blade length ranged from 4.44 cm ('Mojdiški sitnozrni') to 5.58 cm ('Domaći kiseli'), the leaf width varied from 1.56 cm ('Wonderful') to 1.97 cm ('Slatki crveni'), while the LL/LW ratio ranged from 2.45 ('Mojdiški sitnozrni') to 3.18 ('Domaći kiseli'). In addition, petiole length ranged from 0.42 cm ('Hicaznar') to 0.60 cm ('Pastun'), with anthocyanin coloration ranging from medium (42%) to strong intensity (58%), depending on the cultivar. The shape of the leaf blade—specifically, the shape of the apex excluding the tip—was predominantly moderately obtuse (53%), followed by a right angle (37%), and moderately acute (10%) (Table 2).

The variability in the flower characteristics of the studied cultivars is shown in Table 4. The calyx length ranged from 3.30 cm ('Konjski zub') to 4.28 cm ('Medunac'); the calyx width was from 1.36 cm ('Domaći kiseli') to 1.98 cm ('Glavaš'); the petal length was from 2.49 cm to 2.99 cm, and the petal width was from 1.51 cm to 2.42 cm. The form coefficient of the petals ranged between 0.81 and 0.89, while the cultivars were divided into three groups based on the flower surface, with moderately wrinkled petal surfaces predominating (68%). The predominant color of the calyx was medium red in 74% of the studied cultivars and the color of the corolla was orange–red in 79% of cases (Table 2).

Table 1. Qualitative characteristics of the studied pomegranate cultivars according to UPOV (2013).

	Cultivar	**Barski slatki**	**Bokežan**	**Dividiš**	**Domaći kiseli**	**Dubrovački kasni**	**Glavaš**	**Konjski zub**	**Kristal**	**Medunac**	**Mojdiški sitnozrni**	**Pastun**	**Sladun**	**Slatki crveni**	**Slatki tankokorac**	**Šerbetaš**	**Zamorac**	**Hicaznar**	**Granada**	**Wonderful**
Tree	Plant vigor (PV)	5	7	7	7	5	7	3	5	5	5	5	7	7	5	5	7	7	5	5
	Plant growth habit (PGH)	3	3	3	3	3	3	1	1	1	3	3	1	1	3	3	3	3	3	3
	Intensity of the gray color of the main branches (IGC)	2	1	2	2	2	3	3	3	2	1	1	3	2	3	2	2	3	3	2
	Number of one-year-old shoots ending in thorns	2	2	2	2	2	2	2	2	2	2	2	2	2	2	2	2	2	2	2
	Predominant number of leaves per node on young shoots (DNL)	1	3	3	3	3	3	1	1	3	1	2	3	3	1	2	3	3	1	2
	Predominant number of flowers per node (F/Node)	1	1	1	1	1	1	1	1	1	1	1	1	1	1	1	1	1	1	1
	Arrangement of flowers (predominant type) (AF)	2	2	1	2	1	2	1	2	1	1	2	2	2	2	1	1	2	2	2
Leaf	Leaf blade: shape of apex, excluding the tip (LBS)	4	3	4	3	4	2	4	4	4	3	3	3	3	4	4	3	4	4	2
	Leaf blade: intensity of the green color (IGC)	5	5	5	5	5	5	5	5	5	5	5	5	5	5	5	5	5	3	5
	Petiole: anthocyanin coloration (PAC)	7	5	7	5	7	7	5	5	7	7	7	7	7	7	7	5	5	5	5
Flower	Calix color (CC)	3	3	2	3	3	3	2	1	3	2	3	2	3	3	3	3	3	3	3
	Corolla color (CoC)	5	5	5	5	5	3	5	4	5	5	5	3	6	5	5	5	5	5	5
	Petal surface (PS)	3	3	3	3	3	5	3	3	3	1	3	3	5	3	5	1	3	5	3
Fruit	Fruit shape (FS)	4	4	4	1	1	4	4	1	4	4	4	4	4	4	1	4	1	4	4
	Shape of the base (SFB)	1	2	2	2	2	2	2	2	2	2	2	2	2	2	2	2	2	2	2
	Shape of the fruit apex (FSA)	2	3	3	2	2	3	2	2	2	2	2	3	3	2	2	3	2	2	3
	Crown type (CT)	2	2	1	2	3	1	2	2	3	3	2	1	3	1	2	3	2	3	2
	Fruit overcolor (FOC)	4	6	5	6	4	5	4	3	3	3	6	5	3	5	5	4	6	6	6
	Fruit extent of overcolor (FEOC)	7	7	5	7	1	7	5	1	1	1	7	7	7	7	5	7	7	7	7
	Shape in cross-section (SCS)	3	1	2	1	2	2	2	2	1	1	2	1	1	3	3	2	3	1	2

Table 1. *Cont.*

	Cultivar	Barski slatki	Bokežan	Dividiš	Domaći kiseli	Dubrovački kasni	Glavaš	Konjski zub	Kristal	Medunac	Mojdiški sitnozrni	Pastun	Sladun	Slatki crveni	Slatki tankokorac	Šerbetaš	Zamorac	Hicaznar	Granada	Wonderful
Aril	Aril main color (AMC)	6	6	2	5	6	5	3	2	6	5	5	6	4	4	6	5	7	7	7
Seed	Hardness of seed (HS)	2	3	2	3	3	3	2	2	3	3	3	3	2	3	3	3	3	2	2
Juice	Juice color (JC)	5	3	1	3	3	3	2	2	3	2	3	2	4	3	5	3	5	5	5

Table 2. Distribution frequency (%) for the qualitative characteristics of the studied pomegranate cultivars.

Characteristic	Frequency (%)								
	1	2	3	4	5	6	7	8	9
Plant vigor			Weak (5)		Medium (53)		Strong (42)		
Plant growth habit	Upright (37)		Spreading (63)		Weeping (0)				
The intensity of the gray color of the main branches	Light (16)		Medium (47)		Dark (37)				
No. of one-year-old shoots ending in thorns	None or very few (0)	Few (100)	Medium (0)	Many (0)					
Predominant no. of leaves per node	Two (32)	Three (16)	More than three (52)						
Predominant no. of flowers per node	One (100)	Two (0)	Three (0)	More than three (0)					
Arrangement of flowers	Solitary (37)	Inflorescence (63)							
Leaf blade: the shape of the apex, excluding the tip	Strongly acute (0)	Moderately acute (10)	Right angle (37)	Moderately Obtuse (53)	Strongly Obtuse (0)				
Leaf blade: intensity of the green color			Light (5)		Medium (95)		Dark (0)		
Petiole: anthocyanin coloration			Weak (0)		Medium (42)		Strong (58)		
Calyx color	Orange (5)	Orange–red (21)	Medium red (74)	Dark red (0)					

Table 2. *Cont.*

Characteristic	Frequency (%)								
Corolla color	White (0)	Pink (0)	Light orange (11)	Medium Orange (5)	Orange Red (79)	Medium red (5)			
Petal surface	Smooth or slightly Wrinkled (11)		Moderately Wrinkled (68)		Strongly Wrinkled (21)				
Fruit shape	Spheroid (26)	Ellipsoid (0)	Ovoid (0)	Oblate (74)					
Shape of the base	Truncate (5)	Convex (95)	Angular (0)						
Shape of the fruit apex	Convex (0)	Truncate (63)	Necked (37)						
Crown type	Closed; convergent sepals (21)	Semi-opened; right sepals (47)	Opened; divergent sepals (32)	Largely opened (0)					
Fruit overcolor	Orange (0)	Orange red (0)	Pink (21)	Pink red (21)	Medium red (26)	Red purple (32)	Purple (0)	Dark purple (0)	
Fruit extent of overcolor	Very small (21)		Small (0)		Medium (16)		Large (63)		Very large (0)
Shape in cross-section	Round (37)	Round to angular (42)	Angular (21)						
Aril main color	White (0)	Light pink (10)	Medium pink (5)	Dark pink (10)	Light red (26)	Medium red (32)	Dark red (17)		
Juice color	Creamy-bright pink (5)		Pink (21)		Light red (42)	Red (5)	Dark red (26)		
Hardness of seed	Soft (0)	Medium (37)	Hard (63)						

Table 3. Leaf morphological characteristics of the studied cultivars.

Cultivar	Leaf Blade Length (LL; cm)	Leaf Blade Width (LW; cm)	LL/LW Ratio	Leaf Area (LA; cm^2)	Petiole Length (PL; cm)
Barski slatki	4.57 ± 0.67 e	1.81 ± 0.25 de	2.55 ± 0.39 i	6.15 ± 1.54 de	0.46 ± 0.09 gh
Bokežan	4.76 ± 0.67 cde	1.67 ± 0.19 efg	2.87 ± 0.41 e	5.67 ± 1.17 fg	0.48 ± 0.10 efg
Dividiš	5.20 ± 0.65 b	1.69 ± 0.20 ef	3.11 ± 0.43 ab	6.28 ± 1.24 de	0.55 ± 0.11 bcd
Domaći kiseli	5.58 ± 0.87 a	1.76 ± 0.22 def	3.18 ± 0.42 a	7.02 ± 1.70 bc	0.56 ± 0.13 abc
Dubrovački kasni	4.93 ± 0.64 bcd	1.89 ± 0.24 abc	2.63 ± 0.36 hi	6.97 ± 1.54 c	0.45 ± 0.07 gh
Glavaš	4.59 ± 0.74 cde	1.64 ± 0.18 efg	2.82 ± 0.46 ef	5.46 ± 1.19 fg	0.47 ± 0.09 fgh
Konjski zub	4.97 ± 0.66 bcd	1.63 ± 0.17 efg	3.07 ± 0.38 abc	5.92 ± 1.20 ef	0.51 ± 0.08 cdefg
Kristal	4.98 ± 0.49 bc	1.72 ± 0.23 de	2.93 ± 0.41 de	6.27 ± 1.11 de	0.51 ± 0.06 def
Medunac	5.28 ± 0.78 ab	1.92 ± 0.25 ab	2.76 ± 0.34 fg	7.40 ± 1.95 abc	0.56 ± 0.10 abc
Mojdiški sitnozrni	4.44 ± 0.40 cde	1.82 ± 0.16 abcde	2.45 ± 0.27 j	6.07 ± 0.86 def	0.43 ± 0.05 fgh
Pastun	5.51 ± 0.82 a	1.93 ± 0.25 ab	2.87 ± 0.37 e	7.53 ± 1.83 a	0.60 ± 0.10 a
Sladun	5.18 ± 0.81 b	1.76 ± 0.26 de	2.97 ± 0.48 cde	6.45 ± 1.66 d	0.57 ± 0.12 ab
Slatki crveni	4.86 ± 0.77 cd	1.97 ± 0.26 a	2.48 ± 0.30 j	7.08 ± 1.97 bc	0.52 ± 0.09 cde
Slatki tankokorac	4.71 ± 0.56 cde	1.82 ± 0.25 bcd	2.62 ± 0.44 hi	6.30 ± 1.36 de	0.51 ± 0.10 def
Šerbetaš	5.23 ± 0.57 ab	1.92 ± 0.25 ab	2.76 ± 0.38 fg	7.40 ± 1.42 ab	0.55 ± 0.10 bc
Zamorac	5.19 ± 0.81 b	1.68 ± 0.28 ef	3.13 ± 0.48 a	6.31 ± 1.84 de	0.54 ± 0.10 bcd
Hicaznar	4.58 ± 0.62 de	1.72 ± 0.21 de	2.66 ± 0.27 gh	5.72 ± 1.39 f	0.42 ± 0.08 h
Granada	4.79 ± 0.74 cde	1.60 ± 0.22 fg	3.01 ± 0.38 bcd	5.53 ± 1.32 fg	0.46 ± 0.10 fgh
Wonderful	4.67 ± 0.69 cde	1.56 ± 0.21 g	3.02 ± 0.45 bcd	5.28 ± 1.22 g	0.43 ± 0.10 h
Min.	4.44	1.56	2.45	5.28	0.42
Max.	5.58	1.97	3.18	7.53	0.60
Mean	4.95	1.76	2.81	6.36	0.50
SD	0.33	0.12	0.22	0.70	0.05
CV%	6.65	6.94	7.90	10.97	10.69

Values are given as mean ± SD, n = 100. Different lower-case letters in each column indicate a significant difference between the cultivars at $p \leq 0.05$ by the Tukey's test.

3.1.3. Fruit

Fruit characteristics of the observed pomegranate cultivars are listed in Tables 1 and 5. Significant differences between cultivars were observed for all fruit characteristics ($p \leq 0.05$). The greatest variability was observed for fruit shape, crown type, and fruit shape in cross-section for qualitative characteristics, and in peel weight for quantitative characteristics.

The fruit weight differed significantly between the cultivars and ranged from 261.9 g ('Medunac') to 497.97 g ('Slatki tankokorac'). Fruit length and diameter values varied, receptively, between 68.3 mm ('Medunac') and 85.19 mm ('Pastun') and 78.05 mm ('Zamorac') and 98.63 mm ('Slatki tankokorac'), while FL/FD was 0.81–0.90. In terms of fruit shape (FL/FD ratio), 'Pastun', 'Slatki crveni', and 'Granada' were more spherical, while 'Glavaš' was more oblate, as confirmed by the Fruit Form Index (the ratio between the equatorial fruit diameter and the fruit length, excluding the crown). The IFF ranged from 109.4% ('Granada') to 123.4% ('Glavaš').

The majority of cultivars had a convex base shape (95%), while 63% of the cultivars had a truncated fruit apex and 37% had a necked apex. In the cross-section, the cultivars can be divided into three groups: round (37%), round to angular (42%), and angular (21%) (Table 2).

The crown size and crown type also differed between the cultivars. The values for the crown length ranged from 11.22 mm ('Zamorac') to 17.98 mm ('Slatki tankokorac') and the crown index (IC) ranged from 14% ('Zamorac') to 19% ('Medunac') (Table 5).

Peel weight varied between 97.82 g ('Mojdiški sitnozrni') and 259.88 g ('Dividiš'). The peel weight of 'Mojdiški sitnozrni' did not differ significantly from those of 'Dubrovački kasni', 'Kristal', 'Medunac', and 'Zamorac'. The cultivar 'Sladun' had the thickest peel (5.86 mm) but was not significantly different from the cultivars 'Kristal', 'Pastun', 'Slatki tankokorac', and 'Hicaznar'.

Table 4. Flower morphological characteristics of the cultivars studied.

Cultivar	Calyx Length (CL; cm)	Calyx Width (CW; cm)	CL/CW Ratio	Petal Length (PL; cm)	Petal Width (PW; cm)	PL/PW Ratio	Petal Area (PA; cm^2)	Perimeter of Petal (PP; cm)	Petal Coefficient Form (CoP)	Number of Sepals (NS)	Number of Petals (NP)
Barski slatki	3.97 ± 0.31 abc	1.59 ± 0.18 ef	2.52 ± 0.23 b	2.81 ± 0.25 b–f	2.28 ± 0.17 bcd	1.24 ± 0.10 j	4.60 ± 0.61 b	8.01 ± 1.04 b	0.88 ± 0.05 abc	6.91 ± 0.88 bc	7.00 ± 0.93 bc
Bokežan	3.76 ± 0.47 c–g	1.56 ± 0.23 e–h	2.42 ± 0.22 b–e	2.49 ± 0.23 i	1.51 ± 0.12 j	1.65 ± 0.16 a	2.68 ± 0.35 h	6.50 ± 0.51 k	0.78 ± 0.04 k	7.74 ± 1.21 a	7.78 ± 1.67 a
Dividiš	3.84 ± 0.41 c–f	1.75 ± 0.20 c	2.22 ± 0.26 gh	2.71 ± 0.22 fg	2.13 ± 0.24 e–h	1.29 ± 0.13 ghi	4.06 ± 0.71 de	7.65 ± 0.64 fgh	0.87 ± 0.04 cde	6.90 ± 1.06 bc	6.92 ± 1.09 bc
Domaći kiseli	3.76 ± 0.31 c–g	1.36 ± 0.18 j	2.79 ± 0.35 a	2.80 ± 0.33 b–f	1.87 ± 0.22 i	1.50 ± 0.11 b	3.65 ± 0.79 fg	7.50 ± 0.91 hi	0.81 ± 0.04 j	6.82 ± 0.98 bc	6.82 ± 0.95 bcd
Dubrovački kasni	3.89 ± 0.33 b–e	1.75 ± 0.21 c	2.23 ± 0.17 gh	2.85 ± 0.25 b–e	2.20 ± 0.17 df	1.29 ± 0.09 gh	4.42 ± 0.60 bc	7.98 ± 0.61 bc	0.87 ± 0.04 bcd	7.83 ± 1.00 a	7.83 ± 1.00 a
Glavaš	3.78 ± 0.24 c–g	1.98 ± 0.35 a	1.96 ± 0.31 j	2.73 ± 0.38 ef	1.92 ± 0.19 i	1.42 ± 0.12 cd	3.59 ± 0.71 g	7.36 ± 0.88 i	0.83 ± 0.05 i	7.12 ± 1.08 b	7.12 ± 1.08 b
Konjski zub	3.30 ± 0.26 h	1.52 ± 0.11 f–i	2.18 ± 0.22 hi	2.50 ± 0.17 i	2.36 ± 0.17 ab	1.07 ± 0.10 k	4.18 ± 0.48 cde	7.81 ± 0.43 c–f	0.86 ± 0.04 d–g	6.21 ± 0.89 de	6.21 ± 0.82 f
Kristal	3.61 ± 0.37 efg	1.58 ± 0.17 efg	2.30 ± 0.27 efg	2.57 ± 0.29 hi	2.42 ± 0.31 a	1.07 ± 0.08 k	4.43 ± 0.96 bc	8.00 ± 0.85 bc	0.86 ± 0.04 d–g	6.30 ± 0.52 de	6.30 ± 0.52 ef
Medunac	4.28 ± 0.29 a	1.92 ± 0.20 ab	2.25 ± 0.22 fgh	2.77 ± 0.22 def	2.10 ± 0.20 e–h	1.33 ± 0.10 fg	4.15 ± 0.59 cde	7.75 ± 0.58 d–g	0.87 ± 0.06 b–e	8.17 ± 0.79 a	8.11 ± 0.76 a
Mojdiški sitnozrni	3.79 ± 0.20 c–g	1.65 ± 0.10 de	2.30 ± 0.26 efg	2.73 ± 0.36 ef	2.10 ± 0.20 e–h	1.30 ± 0.09 gh	4.06 ± 0.70 de	7.66 ± 0.35 fgh	0.86 ± 0.04 d–g	6.05 ± 0.56 e	6.06 ± 0.57 f
Pastun	4.02 ± 0.54 abc	1.71 ± 0.18 cd	2.37 ± 0.37 c–f	2.93 ± 0.19 ac	2.09 ± 0.17 gh	1.41 ± 0.12 d	4.19 ± 0.50 cde	7.83 ± 0.49 cde	0.86 ± 0.04 de	6.68 ± 0.87 c	6.67 ± 0.85 cde
Sladun	3.84 ± 0.25 c–f	1.64 ± 0.17 de	2.37 ± 0.25 def	2.59 ± 0.27 hi	1.92 ± 0.23 i	1.36 ± 0.11 ef	3.42 ± 0.66 g	7.06 ± 0.69 j	0.86 ± 0.05 efg	6.69 ± 0.91 c	6.72 ± 0.98 bcd
Slatki crveni	4.09 ± 0.37 ab	1.67 ± 0.21 d	2.48 ± 0.34 b	2.86 ± 0.26 bcd	2.06 ± 0.18 h	1.39 ± 0.09 de	4.15 ± 0.64 cde	7.84 ± 0.66 cde	0.84 ± 0.04 fgh	6.80 ± 1.02 bc	6.81 ± 1.03 bcd
Slatki tankokorac	3.70 ± 0.30 d–g	1.51 ± 0.16 ghi	2.47 ± 0.23 bcd	2.53 ± 0.26 i	2.09 ± 0.24 e–h	1.22 ± 0.14 j	3.71 ± 0.70 fg	7.42 ± 0.92 i	0.85 ± 0.08 fgh	6.67 ± 0.70 cd	6.69 ± 0.73 cde
Šerbetaš	3.61 ± 0.34 d–h	1.49 ± 0.15 i	2.46 ± 0.24 bc	2.60 ± 0.26 ghi	2.12 ± 0.20 e–h	1.23 ± 0.08 ij	3.96 ± 0.67 ef	7.54 ± 0.66 ghi	0.87 ± 0.04 bcd	6.68 ± 0.91 c	6.68 ± 0.91 cde
Zamorac	3.95 ± 0.34 a–d	1.89 ± 0.20 b	2.10 ± 0.19 i	2.99 ± 0.25 a	2.06 ± 0.19 h	1.46 ± 0.11 bc	4.28 ± 0.67 cd	7.95 ± 0.61 bcd	0.85 ± 0.04 efg	7.75 ± 0.91 a	7.67 ± 0.93 a
Hicaznar	3.72 ± 0.35 c–g	1.56 ± 0.21 e–i	2.40 ± 0.22 b–f	2.95 ± 0.24 abc	2.34 ± 0.22 abc	1.27 ± 0.06 g–j	5.05 ± 0.80 a	8.49 ± 0.75 a	0.88 ± 0.03 abc	6.22 ± 0.65 de	6.22 ± 0.65 ef
Granada	3.60 ± 0.33 g	1.48 ± 0.20 hi	2.45 ± 0.25 bcd	2.71 ± 0.19 e–h	2.21 ± 0.16 c–f	1.23 ± 0.08 hij	4.38 ± 0.61 bcd	7.86 ± 0.57 c–f	0.89 ± 0.03 a	6.05 ± 0.57 e	6.07 ± 0.59 f
Wonderful	3.53 ± 0.34 gh	1.71 ± 0.16 cd	2.09 ± 0.24 i	2.68 ± 0.23 fgh	2.18 ± 0.20 d–g	1.23 ± 0.08 hij	4.24 ± 0.69 b,e	7.72 ± 0.58 e–h	0.89 ± 0.03 a	6.30 ± 0.64 de	6.42 ± 0.85 def
Min.	3.30	1.36	1.96	2.49	1.51	1.07	2.68	6.50	0.80	6.05	6.06
Max.	4.28	1.98	2.79	2.99	2.42	1.65	5.05	8.49	0.89	7.83	7.83
Mean	3.79	1.65	2.34	2.73	2.10	1.31	4.06	7.68	0.86	6.84	6.85
SD	0.23	0.17	0.19	0.16	0.21	0.14	0.52	0.43	0.02	0.66	0.67
CV%	6.00	10.05	8.25	5.74	10.04	10.99	12.81	5.59	2.90	9.57	9.35

Values are given as mean ± SD, n = 100. Different lower-case letters in each column indicate a significant difference between cultivars at $p \leq 0.05$ by the Tukey's test.

Table 5. Fruit morphological characteristics of the cultivars studied.

Cultivar	Fruit Weight (FWg; g)	Fruit Length (FL; mm)	Fruit Diameter (FD; mm)	FL/FD Ratio	Fruit Form Index (IFF; %)	Crown Length (CrL; mm)	Crown Index (IC; %)	Peel Thickness (PT; mm)	Peel Thickness Index (IPT; %)	Peel Weight (PWg; g)
Barski slatki	440.50 ± 93.33 abc	84.94 ± 6.89 a	95.76 ± 6.73 ab	0.89 ± 0.07 ab	113.1 ± 8.6 cde	16.35 ± 2.95 abc	16.2 ± 2.6 bcd	4.79 ± 1.22 b	5.0 ± 1.3 bc	207.73 ± 53.42 cd
Bokežan	347.23 ± 61.06 b–f	76.26 ± 4.83 a–e	90.32 ± 5.66 abc	0.84 ± 0.03 abc	118.5 ± 4.9 b	16.62 ± 3.48 abc	17.9 ± 3.4 ab	4.31 ± 1.55 bc	4.8 ± 1.7 c	197.77 ± 47.72 c–f
Dividiš	474.86 ± 161.94 ab	83.40 ± 7.63 ab	98.19 ± 9.32 a	0.85 ± 0.05 abc	117.9 ± 6.9 b	17.31 ± 2.72 abc	17.3 ± 2.9 ab	4.45 ± 0.98 b	4.6 ± 1.0 c	259.88 ± 109.85 a
Domaći kiseli	389.44 ± 118.81 a–f	78.54 ± 8.24 a–e	90.92 ± 7.58 abc	0.86 ± 0.04 abc	116.1 ± 5.9 bcd	15.89 ± 2.27 bcd	17.0 ± 3.0 ab	4.49 ± 1.25 b	4.9 ± 1.2 bc	227.11 ± 66.58 a–d
Dubrovački kasni	262.53 ± 41.72 f	69.40 ± 5.14 e	80.28 ± 4.69 d	0.86 ± 0.05 abc	116.0 ± 6.1 bcd	16.07 ± 3.69 a–d	18.8 ± 4.1 a	3.97 ± 1.08 bc	5.0 ± 1.4 bc	126.68 ± 41.16 hi
Glavaš	392.01 ± 93.63 a–f	74.62 ± 7.14 b–e	91.84 ± 7.64 abc	0.81 ± 0.04 c	123.4 ± 6.7 a	16.20 ± 3.86 a–d	17.9 ± 34.0 ab	4.32 ± 1.25 bc	4.7 ± 1.3 c	219.17 ± 60.47 a–d
Konjski zub	404.60 ± 93.95 a–e	78.22 ± 5.94 a–e	93.38 ± 7.34 ab	0.84 ± 0.06 abc	119.6 ± 8.1 ab	15.54 ± 2.29 cd	16.6 ± 2.2 bc	4.37 ± 0.97 b	4.7 ± 1.0 c	154.05 ± 45.98 fgh
Kristal	326.66 ± 52.69 c–f	73.00 ± 4.66 cde	86.84 ± 5.71 bcd	0.84 ± 0.04 abc	119.1 ± 5.5 ab	15.43 ± 4.63 cd	17.4 ± 4.7 ab	5.05 ± 0.69 ab	5.9 ± 0.9 ab	134.18 ± 32.05 ghi
Medunac	261.90 ± 38.50 ef	68.30 ± 4.84 e	81.64 ± 4.56 cd	0.84 ± 0.03 abc	119.7 ± 4.4 ab	16.09 ± 2.66 a–d	19.0 ± 2.7 a	4.31 ± 0.91 bc	5.3 ± 1.0 abc	133.14 ± 24.47 ghi
Mojdiški sitnozrni	277.26 ± 49.72 def	69.74 ± 7.77 de	80.60 ± 9.91 cd	0.87 ± 0.03 abc	115.5 ± 4.1 bcd	12.77 ± 2.14 ef	15.6 ± 3.2 b	3.79 ± 0.98 bc	4.7 ± 0.9 c	97.82 ± 34.23 i
Pastun	491.83 ± 148.79 a	85.19 ± 9.16 a	95.04 ± 8.93 ab	0.90 ± 0.06 ab	111.9 ± 7.3 de	17.33 ± 2.63 abc	17.0 ± 2.7 ab	4.69 ± 0.45 ab	5.0 ± 0.7 bc	245.59 ± 74.21 ab
Sladun	430.67 ± 122.21 abc	81.69 ± 7.65 abc	96.74 ± 7.25 a	0.84 ± 0.04 bc	118.8 ± 6.0 b	14.40 ± 2.35 de	15.0 ± 2.0 cd	5.86 ± 1.10 a	6.1 ± 1.3 a	251.53 ± 67.14 ab
Slatki crveni	446.59 ± 76.74 abc	84.96 ± 8.22 a	94.88 ± 6.42 ab	0.90 ± 0.07 a	112.2 ± 7.8 de	17.81 ± 4.51 a	17.2 ± 3.6 ab	4.16 ± 1.31 b	4.4 ± 1.4 c	198.11 ± 50.28 de

Table 5. *Cont.*

Cultivar	Fruit Weight (FWg; g)	Fruit Length (FL; mm)	Fruit Diameter (FD; mm)	FL/FD Ratio	Fruit Form Index (IFF; %)	Crown Length (CrL; mm)	Crown Index (IC; %)	Peel Thickness (PT; mm)	Peel Thickness Index (IPT; %)	Peel Weight (PWg; g)
Slatki tankokorac	497.97 ± 175.62 a	84.54 ± 10.23 ab	98.63 ± 8.91 a	0.86 ± 0.04 abc	117.2 ± 6.0 bc	17.98 ± 1.68 ab	17.7 ± 2.1 ab	4.63 ± 1.06 ab	4.7 ± 1.0 c	215.19 ± 93.09 bcd
Šerbetaš	413.28 ± 117.93 a–d	81.62 ± 7.52 abc	92.10 ± 9.05 ab	0.89 ± 0.04 ab	112.9 ± 6.0 cde	14.33 ± 2.10 de	15.0 ± 2.0 cd	2.85 ± 1.27 c	3.1 ± 1.3 d	184.46 ± 61.71 def
Zamorac	304.50 ± 89.12 c–f	69.36 ± 7.96 de	78.05 ± 7.43 d	0.89 ± 0.05 abc	112.9 ± 6.1 cde	11.22 ± 2.52 f	14.0 ± 2.9 d	3.41 ± 0.60 bc	4.4 ± 1.1 c	112.26 ± 36.19 hi
Hicaznar	462.65 ± 101.71 abc	84.71 ± 6.21 ab	95.17 ± 7.47 ab	0.89 ± 0.06 ab	112.6 ± 7.8 cde	17.05 ± 2.01 abc	16.8 ± 2.2 ab	4.63 ± 1.54 ab	4.9 ± 1.7 bc	239.53 ± 68.93 abc
Granada	378.47 ± 91.50 a–f	79.94 ± 11.99 a–d	88.72 ± 7.25 bcd	0.90 ± 0.12 a	109.4 ± 10.2 e	16.84 ± 2.09 abc	17.1 ± 2.3 ab	4.05 ± 1.71 bc	4.6 ± 2.1 c	171.09 ± 40.88 efg
Wonderful	392.76 ± 83.05 a–f	78.56 ± 6.57 a–e	91.87 ± 7.07 ab	0.86 ± 0.04 abc	117.1 ± 6.1 bc	16.96 ± 3.14 abc	17.7 ± 2.8 ab	4.23 ± 0.97 b	4.6 ± 1.0 c	221.08 ± 64.90 a–d
Min.	261.90	68.30	78.05	0.81	109.4	11.22	14.0	2.85	3.1	97.82
Max.	497.97	85.19	98.63	0.90	123.4	17.98	19.0	5.86	6.1	259.88
Mean	389.25	78.26	90.58	0.86	116.0	15.90	16.9	4.33	4.8	189.28
SD	75.03	5.99	6.33	0.03	3.54	1.72	1.29	0.62	0.61	50.03
CV (%)	19.27	7.65	6.99	3.07	3.05	10.79	7.62	14.38	12.60	26.43

Values are given as mean ± SD, n = 36. Different lower-case letters in each column indicate significant differences between cultivars at $p \leq 0.05$ by Tukey's test.

The results of the analysis of variance showed significant differences in the chromatic values of the color parameters L*, a*, b*, C*, and h° for the peel color attributes between the cultivars (Table 6). Peel color varied significantly ($p \leq 0.05$), with the highest coefficient of variation for the values a* and h° being 32.34% and 28.56%, respectively.

The fruit color is an important characteristic, especially for consumer preference. The lightness of the peel (L*) varied between 44.06 and 66.86. The cultivars 'Konjski zub', 'Šerbetaš', 'Dividiš', 'Sladun', and 'Dubrovački kasni' had the lightest peel colors, while the peels of 'Wonderful', 'Granada', and 'Bokežan' were the darkest (Figure 1). The a* value ranges from negative values for green to positive values for red, while the b* value ranges from negative values for blue to positive values for yellow. In our study, the a* value represents the red color of the peel and ranges from 11.97 to 47.53, while b* represents the yellow color of the peel and ranges from 22.52 to 34.97. For the introduced cultivars 'Hicaznar', 'Granada', and 'Wonderful' as well as the native cultivar 'Bokežan', the red color of the peel dominates, while for the other native cultivars, the yellow color of the peel dominates. In addition, high variability in peel color within cultivars was observed in all native cultivars, except 'Bokežan', with the coefficient of variation ranging from 36.45% ('Barski slatki') to 94.42% ('Mojdiški sitnozrni'). Chroma (C*) describes the color intensity and hue angle (h°) describes the visual color impression. The cultivars with the higher red coloration had the highest C*, while the cultivars with the lighter peel had the highest h°.

Figure 1. Visual representation of the studied cultivars.

Table 6. Peel color CIE L*a*b* parameters of the studied cultivars.

Cultivar	Fruit Peel Color					Fruit Peel Color Index; FCI
	L*	a*	b*	C*	h°	
Barski slatki	54.57 ± 8.60 gh	38.04 ± 13.87 bcd	29.25 ± 3.88 cd	49.35 ± 8.36 bcd	39.79 ± 14.35 de	1.35 ± 0.15 cd
Bokežan	47.89 ± 8.01 ij	47.53 ± 6.99 a	23.99 ± 3.29 fg	53.43 ± 6.27 a	27.16 ± 5.41 f	1.51 ± 0.12 ab
Dividiš	64.78 ± 12.49 a–d	28.28 ± 17.94 ef	32.48 ± 5.11 ab	46.10 ± 7.97 def	52.52 ± 21.63 bc	1.16 ± 0.25 efg
Domaći kiseli	59.80 ± 12.57 c–g	35.22 ± 15.56 cde	29.71 ± 5.24 bcd	48.06 ± 8.55 bcd	42.91 ± 17.42 cd	1.28 ± 0.22 de
Dubrovački kasni	64.10 ± 5.16 a–d	19.01 ± 13.75 gh	34.97 ± 6.18 a	42.32 ± 3.06 fgh	61.82 ± 20.11 ab	1.12 ± 0.23 g
Glavaš	56.62 ± 8.27 e–h	35.90 ± 13.14 cde	27.95 ± 3.50 de	46.77 ± 8.13 cde	40.13 ± 14.11 de	1.36 ± 0.15 cd
Konjski zub	66.86 ± 4.49 ab	16.73 ± 8.98 gh	32.00 ± 3.98 abc	37.28 ± 2.18 ij	62.77 ± 14.87 ab	1.13 ± 0.18 g
Kristal	60.94 ± 13.81 c–f	25.66 ± 15.32 fg	26.74 ± 7.97 de	40.46 ± 4.69 ghi	48.81 ± 24.40 cd	1.33 ± 0.37 d

Table 6. *Cont.*

Cultivar	Fruit Peel Color					Fruit Peel Color Index; FCI
	L*	a*	b*	C*	h°	
Medunac	62.64 ± 9.42 a–d	28.67 ± 17.01 ef	33.81 ± 5.26 a	47.11 ± 7.08 bcd	52.50 ± 20.43 bcd	1.17 ± 0.22 efg
Mojdiški sitnozrni	61.60 ± 5.25 b–f	11.97 ± 11.30 h	31.96 ± 3.99 abc	36.02 ± 1.97 j	69.72 ± 19.20 a	1.14 ± 0.25 fg
Pastun	56.30 ± 12.70 fgh	33.49 ± 15.42 def	28.43 ± 8.21 d	46.59 ± 7.81 de	42.69 ± 19.80 cd	1.36 ± 0.31 cd
Sladun	64.32 ± 7.03 a–d	31.88 ± 14.30 def	32.21 ± 4.73 abc	47.16 ± 6.90 bcd	47.53 ± 16.65 cd	1.19 ± 0.15 ef
Slatki crveni	59.25 ± 11.33 d–g	31.85 ± 13.23 def	25.73 ± 6.47 ef	42.91 ± 6.55 efg	41.27 ± 18.47 d	1.38 ± 0.27 bcd
Slatki tankokorac	54.25 ± 11.37 gh	35.48 ± 13.76 cde	28.19 ± 4.78 de	46.94 ± 7.39 bcd	40.84 ± 16.03 de	1.39 ± 0.22 bcd
Šerbetaš	65.24 ± 5.94 abc	19.61 ± 12.60 gh	31.85 ± 5.65 abc	39.65 ± 2.84 hij	59.22 ± 19.82 b	1.16 ± 0.23 efg
Zamorac	64.30 ± 7.13 a–d	31.86 ± 13.65 cde	32.19 ± 4.52 abc	46.89 ± 7.90 cde	47.45 ± 16.57 cd	1.19 ± 0.14 ef
Hicaznar	52.12 ± 5.45 hi	42.43 ± 8.28 abc	24.48 ± 2.30 fg	49.21 ± 7.37 bcd	30.72 ± 5.80 ef	1.47 ± 0.06 bc
Granada	44.28 ± 5.45 j	45.34 ± 4.05 ab	22.80 ± 3.09 fg	50.81 ± 4.49 ab	26.66 ± 2.82 f	1.62 ± 0.14 a
Wonderful	44.06 ± 6.73 j	45.07 ± 4.16 ab	22.52 ± 3.19 g	50.49 ± 4.01 abc	26.60 ± 4.00 f	1.63 ± 0.16 a
Min.	44.06	11.97	22.52	36.02	26.60	1.12
Max.	66.86	47.53	34.97	53.43	69.72	1.63
Mean	57.76	31.79	28.84	45.59	45.20	1.31
SD	7.08	10.28	3.84	4.82	12.91	0.16
CV (%)	12.26	32.34	13.31	10.57	28.56	12.21

Values are given as mean ± SD, n = 9. Different lowercase letters in each column indicate a significant difference between cultivars at $p \leq 0.05$ by Tukey's test. Abbreviations: L*—lightness, a*—red–green color spectrum, b*—yellow–blue color spectrum, C*—Chroma, and h°—hue angle.

3.1.4. Aril

The significant differences between aril characteristics and the high variability (>20%) between cultivars for total aril weight, aril weight, number of arils per fruit, seed length, seed width, and seed yield are shown in Table 7.

The arils contain the edible part of the pomegranate fruit, which contains the juice and the seed. The total aril weight varied between 128.76 g ('Medunac') and 282.78 g ('Slatki tankokrac'). Aril yield was highest in 'Mojdiški sitnozrni' (65.49%) compared to all studied cultivars except 'Konjski zub', and 'Zamorac', while it was lowest in 'Sladun' (40.92%).

The weight of the individual aril was between 0.27 g and 0.59 g. The cultivars 'Kristal' (0.59 g), 'Dividiš' (0.58 g), and 'Konjski zub' (0.55 g) had the highest values and showed no difference to 'Slatki tankokorac' (0.52 g). In addition, 'Kristal' had the longest and widest aril (14.58 mm and 10.05 mm), while 'Wonderful' and 'Granada' had the shortest and narrowest (10.34 mm and 7.26 mm and 10.53 and 7.87 mm, respectively). The number of arils in the fruit varied between 336 and 692 in the different cultivars.

The cultivars also differed in seed characteristics with a coefficient of variation of 23.71, 23.77, and 20.27% for seed length, seed width, and seed yield, respectively. The longest seeds had 'Konjski zub' (8.16 mm) and 'Kristal' (8.10 mm) and the shortest 'Domaći kiseli' (6.58 mm) of all other cultivars except for 'Dubrovački kasni'. The widest seed was 'Dividiš' (3.86 mm) compared to all cultivars except for 'Konjski zub', while the narrowest was 'Mojdiški sitnozrni' (2.83 mm). Seed yield ranged from 4.29% to 8.47% and the heaviest seed was 'Zamorac' (0.043 g). In our study, 63% of the cultivars had hard seeds and 37% had medium seeds (Table 2).

The mean values of the aril and juice color parameters (L*, a*, b*, C*, and h°) and significant differences of studied cultivars are shown in Tables 8 and 9.

Table 7. Aril morphological characteristics of the studied cultivars.

Cultivar	Total Arils Weight (TAW; g)	Arils Yield (AY; %)	Aril Weight (AWg; g)	Aril Length (AL; mm)	Aril Width (AW; mm)	Number of Arils in the Fruit (No/AF)	Seed Length (SL; mm)	Seed Width (SW; mm)	Seed Weight (SWg; g)	Seed Yield (SY; %)
Barski slatki	232.62 ± 52.09 a–d	52.89 ± 5.32 ef	0.36 ± 0.08 c–f	11.81 ± 0.90 ef	8.34 ± 0.86 ef	653 ± 158 ab	7.56 ± 0.58 cd	3.67 ± 0.37 b	0.031 ± 0.008 c	8.06 ± 1.63 ab
Bokežan	149.46 ± 29.17 fg	43.34 ± 6.09 jk	0.31 ± 0.03 efg	10.73 ± 0.76 ij	8.14 ± 0.67 fg	482 ± 109 cde	6.89 ± 0.48 g	3.38 ± 0.29 fgh	0.027 ± 0.003 e–h	8.12 ± 1.40 ab
Dividiš	214.98 ± 70.68 a–f	46.01 ± 12.23 hij	0.58 ± 0.12 a	13.62 ± 1.33 b	9.86 ± 1.14 a	410 ± 199 def	7.73 ± 0.76 bc	3.86 ± 0.41 a	0.026 ± 0.006 f–i	4.65 ± 1.79 g
Domaći kiseli	168.99 ± 55.41 d–g	43.41 ± 9.45 jk	0.31 ± 0.06 fg	11.00 ± 0.58 hi	8.49 ± 0.56 de	518 ± 221 cd	6.58 ± 0.45 h	3.27 ± 0.31 h	0.028 ± 0.003 b–g	8.47 ± 1.58 a
Dubrovački kasni	135.85 ± 30.17 g	52.24 ± 9.56 efg	0.36 ± 0.04 c–g	11.96 ± 0.82 c–f	9.00 ± 0.90 c	383 ± 97 ef	6.87 ± 0.65 fgh	3.44 ± 0.36 d–h	0.025 ± 0.003 f–i	6.81 ± 1.11 c–f
Glavaš	172.84 ± 40.81 c–g	44.45 ± 5.51 ijk	0.37 ± 0.07 c–f	11.83 ± 0.73 ef	8.66 ± 0.84 cd	463 ± 129 c–f	7.72 ± 0.60 bc	3.61 ± 0.35 bcd	0.030 ± 0.007 bcd	7.91 ± 2.30 a–d
Konjski zub	250.55 ± 60.71 ab	62.00 ± 6.84 ab	0.55 ± 0.08 a	13.48 ± 1.53 b	9.49 ± 1.09 b	467 ± 146 cde	8.16 ± 0.83 a	3.67 ± 0.48 abc	0.026 ± 0.006 f–i	4.85 ± 1.19 g
Kristal	192.47 ± 38.39 b–f	58.98 ± 7.71 bc	0.59 ± 0.09 a	14.58 ± 1.02 a	10.05 ± 0.88 a	336 ± 93 f	8.10 ± 0.82 a	3.53 ± 0.47 c–f	0.023 ± 0.006 ij	4.29 ± 1.74 g
Medunac	128.76 ± 21.14 g	49.17 ± 4.30 fghi	0.33 ± 0.07 d–g	11.59 ± 0.90 fg	8.96 ± 0.98 c	420 ± 145 def	6.89 ± 0.65 fg	3.41 ± 0.39 fgh	0.023 ± 0.002 hij	7.52 ± 1.05 a–e
Mojdiški sitnozrni	179.44 ± 22.75 b–g	65.49 ± 6.33 a	0.32 ± 0.06 d–g	12.46 ± 0.49 c	8.94 ± 0.65 c	587 ± 123 abc	7.16 ± 0.49 ef	2.83 ± 0.56 i	0.032 ± 0.008 bc	6.36 ± 1.07 ef
Pastun	246.24 ± 79.50 abc	49.93 ± 3.62 fgh	0.39 ± 0.09 c–f	12.24 ± 1.12 cd	8.89 ± 0.97 c	665 ± 228 ab	7.74 ± 0.65 bc	3.60 ± 0.37 b–e	0.030 ± 0.005 b–e	7.88 ± 1.70 abc
Sladun	179.14 ± 75.42 efg	40.92 ± 6.73 k	0.42 ± 0.11 cd	11.30 ± 0.86 gh	8.74 ± 0.90 cd	428 ± 123 def	6.93 ± 0.41 g	3.62 ± 0.27 bc	0.027 ± 0.002 def	6.96 ± 0.85 ce
Slatki crveni	248.49 ± 49.09 ab	55.86 ± 8.35 cde	0.40 ± 0.09 cde	12.01 ± 0.89 de	8.78 ± 0.86 cd	615 ± 143 ab	7.44 ± 0.60 d	3.41 ± 0.33 fgh	0.025 ± 0.005 f–i	5.89 ± 1.03 f
Slatki tankokorac	282.78 ± 89.15 a	57.31 ± 5.24 bcd	0.52 ± 0.12 ab	13.60 ± 1.24 b	9.51 ± 0.87 b	558 ± 228 bc	7.82 ± 0.71 b	3.37 ± 0.38 gh	0.024 ± 0.003 g–j	4.81 ± 1.33 g
Šerbetaš	228.81 ± 64.82 a–e	55.61 ± 5.90 cde	0.45 ± 0.05 bc	12.06 ± 1.29 cde	8.86 ± 1.15 c	483 ± 193 cd	7.46 ± 0.52 cde	3.74 ± 0.33 ab	0.027 ± 0.003 d–g	5.99 ± 0.51 f
Zamorac	192.24 ± 67.73 a–g	62.62 ± 7.31 ab	0.35 ± 0.04 c–g	12.20 ± 1.10 cd	8.88 ± 0.95 c	579 ± 236 abc	7.43 ± 0.51 d	3.61 ± 0.34 bcd	0.043 ± 0.001 a	5.57 ± 0.79 fg
Hicaznar	223.12 ± 54.03 a–f	48.59 ± 6.86 ghi	0.35 ± 0.08 c–g	11.24 ± 0.71 gh	7.39 ± 0.71 h	659 ± 177 ab	7.56 ± 0.46 cd	3.48 ± 0.24 d–g	0.024 ± 0.004 hij	6.76 ± 1.73 ef
Granada	207.39 ± 63.62 b–f	54.28 ± 6.67 de	0.32 ± 0.12 fg	10.53 ± 0.79 jk	7.87 ± 0.71 g	692 ± 174 a	7.72 ± 0.48 bc	3.46 ± 0.33 efg	0.021 ± 0.005 j	7.20 ± 2.68 b–e
Wonderful	171.68 ± 31.47 d–g	44.41 ± 7.08 ijk	0.27 ± 0.04 g	10.34 ± 0.86 k	7.26 ± 0.79 h	641 ± 147 ab	7.17 ± 0.51 e	3.39 ± 0.32 gh	0.023 ± 0.005 ij	8.23 ± 1.92 a
Min.	128.76	40.92	0.27	10.24	7.26	336	6.58	2.83	0.021	4.29
Max.	282.78	65.49	0.59	14.58	10.05	692	8.16	3.86	0.043	8.47
Mean	200.31	51.97	0.40	12.02	8.71	528	7.42	3.49	0.03	6.65
SD	42.05	7.27	0.10	1.17	0.79	108	1.76	0.83	0	1.35
CV (%)	20.99	13.99	24.32	9.70	9.12	20.51	23.71	23.77	17.99	20.27

Values are given as mean ± SD, n = 120. Different lower-case letters in each column indicate a significant difference between the cultivars at $p \leq 0.05$ by the Tukey's test.

Table 8. CIE L*a*b* aril color parameters of the studied cultivars.

Cultivar	Aril Color					Aril Color Index; ACI
	L*	a*	b*	C*	h°	
Barski slatki	33.61 ± 2.41 abc	17.60 ± 2.36 def	6.87 ± 1.64 bc	18.91 ± 2.73 d–g	21.09 ± 2.71 bc	3.05 ± 0.34 c
Bokežan	41.88 ± 3.72 fg	14.45 ± 4.29 a–d	9.45 ± 1.94 f–i	17.50 ± 3.72 c–f	34.36 ± 9.94 ef	2.47 ± 0.29 i
Dividiš	46.68 ± 3.78 h	12.47 ± 4.50 a	10.49 ± 1.29 ij	16.60 ± 3.38 cde	41.92 ± 11.47 g	2.19 ± 0.21 l
Domaći kiseli	44.22 ± 5.01 gh	13.38 ± 4.49 ab	9.69 ± 1.70 g–j	16.83 ± 3.51 cde	37.82 ± 12.55 fg	2.35 ± 0.33 k
Dubrovački kasni	35.55 ± 2.28 cd	18.39 ± 1.47 f	7.51 ± 1.08 cd	19.88 ± 1.62 fg	22.16 ± 2.45 bcd	2.86 ± 0.20 de
Glavaš	41.25 ± 4.53 f	14.02 ± 4.28 abc	10.38 ± 2.00 hij	17.77 ± 3.23 c–g	37.97 ± 11.82 fg	2.45 ± 0.33 jk
Konjski zub	34.77 ± 2.31 bcd	18.63 ± 3.30 f	8.22 ± 1.70 c–f	20.39 ± 3.54 g	23.87 ± 3.38 bcd	3.34 ± 0.40 b
Kristal	40.08 ± 4.64 f	11.91 ± 5.09 a	9.28 ± 1.80 f–i	15.60 ± 3.61 bc	40.62 ± 15.70 fg	3.77 ± 0.30 a
Medunac	40.14 ± 2.59 f	17.71 ± 1.94 ef	7.79 ± 1.22 cde	19.38 ± 1.97 efg	23.78 ± 3.48 bcd	2.84 ± 0.21 de
Mojdiški sitnozrni	41.29 ± 3.22 fg	14.61 ± 2.99 a–e	7.46 ± 1.38 cd	16.51 ± 2.72 cd	27.61 ± 6.58 cde	2.53 ± 0.40 hij
Pastun	41.16 ± 5.05 f	14.73 ± 6.23 a–e	11.13 ± 1.60 j	18.92 ± 4.92 d–g	40.03 ± 13.74 fg	2.64 ± 0.20 fgh
Sladun	39.62 ± 3.32 ef	16.35 ± 2.69 b–f	9.00 ± 1.62 e–h	18.74 ± 2.65 d–g	29.05 ± 5.36 de	2.65 ± 0.25 fgh
Slatki crveni	37.10 ± 1.89 de	16.33 ± 3.55 b–f	8.42 ± 2.05 d–g	18.43 ± 3.85 c–g	27.42 ± 4.49 cde	2.34 ± 0.31 k
Slatki tankokorac	37.06 ± 2.55 de	18.06 ± 2.71 f	8.25 ± 1.64 c–g	19.89 ± 2.98 fg	24.46 ± 3.26 cd	2.60 ± 0.25 ghi
Šerbetaš	35.79 ± 3.23 cd	16.76 ± 2.68 c–f	7.40 ± 1.83 cd	18.37 ± 2.97 c–g	23.68 ± 4.33 bcd	2.77 ± 0.26 def
Zamorac	41.14 ± 5.07 f	14.70 ± 6.26 a–e	11.10 ± 1.63 j	18.90 ± 4.94 d–g	39.89 ± 13.70 fg	2.33 ± 0.32 k
Granada	31.32 ± 1.99 a	17.31 ± 3.40 def	5.43 ± 2.10 b	18.17 ± 3.85 c–g	16.83 ± 3.32 ab	2.75 ± 0.25 efg
Hicaznar	31.68 ± 1.45 a	13.02 ± 1.90 a	2.67 ± 0.94 a	13.30 ± 2.04 ab	11.25 ± 2.52 a	3.79 ± 0.31 a
Wonderful	32.20 ± 1.42 ab	12.34 ± 1.93 a	2.46 ± 0.93 a	12.59 ± 2.06 a	10.93 ± 2.63 a	2.92 ± 0.35 cd
Min.	31.32	11.91	2.46	12.59	10.93	2.19
Max.	46.68	18.63	11.13	20.39	41.92	3.79
Mean	38.08	15.45	7.88	17.65	27.49	2.77
SD	4.42	2.27	2.39	2.14	9.63	0.45
CV (%)	11.61	14.69	30.32	12.12	35.03	16.25

Values are given as mean ± SD, n = 30. Different lowercase letters in each column indicate a significant difference between cultivars for $p \leq 0.05$ by the Tukey's test. Abbreviations: L*—lightness, a*—red–green color spectrum, b*—yellow–blue color spectrum, C*—Chroma, and h°—hue angle.

Table 9. CIE L*a*b* juice color parameters of the studied cultivars.

Cultivar	Juice Color					Juice Color Index: JCI
	L*	a*	b*	C*	h°	
Barski slatki	20.36 ± 0.61 abc	3.47 ± 0.32 ab	3.44 ± 0.17 ghi	4.89 ± 0.20 bc	44.83 ± 3.42 hij	5.35 ± 0.08 bcd
Bokežan	22.51 ± 0.26 def	7.49 ± 0.34 hij	4.37 ± 0.05 j	8.67 ± 0.31 h	30.27 ± 0.89 def	4.80 ± 0.06 g
Dividiš	26.07 ± 0.24 h	5.90 ± 0.19 efg	1.23 ± 0.04 a	6.03 ± 0.19 cde	11.81 ± 0.08 a	5.24 ± 0.02 d
Domaći kiseli	23.47 ± 0.10 fg	7.82 ± 0.29 ij	2.75 ± 0.07 def	8.29 ± 0.29 gh	19.39 ± 0.26 abc	5.06 ± 0.04 f
Dubrovački kasni	23.24 ± 0.10 efg	6.70 ± 0.14 ghi	2.32 ± 0.11 cd	7.10 ± 0.09 efg	19.07 ± 1.16 abc	5.30 ± 0.01 cde
Glavaš	20.30 ± 0.30 abc	5.08 ± 0.31 def	3.75 ± 0.12 i	6.32 ± 0.29 de	36.49 ± 1.46 fgh	5.39 ± 0.03 bcd
Konjski zub	24.72 ± 0.61 gh	5.59 ± 0.29 d–g	1.39 ± 0.25 ab	5.77 ± 0.24 b–e	13.99 ± 2.88 ab	5.44 ± 0.09 b
Kristal	23.94 ± 0.51 fg	2.27 ± 1.01 a	1.01 ± 0.17 a	2.50 ± 0.97 a	26.15 ± 8.95 cde	5.82 ± 0.26 a
Medunac	24.33 ± 0.33 g	8.24 ± 0.80 j	1.86 ± 0.23 bc	8.45 ± 0.73 gh	12.86 ± 2.80 ab	5.10 ± 0.09 f
Mojdiški sitnozrni	21.78 ± 0.28 cde	6.39 ± 1.05 fgh	2.57 ± 0.07 de	6.89 ± 1.01 ef	22.15 ± 2.54 bcd	5.51 ± 0.05 b
Pastun	19.22 ± 1.50 a	4.33 ± 0.40 bcd	3.16 ± 0.30 fgh	5.35 ± 0.49 bcd	36.14 ± 1.13 fgh	5.86 ± 0.23 a
Sladun	21.61 ± 0.24 b–e	4.94 ± 0.49 cde	3.04 ± 0.16 efg	5.80 ± 0.49 b–e	31.73 ± 1.63 efg	5.41 ± 0.09 bcd
Slatki crveni	20.59 ± 0.38 abc	3.46 ± 0.12 ab	3.04 ± 0.47 efg	4.62 ± 0.26 b	41.17 ± 5.00 ghi	5.50 ± 0.17 b
Slatki tankokorac	23.96 ± 0.76 fg	7.38 ± 0.09 hij	2.38 ± 0.15 cd	7.75 ± 0.05 fgh	17.91 ± 1.21 abc	5.11 ± 0.09 ef
Šerbetaš	21.43 ± 0.38 bcd	4.21 ± 0.20 bcd	2.85 ± 0.24 def	5.10 ± 0.06 bcd	34.06 ± 3.50 efg	5.50 ± 0.06 bc
Zamorac	19.20 ± 1.50 a	4.23 ± 0.30 bcd	3.12 ± 0.30 fgh	5.32 ± 0.50 bcd	36.05 ± 1.15 gfh	5.87 ± 0.22 a
Granada	19.18 ± 0.09 a	2.97 ± 0.31 ab	3.68 ± 0.22 hi	4.73 ± 0.33 bc	51.09 ± 2.35 j	5.39 ± 0.10 bcd
Hicaznar	20.07 ± 0.03 ab	3.33 ± 0.07 ab	3.63 ± 0.02 ghi	4.93 ± 0.05 bc	47.49 ± 0.68 ij	5.31 ± 0.20 d
Wonderful	19.28 ± 0.74 a	3.62 ± 0.26 abc	3.81 ± 0.04 ij	5.26 ± 0.16 bcd	46.54 ± 2.15 ij	5.44 ± 0.22 bc
Min.	19.18	2.27	1.01	2.50	11.81	4.80
Max.	26.07	8.24	4.37	8.67	51.09	5.87
Mean	22.00	5.18	2.79	6.03	30.17	5.36
SD	2.11	1.84	0.96	1.60	12.84	0.25
CV (%)	9.63	35.71	34.03	26.78	42.59	5.39

Values are given as mean ± SD, n = 9. Different lowercase letters in each column indicate a significant difference between the aril color characteristics of the cultivars for $p \leq 0.05$ by Tukey's test. Abbreviations: L*—lightness, a*—red–green color spectrum, b*—yellow–blue color spectrum, C*—Chroma, and h°—hue angle.

The aril colors of 'Granada' and 'Hicaznar' were darker (the lowest L* value) compared to other studied cultivars except for 'Wonderful' and 'Barski slatki', while 'Dividiš' and

'Domaći kiseli' had the lightest aril colors. The red–green color spectra (a*) of all studied cultivars indicated the red aril colors. 'Konjski zub' had the highest red color intensity followed by 'Dubrovački kasni', 'Slatki tankokorac', 'Barski slatki', 'Medunac', 'Sladun', 'Slatki crveni', 'Šerbetaš', and 'Granada'. No significant differences were found between these cultivars. 'Kristal', 'Dividiš', 'Wonderful', and 'Granada' had the lowest values for red. In all of the studied cultivars, the yellow–blue color spectrum (b*) was oriented toward yellow, which was the most pronounced in the cultivars 'Pastun' (11.13), 'Dividiš' (10.49), 'Glavaš' (10.38), and 'Domaći kiseli' (9.69), while 'Wonderful' and 'Hicaznar' had the lowest yellow color intensities (2.46 and 2.67, respectively). The lowest color intensity (C*) of the aril was observed in the cultivars 'Wonderful' (12.59) and 'Hicaznar' (13.30), while the C* of 'Konjski zub' was higher compared to the cultivars 'Bokežan', 'Dividiš', 'Domaći kiseli', 'Kristal', 'Mojdiški sitnozrni', 'Hicaznar', and 'Wonderful' (Table 8). According to the h° value, all cultivars belonged to the red–violet group (10.93–41.92). The h° value was the lowest for dark cultivars ('Granada', 'Hicaznar', and 'Wonderful').

In addition, statistical differences were found between the cultivars in the color parameters of the juice, along with a high degree of variability in the red–green (a*) and yellow–blue (b*) color spectra, color intensity (C*), and h° value (Table 9).

The darkest juice (L*) values were found in 'Granada', 'Zamorac', 'Pastun', and 'Wonderful', and did not differ from 'Hicaznar', 'Barski slatki', 'Glavaš', and 'Slatki crveni', while 'Dividiš' had the lightest juice color compared to the other studied cultivars except for 'Konjski zub'. The a* value varied from 2.27 to 8.24. 'Medunac' had the highest a* value but did not differ from Bokežan, Domaći kiseli, and 'Slatki tankokorac'. The C* value varied between 2.50 and 8.67, while h° was between 11.81 and 51.09. According to the value for h°, all cultivars belonged to the red–violet juice color group.

3.2. Juice Yield, Total Soluble Solids, and Total Acidity Content

The high juice yield and the quality characteristics of the juice are very important properties for producers, breeders, and the processing industry. The juice yield (JY), total soluble solid (TSS; °Brix), total acidity (TA; %), and the TSS/TA ratio of the juice of the cultivars studied are shown in Table 10.

Table 10. Juice yield, total soluble solid, total acidity, and TSS/TA ratio of the studied cultivars.

Cultivar	Juice Yield (JY; %)	Total Soluble Solid (TSS; °Brix)	Total Acidity (TA; %)	TSS/TA
Barski slatki	64.3 ± 13.5 ab	15.9 ± 1.2 ab	0.66 ± 0.06 e	24.13 ± 2.29 d
Bokežan	64.3 ± 9.1 ab	16.0 ± 1.7 ab	2.11 ± 0.71 bc	8.29 ± 2.56 e
Dividiš	69.0 ± 6.1 ab	13.8 ± 1.6 d	1.69 ± 0.28 d	8.37 ± 1.49 e
Domaći kiseli	68.5 ± 8.4 ab	16.1 ± 1.2 ab	2.10 ± 0.37 bc	7.88 ± 1.46 e
Dubrovački kasni	65.9 ± 8.2 ab	16.8 ± 1.4 ab	0.54 ± 0.07 e	31.35 ± 3.30 abc
Glavaš	67.2 ± 8.3 ab	16.0 ± 2.5 ab	2.46 ± 0.29 ab	6.60 ± 1.21 e
Konjski zub	67.1 ± 1.6 ab	15.3 ± 1.1 bcd	0.46 ± 0.08 e	34.30 ± 8.20 a
Kristal	71.8 ± 10.0 ab	15.3 ± 1.6 a–d	0.46 ± 0.07 e	33.74 ± 5.53 a
Medunac	69.3 ± 3.8 ab	15.2 ± 2.3 a–d	0.57 ± 0.17 e	27.96 ± 6.18 bcd
Mojdiški sitnozrni	66.4 ± 5.9 ab	12.8 ± 0.49 d	0.49 ± 0.06 e	26.42 ± 3.89 cd
Pastun	61.9 ± 8.9 b	14.1 ± 1.4 cd	1.81 ± 0.31 cd	8.01 ± 1.55 e
Sladun	65.8 ± 7.7 ab	18.9 ± 1.4 ab	0.66 ± 0.08 e	24.24 ± 2.51 d
Slatki crveni	68.5 ± 10.2 ab	16.6 ± 1.3 ab	0.53 ± 0.08 e	31.81 ± 3.71 ab
Slatki tankokorac	73.7 ± 5.9 a	14.1 ± 1.4 cd	0.61 ± 0.11 e	23.47 ± 4.26 d
Šerbetaš	73.4 ± 2.4 a	17.2 ± 1.3 a	0.64 ± 0.11 e	27.25 ± 3.12 cd
Zamorac	68.3 ± 2.6 ab	15.7 ± 0.8 a–d	2.17 ± 0.29 abc	7.38 ± 1.30 e
Hicaznar	67.5 ± 9.0 ab	16.8 ± 1.9 ab	2.10 ± 0.44 bc	8.33 ± 1.89 e
Granada	68.9 ± 9.9 ab	15.7 ± 2.0 abc	2.50 ± 0.51 a	6.63 ± 1.96 e
Wonderful	63.9 ± 9.5 ab	16.8 ± 1.1 ab	2.15 ± 0.34 abc	7.97 ± 1.19 e
Min.	61.9	12.8	0.46	6.60
Max.	73.7	18.9	2.50	34.30

Table 10. *Cont.*

Cultivar	Juice Yield (JY; %)	Total Soluble Solid (TSS; °Brix)	Total Acidity (TA; %)	TSS/TA
Mean	67.7	15.75	1.30	18.64
SD	3.10	1.39	0.82	11.04
CV (%)	4.59	8.80	63.08	59.21

Values are given as mean ± SD, n = 12. Different lower-case letters in each column indicate a significant difference between cultivars at $p \leq 0.05$ by the Tukey's test.

There were significant differences in the juice yield of the studied cultivars, ranging from 61.86% to 73.69%. 'Slatki tankokorac' and 'Šerbetaš' had higher juice yields than 'Pastun'. The total soluble solid content and total titratable acidity varied significantly between 12.8 °Brix ('Mojdiški sitnozrni') and 18.9 °Brix ('Sladun') and between 0.46% ('Konjski zub' and 'Kristal') and 2.50% ('Granada'). The TSS/TA ratio is also an important parameter and plays a major role in the perception of pomegranate flavor, as it creates a balance between sweetness and acidity and determines the taste. Each pomegranate cultivar requires a specific TSS/TA ratio at the time of harvest. In this study, the TSS/TA ratio varied between 6.60 ('Glavaš') and 34.30 ('Konjski zub').

3.3. Relationship between Different Pomegranate Cultivars and their Morphological and Chemical Characteristics

The use of a heat map allows visualizing complex relationships between 19 pomegranate cultivars, clustered based on 44 characters specifically selected for their coefficient of variation exceeding 15%; this approach helps in identifying patterns and understanding their interrelationships or similarities (Figure 2).

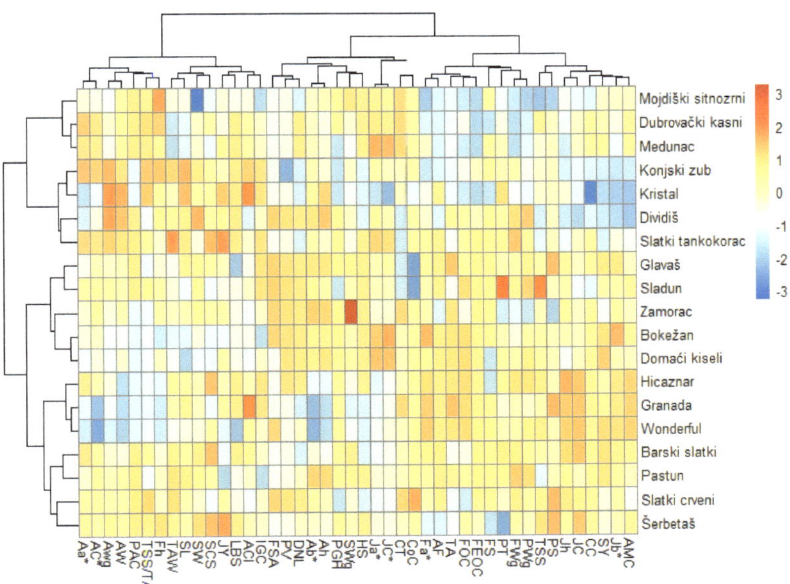

Figure 2. Clustering of pomegranate cultivars based on the morphological parameters of the tree, leaf, flower, fruit, aril, and seed, as well as the chemical parameters of the juice. The rows of the diagram represent the cultivars and the columns denote the analyzed characteristics. For the abbreviations of the analyzed traits, see Tables 1 and 3–10. In addition, the color parameters obtained from the CIE L*a*b* method are represented by the letters L*, a*, b* C*, and h° located behind the label representing the object of analysis, i.e., fruit (F), aril (A) or juice (J). The colors of the cells of the heatmap indicate low (dark blue), medium (white), and high (dark red) values of a particular trait.

The pomegranate cultivars were divided into two clusters, which were further subdivided into subgroups. The first cluster featured the fruit peel h° color value (Fh°) and lower values of the red color of fruit peel (Fa*) and juice b* (Jb*) values. Within this cluster, 'Mojdiški sitonozrni', 'Dubrovački kasni', and 'Medunac' stood out with high red juice color values (Ja*), but also had lower values for fruit and peel weight (FWg and PWg) and fruit extent of overcolor (FEOC). The second sub-cluster was characterized by high aril weight (AWg) and TSS/TA values but low aril main color (AMC), juice b* values, h° values, seed yield (SY), and calix color (CC) values; it consisted of four different cultivars ('Kristal', 'Dividiš', 'Konjski zub', and 'Slatki tankokorac'). The second cluster was generally characterized by higher values for fruit peel red color (Fa*), fruit extent of overcolor (FEOC), juice b* and h° values, and lower values for aril width (AW). A more pronounced aril main color (AMC), fruit peel red color (a*), total acidity (TA), as well as juice h° (Jh°), juice color (JC), and seed yield (SY) characterize the three cultivars ('Wonderful', 'Hicaznar', 'Granada'), forming the sub-cluster in the lower part of the heatmap display. On the other hand, these cultivars have lower values of aril weight (AWg), petiole anthocyanin color (PAC), TSS/TA ratio, and fruit peel h° value. 'Barski slatki', 'Pastun', 'Slatki crveni', and 'Šerbetaš' have higher total aril weight (TAW) and seed length (SL), along with lower juice color (JC) values and intensity of red color in the juice (a*). 'Zamorac', 'Domaći kiseli', and 'Bokežan' are cultivars that, in addition to the cluster characteristics mentioned above, are distinguished by lower petiole anthocyanin coloration and TSS/TA ratio, and a higher predominant number of leaves per node (DNL), aril h°, aril b* value, and seed weight (SWg) (Figure 2).

In addition, the morphological characteristics of the tree, flower, fruit, aril, and seed, as well as the chemical properties of the juice, could be divided into three categories by cluster analysis (Figure 2). The intensity of the red color of the aril (Aa*) was closely related to the weight of the aril (AWg) and the taste characteristic of the juice (TSS/TA). Juice yield (JY) was related to total aril weight (TAW) and seed characteristics (SL, SW, SCS). Tree characteristics, i.e., plant vigor, plant growth habit, and predominant number of leaves per node on young shoots (PV, PGH, and DNL, respectively), were clustered together; they are closely related to corolla color (CoC) and crown type (CT). The acidity of the juice (TA), the red color of the fruit peel (Fa*), fruit overcolor, and fruit extent of overcolor (FOC and FEOC, respectively) were grouped together (Figure 2).

According to the correlation analysis of the analyzed characteristics (Supplementary Materials, Table S1), the fruit weight (FWg) was generally proportional to the total aril weight (TAW), while the juice yield (JY) was proportional to aril weight (AWg). Seed hardness is a genetic characteristic of the cultivar, but consumers prefer soft seeds for fresh consumption, while the processing industry uses cultivars with hard seeds. Seed hardness is significantly negatively correlated with seed length (SL), while it is positively correlated with the intensity of the red color of the juice (Ja*). There is no significant correlation between the intensity of the red color of the fruit peel (Fa*) and the red color of the aril or juice (Aa*, Ja*), while FEOC and FOC are significantly positively correlated with acidity and the red color of the fruit peel (TA and Fa*).

4. Discussion

4.1. Morphological Characteristics

The results of this study showed considerable variability between the cultivars studied in terms of qualitative and quantitative characteristics (Tables 1–10, Figure 1). Narzary et al. [14] found that the leaf sizes varied between 2 and 11 cm in length and 1 and 3 cm in width. According to these data, the cultivars in our study belong to the group of cultivars with medium leaf lengths and widths. A significant difference between cultivars was also observed in the flower characteristics (Table 4). All studied cultivars had long petals, except for 'Bokežan', whose petals were medium-sized. The average leaf length and width of the studied cultivars were similar to Spanish pomegranates [15], while calyx and petals were longer and wider than in Spanish cultivars. Although significant diversity was found among cultivars, the coefficients of variation (CVs) among cultivars were lower, with 6.65–10.97% for leaf characteristics and 2.90–12.81% for flower characteristics, suggesting that these characteristics have less discriminating power.

The morphological characteristics examined showed considerable differences between the cultivars in the ex situ collection. Fruit weight and peel weight showed coefficients of variation of more than 15% (Table 5). Audergon (1987), cited in Mansour et al. [16], considered values between 15 and 20% as medium and above 20% as significant, indicating a large variability related to the studied traits. The fruit weight of the pomegranate is considered the most commonly used measure for identifying some cultivars. Fruit size and peel color are important characteristics that attract the attention of consumers in the market. Consumer preferences show that Indian consumers [17] value small- to medium-sized fruits more, while in Croatia, medium-to-large fruits are preferred [18]. Studies by Parashuram et al. [17] in India, Mansour et al. [16] in Lebanon, Chen et al. [19] in China, Khadivi and Arab [20] in Iran, and Ferrara et al. [21] in Italy found similar values for fruit weight, while Tapia-Campos et al. [22] in Mexico reported slightly lower values for the fruit weights of different pomegranate cultivars. The fruits of the 'Glavaš' cultivar were significantly smaller than in earlier studies [18,23]. The fruit weights of the cultivars 'Barski slatki', 'Konjski zub', 'Sladun', and 'Šerbetaš' were higher (and for 'Dividiš'—lower) than the fruit weights of the same cultivars grown in a production plantation under agroecological conditions in Metković, Croatia [24]. This was most likely due to different environmental conditions, as the study by Ghasemi-Soloklui [25] found that the change in the initial climate of pomegranate cultivars affects the weight of the fruit, aril, and peel. We divided the cultivars into three groups according to fruit weight: cultivars with medium ('Dubrovački kasni', 'Medunac', 'Mojdiški sitnozrni'), large ('Bokežan', 'Domaći kiseli', 'Glavaš', 'Kristal', 'Zamorac', 'Granada', and 'Wonderful'), and very large ('Barski slatki', 'Dividiš', 'Konjski zub', 'Pastun', 'Sladun', 'Slatki crveni', 'Slatki tankokorac', 'Šerbetaš', and 'Hicaznar') fruits. According to the UNECE international standards [26], 9 of the investigated cultivars belong to class B by weight, while 16 cultivars belong to class A by diameter. Fruit size is a genetic characteristic of the cultivar, but can vary considerably depending on climatic conditions, year, and cultivation technique. In the present study, the fruit weight variability was more genetically determined, as the ex situ collection and fruit testing took place at one location. The peel weight, crown type, fruit shape, fruit shape in cross-section, total aril weight, aril weight, number of arils per fruit, seed length and width, seed yield, total acidity, and TSS/TA value were characteristics that had the greatest discriminating power.

The fruit length/diameter ratio of Italian native pomegranates [27] was similar to the FL/diameter ratio in our study. Martinez et al. [28] used the Fruit Form Index (IF), which is the ratio between the equatorial fruit diameter and the fruit length excluding the crown. In our study, the IFF was between 109.4% ('Granada') and 123.4% ('Glavaš'), while in Spain it was between 108.7% and 115.1% [28].

There were significant differences in the shape of the fruit and the shape of the fruit cross-section of the cultivars ($p < 0.001$ and $p = 0.033$, receptively). Most cultivars had an oblate shape (74%), a semi-open crown (47%), and a round to angular shape in the cross-section (42%) (Table 2). The crown index (IC) ranged from 14% ('Zamorac') to 19% ('Medunac'). The cultivars in our study have higher ICs than the cultivars from Morocco (11.7–14.9%) or Spain (15.0–18.9%) [28]. Fruits with a higher IC are prone to crown breakage. We found three types of crowns: closed-convergent sepals (21%), semi-open right sepals (47%), and open divergent sepals (32%) (Table 2). The advantage of a closed crown is that the sepals break less, while an open crown is preferable due to easier disease control. A significant difference was observed in peel weight and thickness (Table 5). The thickness of the peel is influenced by the genotype and the cultivation practices. Peel thickness ranged between 2.85 mm and 5.86 mm, while the index of peel thickness (IPT) varied between 3.1% and 5.9%. This is similar to the Turkish [29] and Spanish cultivars [28]. Consumers generally prefer pomegranate fruit with a thin peel, as this causes less fruit waste and is easy to clean. However, the thinner peel of the fruit dries out faster and the external appearance of the fruit often repels consumers, making it unattractive in the market, even if the quality of the arils has not changed. In a previous study by Radunić et al. [24] conducted in a production orchard in Metković, the total aril weights of 'Barski slatki' and 'Dividiš were higher and weights of 'Konjski zub' and 'Šerbetaš' were lower, indicating that they were under the influence of the environment and the year. The results of the aril yield (Table 7) were consistent with the aril yield results of cultivars in Morocco [28], Iran [30], Italy [27], and Turkey [31]. Cultivars with a higher number of arils in the fruit had a lower individual aril weight. Our results, 336 to 692 arils per fruit (Table 7), are slightly higher than in reports from Oman [32], but significantly higher than reports from Turkey [31]. The number of arils in the fruit depends on the percentage of successfully fertilized ovules [33], as each aril originates from one ovule.

The coefficients of variation for seed length and seed width were 23.71 and 23.77%, respectively (Table 7). The longest seeds had 'Konjski zub' and 'Kristal', and the shortest 'Domaći kiseli' compared to all cultivars, except 'Dubrovački kasni'. The widest seed was 'Dividiš', while the narrowest was 'Mojdiški sitnozrni'. The cultivars differed in seed hardness ($p = 0.032$). A total of 63% had hard seeds and 37% had medium seeds (Table 2) People prefer pomegranate fruits with soft seeds for fresh consumption [19]. In general, the wild genotypes have lower aril yield and higher seed hardness [32]

4.2. Color of the Fruit Peel, Arils, and Juice

The color perceived by the human eye is easily influenced by individual differences and environmental factors. Colorimeters can accurately evaluate colors [34] by using different color spaces and determining a quantitative color value [35]. Although there are many different color spaces, the CIE L*a*b* color space is most commonly used for food because it has a uniform color distribution and the color perception is closest to the human eye [35]. Appearance influences consumer behavior; in particular, the red color and size are considered the most important external quality parameters for pomegranates [8,36]. The color of the pomegranate peel varies from yellow, green, or pink to deep red or indigo, to completely red [37], depending on the characteristics of the genotype, and is influenced by climate, fertilization, irrigation, and many factors during ripening. The pomegranate is known for its attractive color, with high color variability between cultivars from different collections around the world [38]. Color parameters in our study—the yellow–blue (b*) value of aril and juice, the C* value of juice, the red–green (a*) value of fruit and juice, and the h° value of fruit, aril, and juice—were characteristics that had a coefficient of variation above 20 (see Tables 6, 8 and 9). The fruit color is related to the accumulation of chlorophyll, carotenoids, anthocyanins, and other pigments [34].

According to the literature, anthocyanin accumulation in plants is sensitive to environmental conditions such as sunlight, temperature, and altitude, but fruit maturity, canopy position, and cultivar also have significant impacts on some qualitative characteristics of fruit, including color development [39–41]. It is known that high temperatures inhibit the synthesis of anthocyanins. The position of the canopy influences the quality characteristics of the fruit. Shaded fruits have a greener base color than unshaded fruits [42,43], and the results of our study show variability in the color of the peel as well as the arils and juice. According to the color of the peel, the cultivars studied were divided into two groups: cultivars whose fruits were uniformly red–purple-colored over the entire upper surface (coefficient of variation < 20%), which included the commercial cultivars 'Hicaznar', 'Granada', and 'Wonderful' as well as the native cultivar 'Bokežan', as well as a group of cultivars with a dominant yellow color spectrum of the peel and a more or less pronounced overcoloring on the sunny side of the fruit. Based on the overcoloring of the peel, cultivars were divided into five subgroups: the first subgroup: 'Domaći kiseli' and 'Pastun' had a red–purple peel overcolor; the second subgroup: 'Barski slatki', 'Dubrovački kasni', 'Medunac', 'Sladun', and 'Zamorac' had a pink–red peel overcolor; the third subgroup: 'Konjski zub', 'Kristal', and 'Slatki crveni' had a pink peel overcolor; the fourth subgroup:

'Dividiš', 'Glavaš', 'Slatki tankokorac', and 'Šerbetaš' had a red peel overcolor, and the fifth subgroup, 'Mojdiški sitnozrni', had a yellow–green peel color.

While our data on the color characteristics of the juice were consistent with those of Tarantino et al. [44] the values for a* and b* of the peel, aril, and juice were lower in the cultivar 'Wonderful' than in the same cultivars described by Passafiume et al. [45]. Carreno et al. [12] propose a color index for red grapes, which is best suited for evaluating the color of red table grapes and can be used for the objective evaluation of their external color. In the same study, the average color index value was 1.55 for yellow, 2.49 for pink, 3.66 for red, 4.75 for purple, and 5.57 for dark purple. According to our data, a pomegranate cultivar with an optimal commercial color should have a value of 1.50 for the peel, 2.70 for the aril, and 5 for the juice. The significant differences in the color attributes of pomegranate cultivars can be used as indicators of maturity indices for harvest management and classifying harvested fruit into grades.

There was no relationship between the red peel color and the red aril color (Tables 6 and 8 and Supplementary Table S1), which is consistent with other studies [32,37,46]. The 'Granada', 'Hicaznar', 'Wonderful', and 'Bokežan' cultivars had more pronounced red colors and much more attractive appearances, which certainly affected the visual preferences of customers in the markets.

The juice yield in our study was between 73.69% and 61.86% (Table 10) and was higher than in the study by Martinez et al. [28]. The juice yield is influenced by the cultivar, the technology, and the juice extraction method.

4.3. *Juice Yield, Total Soluble Solids, and Total Acidity Content*

The quality of pomegranate fruit can be assessed based on external characteristics such as shape, size, and color [37,47], as well as internal quality traits. Although external characteristics may not always determine the ideal harvest time, crucial internal factors such as aril color, total soluble solid content, and titratable acidity are paramount for optimal harvest maturity [37,47–51]. The chemical composition of the fruit is influenced by the cultivar, growing region, climate, maturity, growing practices, and storage conditions. In our study, the values for total soluble solids ranged from 12.8 °Brix ('Mojdiški sitnozrni') to 18.9 °Brix ('Sladun'), and for acidity from 0.46% ('Konjski zub' and 'Kristal') to 2.50% ('Granada') (Table 10). The coefficient of variance for total acidity (63.08%) indicates a significant difference between the cultivars studied, which serves as a good discriminating feature (Table 10), with the values obtained in line with other studies [27,28,30,49,52]. Chace et al. [48] reported that a TA value $< 1.85\%$, a TSS $\geq 17\%$, and a TSS/TA ratio between 11 and 16 is recommended for the 'Wonderful' cultivar grown in California. In our study, the TA value (2.15%, Table 10) of 'Wonderful' was higher than in previous studies and belonged to the sour cultivar group under our agroecological conditions. According to the classification by Onur and Kaska [53] and Kader [47], 'Barski slatki', 'Dubrovački kasni', 'Konjski zub', 'Kristal', 'Medunac', 'Mojdiški sitnozrni', 'Sladun', 'Slatki crveni', 'Slatki tankokorac', and 'Šerbetaš' belong to the group of sweet cultivars; 'Dividiš' and 'Pastun' belong to the sour–sweet cultivars, while 'Bokežan', 'Domaći kiseli', 'Glavaš', and 'Zamorac', and the introduced cultivars 'Hicaznar', 'Granada', and 'Wonderful' are considered sour cultivars. Our growing area is further north than the growing area of the introduced cultivar, which affects higher overall acidity. The sweet and sour–sweet cultivars are suitable for fresh consumption. The TSS/TA ratio plays an important role in determining fruit quality and ripeness and is a common parameter for determining the quality of pomegranate fruit [30,50]. In our study, the TSS/TA ratio varied between 6.60 ('Glavaš') and 34.30 ('Konjski zub'). We found that nine of the cultivars studied had a TSS/TA ratio of less than 9 (Table 10). Among them, commercial cultivars are introduced that do not achieve a harmonious ratio between TSS and TA under our agroecological conditions. Chace et al. [48] reported that fruits with an MI (TSS/TA ratio) of 12 are better accepted.

Categorizations or groupings of cultivars are usually based on only one characteristic, usually the fruit, seeds, or juice. Looking at the whole picture, including an extremely large number of parameters (44 parameters with medium and high coefficients of variability out of a total of 81 analyzed) led to clearer groupings (Figure 2). The cultivars were divided into two major clusters: (1) cultivars with more pronounced green and yellow colors in the fruit peel, larger arils with lighter colors, and juice; and (2) cultivars with darker red colors in the peel, aril, and juice. Further clusters of cultivars were distinguished as subgroups, which can help in selecting assortments when planning plantations and also in determining their intended use, whether for fresh consumption or processing in the industry.

5. Conclusions

By evaluating the morphological and some fruit chemical characteristics of pomegranate cultivars, significant diversity was determined. Plant vigor, plant growth habit, the predominant number of leaves per node on young shoots, crown type, fruit shape, fruit shape in cross-section, peel weight, total aril weight, aril weight, number of arils per fruit, seed length and width, seed yield, total acidity, TSS/TA ratio, color parameters of the peel, arils, and juice showed high variability, indicating their great discriminating power in determining the diversity of pomegranate cultivars. The database created provides a solid basis for further research to determine the genetic and chemical diversity of pomegranate cultivars and their potential for sustainable use. In addition, the selected number of characteristics needed to describe the cultivars will facilitate easier and faster evaluations in our subsequent studies as well as in other collection plantations.

Supplementary Materials: The following supporting information can be downloaded at https://www.mdpi.com/article/10.3390/horticulturae10060563/s1, Table S1. Correlation matrix of the pomegranate's physical and chemical properties.

Author Contributions: Conceptualization and methodology, M.R.; software and data analysis M.J.Š.; investigation, resources, and data curation, M.R.; writing—original draft preparation and review and editing, M.R., J.G. and M.J.Š.; funding acquisition, M.R. and M.J.Š. All authors have read and agreed to the published version of the manuscript.

Funding: This research was funded by the Croatian National Program for Conservation and Sustainable Use of Plant Genetic Resources for Food and Agriculture 2017–2020 (01-181/2-17), Split-Dalmatia County, (011-352/2-15) and by the project INOMED-2I (09-207/1-23) granted by the European Union—"NextGenerationEU".

Data Availability Statement: The data presented in this study are available in the article.

Acknowledgments: We are thankful to Jakša Rošin, Radojka Plećaš, and Sandra Đirlić for the technical assistance.

Conflicts of Interest: The authors declare no conflicts of interest.

References

1. Evreinoff, V.A. Contribution à l'étude Du Grenadier. *J. Agric. Trop. Bot. Appl.* **1957**, *4*, 124–138. [CrossRef]
2. Teixeira da Silva, J.A.; Rana, T.S.; Narzary, D.; Verma, N.; Meshram, D.T.; Ranade, S.A. Pomegranate Biology And Biotechnology: A Review. *Sci. Hortic.* **2013**, *160*, 85–107. [CrossRef]
3. Sarkhosh, A.; Yavari, A.M.; Zamani, Z. *The Pomegranate: Botany, Production and Uses*; CABI: Wallingford, UK, 2020.
4. Mars, M. La Culture Du Grenadier (*Punica granatum* L.) et Du Figuier (*Ficus carica* L.) En Tunisie. *Cah. Options Méditerranéennes* **1995**, *13*, 85–95.
5. Still, D.W. Pomegranates: A Botanical Perspective. In *Pomegranates: Ancient Roots to Modern Medicine*; Seeram, N., Schulman, R., Heber, D., Eds.; Taylor & Francis: Oxfordshire, UK, 2006; p. 243.
6. Hawkes, J.G. Genetic Conservation of World Crop Plants. *Biol. J. Linn. Soc.* **1991**, *43*, 11–22. [CrossRef]
7. Migicovsky, Z.; Warschefsky, E.; Klein, L.L.; Miller, A.J. Using Living Germplasm Collections to Characterize, Improve, and Conserve Woody Perennials. *Crop Sci.* **2019**, *59*, 2365–2380. [CrossRef]
8. Celik, A.; Ercisli, S. Some Physical Properties of Pomegranate Cv. Eksinar. *Int. Agrophysics* **2009**, *23*, 295–298.
9. Kottek, M.; Grieser, J.; Beck, C.; Rudolf, B.; Rubel, F. World Map of the Köppen-Geiger Climate Classification Updated. *Meteorol. Z.* **2006**, *15*, 259–263. [CrossRef] [PubMed]

10. UPOV. *Guidelines for the Conduct of Tests for Distinctness, Uniformity and Stability. Pomegranate (Punica granatum L.)*; TG/PGRAN/3; UPOV: Geneva, Switzerland, 2012.
11. Tarhan, S. Selection of Chemical and Thermal Pretreatment Combination for Plum Drying at Low and Moderate Drying Air Temperatures. *J. Food Eng.* **2007**, *79*, 255–260. [CrossRef]
12. Carreño, J.; Martínez, A.; Almela, L.; Fernández-López, J.A. Proposal of an Index for the Objective Evaluation of the Colour of Red Table Grapes. *Food Res. Int.* **1995**, *28*, 373–377. [CrossRef]
13. AOAC. *Official Methods of Analysis*, 16th ed.; AOAC: Washington, DC, USA, 1995.
14. Narzary, D.; Yazdanbakhsh, N.; Rana, T. Taxonomy, Botany and Physiology. In *The Pomegranate: Botany, Production and Uses*; CAB International: Wallingford, UK, 2021; pp. 15–58.
15. Martinez-Nicolas, J.J.; Melgarejo, P.; Legua, P.; Garcia-Sanchez, F.; Hernández, F. Genetic Diversity of Pomegranate Germplasm Collection from Spain Determined by Fruit, Seed, Leaf and Flower Characteristics. *PeerJ* **2016**, *4*, e2214. [CrossRef]
16. Mansour, E.; Ben Khaled, A.; Haddad, M.; Abid, M.; Bachar, K.; Ferchichi, A. Selection of Pomegranate (*Punica granatum* L.) in South-Eastern Tunisia. *Afr. J. Biotechnol.* **2011**, *10*, 9352–9361. [CrossRef]
17. Parashuram, S.; Singh, N.V.; Gaikwad, N.N.; Corrado, G.; Roopa Sowjanya, P.; Basile, B.; Devaraja, N.S.; Chandra, R.; Babu, K.D.; Patil, P.G.; et al. Morphological, Biochemical, and Molecular Diversity of an Indian Ex Situ Collection of Pomegranate (*Punica granatum* L.). *Plants* **2022**, *11*, 3518. [CrossRef] [PubMed]
18. Gadže, J.; Radunić, M.; Petric, I.V.; Ercisli, S. In Vitro Pollen Viability, Germination and Pollen Tube Growth in Some Pomegranate (*Punica granatum* L.) Cultivars from Croatia and Bosnia and Herzegovina. *Acta Sci. Pol. Hortorum Cultus* **2011**, *10*, 297–305.
19. Yan-Hui, C.C.; Hui-Fang, G.G.; Sa, W.W.; Xian-Yan, L.L.; Qing-Xia, H.H.; Zai-Hai, J.J.; Ran, W.W.; Jin-Hui, S.S.; Jiang-Li, S.S. Comprehensive evaluation of 20 pomegranate (*Punica granatum* L.) cultivars in China. *J. Integr. Agric.* **2022**, *21*, 434–445. [CrossRef]
20. Khadivi, A.; Arab, M. Identification of the Superior Genotypes of Pomegranate (*Punica granatum* L.) Using Morphological and Fruit Characters. *Food Sci. Nutr.* **2021**, *9*, 4578–4588. [CrossRef] [PubMed]
21. Ferrara, G.; Cavoski, I.; Pacifico, A.; Tedone, L.; Mondelli, D. Morpho-Pomological and Chemical Characterization of Pomegranate (*Punica granatum* L.) Genotypes in Apulia Region, Southeastern Italy. *Sci. Hortic.* **2011**, *130*, 599–606. [CrossRef]
22. Tapia-Campos, E.; Castañeda-Saucedo, M.C.; del Pilar Ramirez-Anaya, J.; Alarcón-Dominguez, K.; Valdés-Miramontes, E.H.; Núñez-Maciel, O. Physical-Chemical Characterization of Fourteen Pomegranate Genotypes of Southern Jalisco, Mexico. *Sci. Hortic.* **2016**, *199*, 163–169. [CrossRef]
23. Ugarković, J.; Radunić, M.; Kozina, A.; Čmelik, Z. Osobine Sorata Šipka (*Punica granatum* L.) Glavaš i Paštrun. *Pomol. Croat.* **2009**, *15*, 87–94.
24. Radunić, M.; Jukić Špika, M.; Goreta Ban, S.; Gadže, J.; Díaz-Pérez, J.C.; Maclean, D. Physical and Chemical Properties of Pomegranate Fruit Accessions from Croatia. *Food Chem.* **2015**, *177*, 53–60. [CrossRef]
25. Ghasemi-Soloklui, A.A.; Kordrostami, M.; Gharaghani, A. Environmental and Geographical Conditions Influence Color, Physical Properties, and Physiochemical Composition of Pomegranate Fruits. *Sci. Rep.* **2023**, *13*, 15447. [CrossRef]
26. UNECE. *Unece Standard FFV-64 Concerning the Marketing and Commercial Quality Control of Pomegranates 2023 Editi-On*; 2023. Available online: https://unece.org/sites/default/files/2023-12/ECE_CTCS_WP.7_2022_06_E.pdf (accessed on 11 April 2024).
27. Cristofori, V.; Caruso, D.; Latini, G.; Dell'Agli, M.; Cammilli, C.; Rugini, E.; Bignami, C.; Muleo, R. Fruit Quality of Italian Pomegranate (*Punica granatum* L.) Autochthonous Varieties. *Eur. Food Res. Technol.* **2011**, *232*, 397–403. [CrossRef]
28. Martínez, J.J.; Hernández, F.; Abdelmajid, H.; Legua, P.; Martínez, R.; El Amine, A.; Melgarejo, P. Physico-Chemical Characterization of Six Pomegranate Cultivars from Morocco: Processing and Fresh Market Aptitudes. *Sci. Hortic.* **2012**, *140*, 100–106. [CrossRef]
29. Erkan Mustafa, D.A. Pomegranate/Roma—Punica Granatum. In *Exotic Fruits*; Rodrigues, S., de Oliveira Silva, E., de Brito, E.S., Eds.; Academic Press: Cambridge, MA, USA, 2018; pp. 355–361.
30. Tehranifar, A.; Zarei, M.; Esfandiyari, B.; Nemati, Z. Physicochemical Properties and Antioxidant Activities of Pomegranate Fruit (*Punica granatum*) of Different Cultivars Grown in Iran. *Hortic. Environ. Biotechnol.* **2010**, *51*, 573–579.
31. Durgaç, C.; Özgen, M.; Şimşek, Ö.; Kaçar, Y.A.; Kiyga, Y.; Çelebi, S.; Gündüz, K.; Serçe, S. Molecular and Pomological Diversity among Pomegranate (*Punica granatum* L.) Cultivars in Eastern Mediterranean Region of Turkey. *Afr. J. Biotechnol.* **2008**, *7*, 1294–1301.
32. Al-Said, F.A.; Opara, L.U.; Al-Yahyai, R.A. Physico-Chemical and Textural Quality Attributes of Pomegranate Cultivars (*Punica granatum* L.) Grown in the Sultanate of Oman. *J. Food Eng.* **2009**, *90*, 129–134. [CrossRef]
33. Wetzstein, H.Y.; Zhang, Z.; Ravid, N.; Wetzstein, M.E. Characterization of Attributes Related to Fruit Size in Pomegranate. *HortScience* **2011**, *46*, 908–912. [CrossRef]
34. Guo, Y.; Bai, J.; Duan, X.; Wang, J. Accumulation Characteristics of Carotenoids and Adaptive Fruit Color Variation in Ornamental Pepper. *Sci. Hortic.* **2021**, *275*, 109699. [CrossRef]
35. Markovic, I.; Ilic, J.; Markovic, D.; Simonovic, V.; Kosanic, N. Color Measurement of Food Products Using CIE L * a * b * and RGB Color Space. *J. Hyg. Eng. Des.* **2013**, *4*, 50–53.
36. Alighourchi, H.; Barzegar, M. Some Physicochemical Characteristics and Degradation Kinetic of Anthocyanin of Reconstituted Pomegranate Juice during Storage. *J. Food Eng.* **2009**, *90*, 179–185. [CrossRef]
37. Holland, D.; Hatib, K.; Bar-Ya'akov, I. Pomegranate: Botany, Horticulture, Breeding. In *Horticultural Reviews*; Janick, J., Ed.; John Wiley & Sons: Hoboken, NJ, USA, 2009; Volume 35.

38. Ben-Simhon, Z.; Judeinstein, S.; Trainin, T.; Harel-Beja, R.; Bar-Yaakov, I.; Borochov-Neori, H.; Holland, D. A "White" Anthocyanin-Less Pomegranate (*Punica granatum* L.) Caused by an Insertion in the Coding Region of the Leucoanthocyanidin Dioxygenase (LDOX; ANS) Gene. *PLoS ONE* **2015**, *10*, e0142777. [CrossRef]
39. Gil, M.I.; García-Viguera, C.; Artés, F.; Tomás-Barberán, F.A. Changes in Pomegranate Juice Pigmentation during Ripening. *J. Sci. Food Agric.* **1995**, *68*, 77–81. [CrossRef]
40. Khayyat, M.; Barati, Z.; Aminifard, M.H.; Samadzadeh, A. Anthocyanin Accumulation and Color Development in Seedless Barberry (*Berberis vulgaris* L.) Fruits: The Role of Altitude and Sun Light—The Preliminary Results. *Int. J. Fruit Sci.* **2020**, *20*, S955–S968. [CrossRef]
41. Candir, E.E.; Ozdemir, A.E.; Kaplankiran, M.; Toplu, C. Physico-Chemical Changes during Growth of Persimmon Fruits in the East Mediterranean Climate Region. *Sci. Hortic.* **2009**, *121*, 42–48. [CrossRef]
42. Lewallen, K.S.; Marini, R.P. Relationship between Flesh Firmness and Ground Color in Peach as Influenced by Light and Canopy Position. *J. Am. Soc. Hortic. Sci.* **2003**, *128*, 163–170. [CrossRef]
43. Kviklys, D.; Viškelis, J.; Liaudanskas, M.; Janulis, V.; Laužikė, K.; Samuolienė, G.; Uselis, N.; Lanauskas, J. Apple Fruit Growth and Quality Depend on the Position in Tree Canopy. *Plants* **2022**, *11*, 196. [CrossRef] [PubMed]
44. Tarantino, A.; Difonzo, G.; Disciglio, G.; Frabboni, L.; Paradiso, V.M.; Gambacorta, G.; Caponio, F. Fresh Pomegranate Juices from Cultivars and Local Ecotypes Grown in Southeastern Italy: Comparison of Physicochemical Properties, Antioxidant Activity and Bioactive Compounds. *J. Sci. Food Agric.* **2022**, *102*, 1185–1192. [CrossRef] [PubMed]
45. Passafiume, R.; Perrone, A.; Sortino, G.; Gianguzzi, G.; Saletta, F.; Gentile, C.; Farina, V. Chemical–physical Characteristics, Polyphenolic Content and Total Antioxidant Activity of Three Italian-Grown Pomegranate Cultivars. *NFS J.* **2019**, *16*, 9–14. [CrossRef]
46. Borochov-Neori, H.; Judeinstein, S.; Tripler, E.; Harari, M.; Greenberg, A.; Shomer, I.; Holland, D. Seasonal and Cultivar Variations in Antioxidant and Sensory Quality of Pomegranate (*Punica granatum* L.) Fruit. *J. Food Compos. Anal.* **2009**, *22*, 189–195. [CrossRef]
47. Kader, A.A. Postharvest Biology and Technology of Pomegranates. In *Pomegranates: Ancient Roots to Modern Medicine*; Heber, D., Schulman, R.N., Seeram, N.P., Eds.; CRC Press Taylor & Francis group: Boca Raton, FL, USA, 2006; pp. 211–220.
48. Chace, E.M.; Church, G.G.; Poore, H.H. The Wonderful Variety of Pomegranate. *USDA Circ* **1981**, *98*, 15.
49. Ben-Arie, R.; Segal, N.; Guelfat-Reich, S. The Maturation and Ripening of the 'Wonderful' Pomegranate. *J. Am. Soc. Hortic. Sci.* **2022**, *109*, 898–902. [CrossRef]
50. Martínez, J.J.; Melgarejo, P.; Hernández, F.; Salazar, D.M.; Martínez, R. Seed Characterisation of Five New Pomegranate (*Punica granatum* L.) Varieties. *Sci. Hortic.* **2006**, *110*, 241–246. [CrossRef]
51. Fawole, O.A.; Opara, U.L. Changes in Physical Properties, Chemical and Elemental Composition and Antioxidant Capacity of Pomegranate (Cv. Ruby) Fruit at Five Maturity Stages. *Sci. Hortic.* **2013**, *150*, 37–46. [CrossRef]
52. Ozgen, M.; Durgaç, C.; Serçe, S.; Kaya, C. Chemical and Antioxidant Properties of Pomegranate Cultivars Grown in the Mediterranean Region of Turkey. *Food Chem.* **2008**, *111*, 703–706. [CrossRef]
53. Onur, C.; Kaska, N. Selection of Pomegranate of Mediterranean Region. *Turk. J. Agric. For. D2* **1985**, *9*, 25–33.

MDPI AG
Grosspeteranlage 5
4052 Basel
Switzerland
Tel.: +41 61 683 77 34

Horticulturae Editorial Office
E-mail: horticulturae@mdpi.com
www.mdpi.com/journal/horticulturae